MODERN COLLEGE ALGEBRA

MODERN COLLEGE ALGEBRA

Herman R. Hyatt
James N. Hardesty
Los Angeles Pierce College

Scott, Foresman and Company

Glenview, Illinois

Dallas, Tex. ● Oakland, N. J.

Palo Alto, Cal. ● Tucker, Ga. ● Brighton, England

Library of Congress Catalog Card Number: 73-89378

ISBN: 0-673-07831-0

AMS 1970 Subject Classification: 98-A10

PREFACE

Typically, a college algebra course is designed to provide needed algebraic skills, problem-solving techniques, and some degree of mathematical maturity to students with varied backgrounds and different plans for their future studies. Some will go on in mathematics, the physical sciences, or engineering; some will continue into the social or behavioral sciences; still others will go on into business administration. To meet this diversity of needs, a college algebra course should offer flexibility of choice as well as a comprehensive number of topics from which to choose. In the opinion of the authors, the following features of this text provide such flexibility and breadth of choice.

Diagnostic Test The text opens with a diagnostic test designed to assist students to pinpoint weaknesses in preparation. Answers to all test questions are provided in the answer section of the book. Additionally, test questions are keyed by number to specific sections in introductory Chapter 0. Thus, Chapter 0 serves as a review chapter, as well as a reference source of elementary algebra. To meet the needs of a given class, all, part, or none of Chapter 0 may be assigned.

Exercise Sets Each section in each chapter is designed to be covered in a day's lecture and study. Exercise sets at the end of each section are carefully graded and arranged in **(A)** and **(B)** sections, as appropriate. The **(A)** exercises help to sharpen manipulative skills and the understanding of basic concepts, as well as providing sufficient material for a basic but minimum course. The **(B)** exercises offer more challenging problems and/or enrichment material. Most exercises are keyed by number to specific examples in the text, which helps the student to "get started" on the assigned work. This feature also permits the use of this text on a *self-study* basis.

Examples Hundreds of examples in the text illustrate how to solve typical problems and also serve to clarify new concepts as they are introduced. The examples are sufficiently detailed to further encourage the use of the book for self-study.

Rigor The wide variation in the background and needs of students enrolled in college algebra courses precludes a strong emphasis on rigor and proof. However, theorems, axioms, and definitions are carefully stated and identified as such. Several exercise sets include **(P)** sections in which consideration is given to proofs of some theorems. Students are encouraged to participate in the construction of such proofs by being given "semi-programmed" proofs or by being given a substantial hint. Because of the **(P)** sections, very few proofs are included in the text proper. The concepts of equivalent equations, equivalent inequalities, and equivalent systems are carefully integrated into the exposition in the text so that, to some degree, students may learn to appreciate some of the theory that underlies the methods they are studying.

Self-Tests Each chapter concludes with a Self-Test for which all answers are given in the answer section. The problems in each Self-Test are keyed by number to specific sections of the chapter. Thus, the Self-Test may be taken as a "test" or may be used as a chapter review.

Computer Programs Where applicable to material being studied, certain Exercise Sets include (C) sections in which computer programs in BASIC computer language are presented. (These sections may be omitted without loss of continuity.) Instructors may wish to expand upon this material by suggesting a text on BASIC programming as a supplement to the course.

Supplements Fourteen supplementary sections are strategically positioned throughout the text, ranging from a review of *computations with base 10 logarithms* to such calculus preparatory topics as *partial fractions* and *the average rate of change of a function.* (Any or all of these supplements may be omitted without disturbing the continuity of the text.) Although the supplements are generally short, each includes examples and an exercise set, thus expanding the choice of topics available to a given class.

Graphing Because competency in graphing is significantly important in helping students to both *do* and *understand* mathematics, a thorough development of techniques of graphing is included. The concept of *translation* is introduced early for the study of conic sections. Thereafter, translations are routinely applied to graphs other than those of conic sections. *Sign-graph* methods are used to solve inequalities. Characteristic properties that aid in graphing a function and determining whether the function is one-to-one (indicating the existence of an inverse function) are introduced early. Supplement D (Related Graphs) demonstrates how to sketch more "complicated" graphs from "simpler" graphs by applying techniques such as taking reciprocals, multiplying by a constant, addition of ordinates, etc. Supplement K (Solving Open Sentences Graphically) displays the use of graphing for solving certain types of equations and inequalities.

Matrices Matrices are first introduced in Chapter 5 for the purpose of solving linear systems. We return to matrices in Chapter 7 where realistic examples are used to motivate the introduction of the rather abstract operations on matrices. The theory of matrices is considered, but only to the extent needed to enable the student to feel "comfortable" with the concept and to operate competently with this extremely useful working tool.

Exponential and Logarithmic Functions Exponential and logarithmic functions are carefully developed, including a section on natural logarithms (base e). A separate section is devoted to applications of these functions. Problems that students are likely to encounter in other courses are chosen from sources such as physics, chemistry, mathematics of finance, etc.

The authors wish to thank Ms. Pamela Conaghan of Scott, Foresman and Company for her assistance and advice in preparing the manuscript for production. Further thanks are due to Mr. Howard R. Hyatt for his help and cooperation in typing the manuscript.

Herman R. Hyatt
James N. Hardesty

CONTENTS

DIAGNOSTIC TEST

A course in College Algebra presupposes some knowledge of a basic amount of elementary algebra. The following test includes exercises that can help you to locate possible weaknesses or deficiencies in your own preparation for such a course. Each set of exercises is keyed to a corresponding section in the introductory Chapter 0. Thus, should you discover that you are uncertain as to your understanding of a particular concept or technique, you may wish to study the appropriate section of Chapter 0. At the very least, an awareness of what material is offered in Chapter 0 is desirable so that you can refer to the chapter later on in your studies, should the need arise.

Chapter 0   Diagnostic Test

[0.1] *Briefly describe the meaning of each expression.*

1. $12 \in \{10, 12, 14\}$ 2. $x, y \in R$

3. List the members of $\{2n + 3: n = 1, 2, 3, 4\}$.

4. Given $A = \{3, 6, 9, 12\}$ and $B = \{2, 4, 6, 8, 10, 12\}$. List the members of   a. $A \cup B$;   b. $A \cap B$.

[0.2] *Name the real number axiom that best justifies each equation.*

5. $xy = yx$ 6. $y + z = z + y$ 7. $2(3x) = (2 \cdot 3)x$

8. $5 + (-5) = 0$ 9. $5 \cdot \frac{1}{5} = 1$ 10. $2(x + y) = 2x + 2y$

[0.3] 11. Specify (in words) the hypothesis and the conclusion of the theorem "If $x \in R$, then $x^2 \geq 0$."

12. Supply a reason that best justifies each step in the following proof that $(-3) + (-6) = -(3 + 6)$.

   a. $(-3) + (-6) = (-3) + (-6) + 0$

   b. $\qquad = (-3) + (-6) + \big((3 + 6) + [-(3 + 6)]\big)$

   c. $\qquad = [(-3) + 3] + [(-6) + 6] + [-(3 + 6)]$

   d. $\qquad = 0 + 0 + [-(3 + 6)]$

   e. $(-3) + (-6) = -(3 + 6)$

[0.4] 13. Graph each set on a number line. Specify whether the graph is an *open* or a *closed* interval; an *unbounded open* or an *unbounded closed* interval.

   a. $\{x: -6 \leq x \leq 4\}$ b. $\{x: -6 < x < 4\}$

   c. $\{x: x < -6\}$ d. $\{x: x \geq 4\}$

14. Graph each intersection and specify the resulting interval in set notation.

   a. $\{x: x \geq -6\} \cap \{x: x \leq 4\}$ b. $\{x: x > -6\} \cap \{x: x < 4\}$

15. Which of the following generalizations is false?
   a. $|x-y| = |y-x|$
   b. $|x+y| = |x| + |y|$
   c. $|xy| = |x| \cdot |y|$
   d. $\left|\dfrac{x}{y}\right| = \dfrac{|x|}{|y|}$

[0.5] *Given* $i = \sqrt{-1}$, *compute:*
   16. $(3+2i) - (6-7i)$
   17. $(2-i)(4+3i)$
   18. $4i \cdot 4i \cdot 4i \cdot 4i$

[0.6] *Simplify:*
   19. $27^{2/3}$
   20. $\left(\dfrac{-32x^4}{4x}\right)^{1/3}$

Find the indicated root of each number.
   21. 49; square root
   22. -125; cube root

[0.7] *Simplify:*
   23. $\sqrt{75x} + \sqrt{108x}$
   24. $(2+\sqrt{3})(5-2\sqrt{3})$
   25. Rationalize the denominator: $\dfrac{x-\sqrt{3}}{x+\sqrt{3}}$.

[0.8] *Compute and express the result in* $a+bi$ *form.*
   26. $\sqrt{-3}\left(\sqrt{-12} + \sqrt{-3}\right)$
   27. $\dfrac{3+i}{4-i}$

[0.9] 28. Given $A(x) = 5x^3 + 3x^2 - 7x - 1$. Find: $A(0)$; $A(-2)$; $A(-x)$; $A(i)$.
   29. Simplify: $(2x^2 - 5x) - 3(x^2 + 3x + 1) + (3x - 2)(x^2 + 4x - 7)$.

[0.10] *Factor completely (if possible) over the set of integers.*
   30. $15x^4 + 57x^3 - 12x^2$
   31. $25x^2 - 144y^2$
   32. $8y^3 - 125$
   33. $x^3 + y^3$
   34. $3x^3 + xy - 6x^2y - 2y^2$
   35. $x^2 + 9$
   36. Factor over the set of rational numbers: $x^2 - \frac{8}{9}x + \frac{16}{81}$.
   37. Factor over the set of complex numbers: $x^2 + 9$.

[0.11] 38. Reduce to lowest terms: $\dfrac{x^2 + 2x - 35}{x^2 + 10x + 21}$.

   39. Simplify: $\left(\dfrac{x+2}{x-4} + \dfrac{x-4}{x+2}\right) \cdot \dfrac{x^2 - 2x - 8}{x^2 - 4x + 12}$.

Write each rational expression in the form $Q(x) + \dfrac{R(x)}{B(x)}$, *where* $Q(x)$ *is the quotient,* $R(x)$ *is the remainder and* $B(x)$ *is the divisor.*
   40. $\dfrac{4x^3 + 2x^2 - 3x - 8}{x^2 + 4x + 5}$
   41. $\dfrac{x^3 - 125}{x - 5}$

[0.12] 42. Graph the set of points: $\{(3, 4), (4, -3), (-3, 4), (-4, -3)\}$.
   43. Describe a property that is a characteristic of all points that are to the right of the y-axis and also below the x-axis.
   44. Graph the set of all points with: a. x-coordinate 3; b. y-coordinate -3.

PRELIMINARY CONCEPTS

A primary concern of mathematicians is the study of mathematical systems. In general, a mathematical system consists of certain *undefined elements*; a set of *defined elements*; a set of assumptions about the elements, called *axioms*; a set of statements called *theorems* that are logical consequences of the definitions and axioms of the system. Underlying our study in this book is the mathematical system called the *Real Number System*, so named because the undefined elements of the system are the real numbers.

0.1 SETS

A **set** is a collection of distinct objects each of which is said to be *a member of*, or *an element of*, the set. Capital letters such as A, N, R, etc., may be used to name sets.

One method of specifying a set is that of listing its members between braces { }. Thus, using N to name the set of *natural numbers*, we may write

$$N = \{1, 2, 3, \ldots\}.^*$$

Because the members of a set must be distinct, we do not list the same member more than once. Furthermore, the arrangement in which the members are listed is usually of no significance. Thus, the set of digits of the number 11,005 can be specified as {0, 1, 5}, {1, 5, 0}, etc.

The symbol $\in$ is used to denote membership in a set. For example, $5 \in \{0, 1, 5\}$ means that 5 is a member of {0, 1, 5}. Notation such as $5 \in N$ means that 5 is a member of set N, or that 5 is a natural number.

A symbol such as one of the lower-case letters a, n, x, . . . may be used to represent, or name, an unspecified member of a given set, in which case the symbol is called a **variable** and the given set is its **replacement set**. Thus, using x to name any natural number, we may write $x \in N$.

---

*The symbol . . ., called *ellipsis*, is used to indicate omitted words or phrases. It may also be used to mean "and so forth."

Then x is a variable and its replacement set is the set of natural numbers. If the replacement set of a symbol includes exactly one member, then the symbol is called a **constant**. If, for example, $n \in \{3\}$, then n is a constant.

A second method of specifying sets, called **set-builder notation**, is of the form

$$\{x: \text{a condition (or conditions) on } x\},$$

which is read "the set of all elements x such that x satisfies some condition (or conditions)." Note that the colon : is read "such that." As an example, the notation $\{n: n \in N\}$, read "the set of all elements n such that n is a natural number," is another method of specifying the set of natural numbers. Note that n is a variable with replacement set N.

Two sets A and B are said to be **equal sets** if they include the same members. We may write $A = B$. Thus,

$$\{0, 1, 5\} = \{1, 5, 0\} \qquad \text{and} \qquad \{1, 2, 3, \ldots\} = \{n: n \in N\}.$$

We sometimes use the slant bar / to indicate *negation*. Thus, in the notation

$$\{0, 1, 5\} \neq \{0, 1, 6\} \qquad \text{and} \qquad 3 \notin \{0, 1, 5\},$$

the symbols $\neq$ and $\notin$ are read "is not equal to" and "is not a member of," respectively.

If every member of a set A is also a member of a set B, then A is called a **subset** of B and we write

$$A \subseteq B.$$

If set A is a subset of set B and B includes at least one member that is not in A, then A is a **proper subset** of B and we write

$$A \subset B.$$

Any set is considered to be a subset (but not a proper subset) of itself. Thus,

$$\{0, 1\} \subset \{0, 1, 5\} \text{ or } \{0, 1\} \subseteq \{0, 1, 5\}, \qquad \text{but} \qquad \{0, 1\} \not\subset \{0, 1\}$$

The set that includes no members is called the **empty set** or the **null set**, and is denoted by the symbol $\emptyset$. The empty set $\emptyset$ is considered to be a subset of any other set.

Two sets A and B are said to be in **one-to-one correspondence** if there is a pairing of the members of A and B such that to each member of A there is associated exactly one member of B, and to each member of B there is associated exactly one member of A. If the members of a given set can be put into one-to-one correspondence with the first n members of the set N of natural numbers, then the given set is said to be a **finite set**. A set that is not finite is an **infinite set**. For example, the set E of letters of the English alphabet is a finite set because the members of E can be put into a one-to-one correspondence with the set of natural numbers $\{1, 2, 3, \ldots, 25, 26\}$. The set N itself is an example of an infinite set because there is no last natural number.

Given sets A and B, the set of all elements that belong *either* to A *or* to B or to both A and B is called the **union** of A and B, denoted by $A \cup B$. Symbolically,

$$A \cup B = \{x: x \in A \text{ or } x \in B\}.$$

The set of all elements that belong to both A *and* B is called the **intersection** of A and B, denoted by $A \cap B$. Symbolically,

$$A \cap B = \{x: x \in A \text{ and } x \in B\}.$$

If A and B have no members in common, that is, if the intersection of A and B is empty, then A and B are said to be **disjoint** sets, indicated by $A \cap B = \emptyset$. Given $A = \{1, 2, 3\}$, $B = \{3, 4, 5\}$ and $C = \{6, 7, 8\}$, you will find it instructive to verify that

$$A \cup B = \{1, 2, 3, 4, 5\}, \qquad A \cap B = \{3\} \qquad \text{and} \qquad A \cap C = \emptyset.$$

It is sometimes helpful to visualize certain relations between sets by means of figures called **Venn diagrams**, as in Figure 0.1-1.

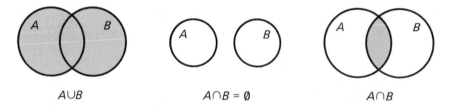

$A \cup B$ $\qquad\qquad$ $A \cap B = \emptyset$ $\qquad\qquad$ $A \cap B$

Figure 0.1-1

In this text, unless specified otherwise, we use "number" to mean "real number," that is, a member of the set R of **real numbers**. Certain familiar subsets of R, named and listed below, will frequently be referred to.

1. The set N of *natural numbers*: $N = \{1, 2, 3, \ldots\}$.
2. The set W of *whole numbers*: $W = \{0, 1, 2, 3, \ldots\}$.
3. The set J of *integers*: $J = \{\ldots, -3, -2, -1, 0, 2, 3, \ldots\}$.
4. The set Q of *rational numbers*: $Q = \left\{\dfrac{m}{n}: m, n \in J, n \neq 0\right\}$.
5. The set H of *irrational numbers*: $H = \{x: x \in R \text{ and } x \notin Q\}$.

Note that a number is a rational number if it can be expressed as a *ratio* (fraction) of two integers. Thus, $\sqrt{16}$ is a rational number because it is equal to $\frac{4}{1}$. On the other hand, it can be shown that $\sqrt{2}$ cannot be expressed as a ratio of two integers and so $\sqrt{2}$ is an irrational number.

Exercise Set 0.1

Briefly describe the meaning of each expression.

1. $3 \in \{1, 2, 3\}$ 2. $6 \notin \{1, 2, 3\}$ 3. $16 \in N$

4. $17 \in Q$ 5. $x \in R$ 6. $n \in J$

7. $a, b \in N$ 8. $x, y \in R$ 9. $\sqrt{5} \notin Q$

10. $\frac{2}{3} \notin J$ 11. $J \subset R$ 12. $N \subseteq W$

13. $H \cap Q = \emptyset$ 14. $H \cup Q = R$

Given $A = \{1, 2, 3, 4, 5\}$, $B = \{2, 4, 6, 8, 10\}$ and $C = \{1, 3, 5, 7, 9\}$.
List the members in each of the following sets.

15. $A \cup B$ 16. $B \cup C$ 17. $B \cap C$

18. $A \cap C$ 19. $A \cup (B \cup C)$ 20. $(A \cap B) \cap C$

21. $A \cap (B \cup C)$ 22. $A \cup (B \cap C)$

For each pair of sets A and B, show that $A = B$.

23. $A = \{2n: n = 1, 2, 3, 4\}$; $B = \{2, 4, 6, 8\}$

24. $A = \{2n + 1: n = 1, 2, 3, 4\}$; $B = \{3, 5, 7, 9\}$

25. $A = \{2x - 1: x = 1, 2, 3, \ldots\}$; $B = \{1, 3, 5, \ldots\}$

26. $A = \{3r: r = 1, 2, 3, \ldots\}$; $B = \{3, 6, 9, \ldots\}$

List the members of each of the following sets. (The word "factor" refers to both positive and negative integer factors. Thus, the set of factors of 9 is $\{\pm 9, \pm 3, \pm 1\}$.)

27. $\{p: p \text{ is a factor of } 6\}$

28. $\{p: p \text{ is a factor of } 25\}$

29. $\{q: q \text{ is a factor of } 8\}$

30. $\{q: q \text{ is a factor of } 15\}$

31. $\left\{\frac{p}{q}: p \text{ is a factor of } 6, \ q \text{ is a factor of } 8\right\}$

32. $\left\{\frac{p}{q}: p \text{ is a factor of } 25, \ q \text{ is a factor of } 15\right\}$

Draw a Venn diagram to illustrate each of the following sets.

33. $A \cup B$, if $A \cap B \neq \emptyset$ 34. $A \cup B$, if $A \cap B = \emptyset$

35. $A \cap B$, if $A \not\subset B$ 36. $A \cap B$, if $A \subset B$

37. $A \cup (B \cap C)$ 38. $A \cap (B \cup C)$

0.2   REAL NUMBER AXIOMS

In mathematics, a declaratory sentence that can be classified as either true or false, but not both, is called a **statement**. Many sentences that involve one (or more) variable(s) cannot be classified as true or false until the variable(s) is replaced by a member of its replacement set.

Such sentences are called **open sentences**. An open sentence involving the equality symbol = is called an **equation**. Thus, the open sentence $x + 2 = 5$ is an equation. The expressions to the left and to the right of the equality symbol are referred to as the *left member* and the *right member*, respectively. If the replacement set for x is the set R of real numbers, and we replace x by 3, we obtain the true statement $3 + 2 = 5$. If we replace x by 6, we obtain the false statement $6 + 2 = 5$.

In the real number system, certain relations between real numbers, and operations on real numbers, are based upon assumptions called **axioms**. Generally, these axioms are given in the form of open sentences that are taken as true statements for all acceptable replacements for the variable. In the following listings of axioms, the phrase "For each $a, b, c, \in R$" is used to mean "The set R of real numbers is the replacement set for the variables a, b, and c." The = symbol may be read "is equal to," or "is the same as," or "is another name for." The first set of axioms states properties of the **equality relation** between real numbers.

Equality Axioms   For each $a, b, c \in R$:

E-1 (Reflexivity) $a = a$.

E-2 (Symmetry)   If $a = b$, then $b = a$.

E-3 (Transitivity)   If $a = b$ and $b = c$, then $a = c$.

E-4 (Substitution)   If $a = b$, then a may replace b and b may replace a in a sentence.

In the set R of real numbers, the operation of *addition* assigns to each pair of real numbers a and b a unique real number called their **sum** and denoted by $a + b$. The operation of *multiplication* assigns to each pair of real numbers a unique real number called their **product** and denoted by $a \times b$, or $a \cdot b$, or simply by ab. The fact that the sum (or the product) of two real numbers is again a real number is referred to as the **closure property** of addition (or multiplication). Properties of the operations of addition and multiplication are given by the next axioms.

Addition Axioms   For each $a, b, c \in R$:

R-1 (Commutativity) $a + b = b + a$.

R-2 (Associativity) $a + (b + c) = (a + b) + c$.

R-3 (Identity)   There exists a real number 0, called the *additive identity element*, such that $a + 0 = a$ and $0 + a = a$.

R-4 (Inverse)   For each $a \in R$ there exists a unique real number $-a$ called the *additive inverse* of a such that $a + (-a) = 0$ and $(-a) + a = 0$.

We sometimes refer to $-a$ as the *negative* of a or the *opposite* of a.

Multiplication Axioms   For each $a, b, c \in R$:

R-5 (Commutativity) $ab = ba$.

R-6 (Associativity) $a(bc) = (ab)c$.

R-7 (Identity)   There exists a real number 1, $1 \neq 0$, called the *multiplicative identity element*, such that $a \cdot 1 = a$ and $1 \cdot a = a$.

R-8 (Inverse)   For each $a \in R$, $a \neq 0$, there exists a unique real number $\frac{1}{a}$, called the *multiplicative inverse* of a such that $a \cdot \frac{1}{a} = 1$ and $\frac{1}{a} \cdot a = 1$.

We sometimes refer to $\frac{1}{a}$ as the *reciprocal* of a. Note that, according to Axiom R-8, the number zero has no reciprocal.

Distributive Axiom   For each $a, b, c \in R$:

R-9 (Distributivity) $a(b+c) = ab + ac$ and $(a+b)c = ac + bc$.

By Axiom R-2, the numbers $a+(b+c)$ and $(a+b)+c$ are not different. Hence, we adopt the convention that $a+b+c$ is the real number such that

$$a + b + c = a + (b + c) = (a + b) + c,$$

a convention that is also applicable to sums of more than three numbers. In sums such as $a+b+c$, the numbers a, b and c are called **terms**. By Axiom R-6, the numbers $a(bc)$ and $(ab)c$ are not different. Hence, we adopt the convention that abc is the real number such that

$$abc = a(bc) = (ab)c,$$

a convention that is also applicable to products of more than three numbers. In products such as abc, the numbers a, b and c are called **factors**.

The operations of *subtraction* and *division* in R are defined in terms of addition and multiplication, respectively.

Definition 0.1   For all $a, b \in R$, the **difference** between a and b, denoted by $a - b$, is given by $a - b = a + (-b)$.

Definition 0.2   For all $a, b \in R$, $b \neq 0$, the **quotient** of a and b, denoted by $\frac{a}{b}$ (or by $a \div b$, or by a/b), is given by $\frac{a}{b} = a \cdot \frac{1}{b}$.

Essentially, Definition 0.1 asserts that the difference between two numbers is the same as the sum of the first number and the additive inverse of the second; Definition 0.2 asserts that the quotient of two numbers is the same as the product of the first number and the multiplicative inverse of the second.

The set R of real numbers may be viewed as the union of three disjoint sets: the set R^+ of **positive** numbers; the set R^- of **negative** numbers; the set $\{0\}$ including only zero. We indicate that a is a positive number ($a \in R^+$) by writing $a > 0$, where the symbol $>$ is read "is greater than." We indicate that a is a negative number ($a \in R^-$) by writing $a < 0$, where the symbol $<$ is read "is less than." The number 0 is neither positive nor negative. The next set of axioms states properties of the **is greater than** relation between real numbers.

Order Axioms   For each $a, b, c \in R$:

O–1 (Trichotomy) Exactly one of the following relations is true:
$$a > b \quad \text{or} \quad a = b \quad \text{or} \quad b > a.$$

O–2 (Transitivity of order) If $a > b$ and $b > c$, then $a > c$.

O–3 (Closure in R^+) If $a > 0$ and $b > 0$, then $a + b > 0$ and $ab > 0$.

The "is less than" relation can be defined in terms of the "is greater than" relation.

Definition 0.3   If $a, b \in R$, then a **is less than** b, denoted by $a < b$, if and only if $b > a$.

Applying Definition 0.3, the Trichotomy Axiom (O–1) can be interpreted as "Given any two real numbers a and b: a is either greater than b, or a is equal to b, or a is less than b." Axiom O–2 is valid if the $>$ symbol is replaced throughout by the $<$ symbol. Whenever such a replacement is made, we shall still refer to the resulting statement as Axiom O–2. Axiom O–3 simply asserts that the sum or the product of two positive numbers is a positive number.

The notation $a \geq 0$, read "a is greater than or equal to zero," is used to mean that a is nonnegative. The notation $a \leq 0$, read "a is less than or equal to zero," is used to mean that a is nonpositive. Open sentences that involve any of the symbols $>$, $<$, $\geq$, or $\leq$ are called **inequalities**.

Exercise Set 0.2
Cite the equality axiom E–1 to E–4 that best justifies each sentence.
1. $3x + 7 = 3x + 7$ 2. If $3x + 7 = 5$, then $5 = 3x + 7$.
3. If $x = 4$ and $3x + 7 = 19$, then $3 \cdot 4 + 7 = 19$.
4. If $x = 4$ and $4 = 3x + 7$, then $x = 3x + 7$.

Cite the axiom R–1 to R–9 that best justifies each sentence.
5. $4x + 0 = 4x$ 6. $3(xy) = (3x)y$

7. $2x + 5 = 5 + 2x$ 8. $(2x) \cdot \left(\dfrac{1}{2x}\right) = 1$

9. $3x = 3x \cdot 1$

10. $4(x+2) = 4x + 4 \cdot 2$

11. $3x + [-(3x)] = 0$

12. $x + (2+y) = (x+2) + y$

13. $3x + 5x = (3+5)x$

14. $2xy = 2yx$

Cite the order axiom O-1 to O-3 that best justifies each sentence.

15. If $2x < 3+x$ and $3+x < 7$, then $2x < 7$.

16. If $x \neq 3$ then $x > 3$.

17. If $x > 0$ then $3x > 0$.

18. If $y > 0$ then $5 + y > 0$.

19. If $x \in R$, then $x > 0$ or $x = 0$ or $x < 0$.

20. If $x \in R$, then $x \in R^+$ or $x \in R^-$ or $x = 0$.

0.3   REAL NUMBER THEOREMS

This section provides a listing of many of the basic real number theorems that are logical consequences of the axioms and definitions introduced in the preceding sections. Each such theorem is numbered so that it can be referred to when necessary. Proofs of theorems in this book are sometimes given in the text proper to serve as examples. Most often, proofs of theorems are considered in the exercise set at the end of the section in which the theorems are stated. Such theorems, and the corresponding parts of the exercise sets, are identified by the symbol **(P)**. Proofs for theorems not so marked are omitted.

Transformation Theorems   Most theorems in mathematics are given in the form

"*If p, then q,*"

where the statements p and q are called the **hypothesis** and the **conclusion**, respectively. Sometimes a theorem is stated in the form

"*p if and only if q,*"

which is understood to mean the two theorems

1. *If q, then p,*   and   2. *If p, then q.*

The first two theorems below are in *if and only if* form.

Theorem 0.1   If $a, b, c \in R$, then $a + c = b + c$ if and only if $a = b$.

Part 1 If $a = b$, then $a + c = b + c$.

Proof

1. $a, b, c \in R$; $a = b$
2. $a + c = a + c$
3. $a + c = b + c$

1. Hypothesis
2. Axiom E-1 (reflexivity)
3. Axiom E-4 (substitution) from (1) into (2)

Part 2 If $a+c=b+c$, then $a=b$.
Proof

1. $a, b, c \in R$; $a+c=b+c$	1. Hypothesis
2. $(a+c)+(-c)=(b+c)+(-c)$	2. Part 1 of Theorem 0.1
3. $a+[c+(-c)]=b+[c+(-c)]$	3. Axiom R-2 (associativity)
4. $[c+(-c)]=0$	4. Axiom R-4 (inverse)
5. $a+0=b+0$	5. Axiom E-4 (substitution) from (4) into (3)
6. $a+0=a$; $b+0=b$	6. Axiom R-3 (identity)
7. $a=b$	7. Axiom E-4 (substitution) from (6) into (5)

The preceding proof is in a *two-column format*, where statements and reasons are listed in columns and numbered to correspond.

Theorem 0.2 If $a, b, c \in R$ and $c \neq 0$, then $ac=bc$ if and only if $a=b$. **(P)***

A proof for Theorem 0.2 can be constructed by using the proof of Theorem 0.1 as a guide, together with Axioms R-6, R-7 and R-8.

Additive Inverse and Zero Theorems The proof of the next theorem is in a paragraph format; the symbol □ indicates the end of the proof.

Theorem 0.3 (Double Negative Theorem) If $a \in R$, then $-(-a)=a$.

Proof Because a is a real number, by Axiom R-4 (inverse) there exists a unique real number $-a$ such that

$$a+(-a)=0. \qquad \langle 1 \rangle$$

Because $-a$ is a real number, by Axiom R-4 there exists a unique real number $-(-a)$ such that

$$-(-a)+(-a)=0. \qquad \langle 2 \rangle$$

Equation $\langle 1 \rangle$ asserts that a is the additive inverse of $-a$; equation $\langle 2 \rangle$ asserts that $-(-a)$ is the additive inverse of $-a$. By the uniqueness of the additive inverse, it follows that a and $-(-a)$ must name the same number. That is,

$$-(-a)=a. \quad □$$

Theorem 0.4 If $a, b, \in R$, then $(-a)+(-b)=-(a+b)$. **(P)**

Theorem 0.5 If $a \in R$, then $a \cdot 0=0$. **(P)**

Theorem 0.6 If $a, b \in R$, and $ab=0$, then $a=0$ or $b=0$.

*The symbol **(P)** following the statement of a theorem means that a proof of that theorem is considered in the very next exercise set.

Theorem 0.7   If $a, b \in R$, then

    I $(-1)a = -a$
    II $(-a)(b) = -ab$
    III $(-a)(-b) = ab.$ **(P)**

Theorems about Fractions   Given a fraction such as $\frac{a}{b}$, recall that a and b are the *terms* of the fraction, a is the *numerator* and b is the *demoninator*. The following theorems review properties of real numbers in fraction form.

Theorem 0.8   If $a, b, c, d \in R$, $b \neq 0$ and $d \neq 0$, then

$$\frac{a}{b} = \frac{c}{d} \quad \text{if and only if} \quad ad = bc. \quad \textbf{(P)}$$

The familiar techniques of "raising a fraction to higher terms" or "reducing a fraction to lower terms" depend upon Theorem 0.9.

Theorem 0.9 (Fundamental Theorem of Fractions)   If $a, b, c \in R$, $b \neq 0$ and $c \neq 0$, then $\frac{a}{b} = \frac{ac}{bc}$.   **(P)**

The next three theorems provide methods for finding sums, products and quotients of fractions.

Theorem 0.10   If $a, b, c \in R$ and $c \neq 0$, then $\frac{a}{c} + \frac{b}{c} = \frac{a+b}{c}$.   **(P)**

Theorem 0.11   If $a, b, c, d \in R$, $b \neq 0$ and $d \neq 0$, then $\frac{a}{b} \cdot \frac{c}{d} = \frac{ac}{bd}$.   **(P)**

Theorem 0.12   If $a, b, c, d \in R$, $b \neq 0$, $c \neq 0$ and $d \neq 0$, then

$$\frac{a}{b} \div \frac{c}{d} = \frac{ad}{bc}.$$

A method for finding the difference between fractions follows the next "sign change" theorem.

Theorem 0.13   If $a, b \in R$ and $b \neq 0$, then $\frac{-a}{b} = \frac{a}{-b} = -\frac{a}{b}$.

Theorem 0.14   If $a, b, c \in R$ and $c \neq 0$, then $\frac{a}{c} - \frac{b}{c} = \frac{a-b}{c}$.   **(P)**

Theorems about Order   To construct proofs for theorems concerning order relations, we introduce a definition that relates "is greater than" with "is equal to."

Definition 0.4   If $a, b \in R$, then $a > b$ if and only if there is a positive number p such that $b + p = a$.

For example, because $-4 + 2 = -2$, it follows that $-2 > -4$.

Definition 0.4 implies that, for any nonzero real number r, either r is positive and $-r$ is negative, or $-r$ is positive and r is negative. To see this, consider first the case that $r > 0$. By Axiom R-4 (inverse), we have

$$(-r) + r = 0.$$

By Definition 0.4, it follows that $0 > -r$. That is, $-r$ is negative. A similar argument establishes that r is negative when $-r$ is positive.

Theorem 0.15   If $a, b, c \in R$, then $a + c > b + c$ if and only if $a > b$.

Part 1 If $a > b$, then $a + c > b + c$.
Proof Because $a > b$, by Definition 0.4 there is a positive number p such that $a = b + p$. By Theorem 0.1, together with commutativity and associativity of addition, it follows that

$$a + c = (b + p) + c$$
$$a + c = (b + c) + p. \qquad\qquad \langle 3 \rangle$$

From $\langle 3 \rangle$, we see that $a + c$ is equal to the sum of $(b + c)$ and a positive number p. Hence, by Definition 0.4, we have that

$$a + c > b + c. \quad \square$$

Part 2 If $a + c > b + c$, then $a > c$. (See Exercise 0.3-XIII.)

Theorem 0.16   For $a, b, c \in R$ and $a > b$:

I   If $c > 0$, then $ac > bc$.
II   If $c < 0$, then $ac < bc$.   **(P)**

Inequalities such as $ac > bc$ and $ac < bc$ of Theorem 0.16, where the order symbols are reversed, are said to be *opposite in sense*. Thus, Part II of the theorem may be stated as "If each member of an inequality is multiplied by a negative number, then the resulting inequality is opposite in sense."

For the last theorem of this section you should recall that the symbol a^2 means $a \cdot a$. It is not difficult to show that *any nonzero real number a^2 is positive*. First, if $a > 0$ then, by Axiom O-3, $a^2 > 0$. Second, if $a < 0$ then $-a > 0$ and $(-a)(-a) > 0$. But, by Theorem 0.7-III, $(-a)(-a) = a \cdot a = a^2$ and we see that in both cases $a^2 > 0$. These results are formally stated next.

Theorem 0.17   If $a \in R$ and $a \neq 0$, then $a^2 > 0$.

The Real Number System   Some of the beginning structure of the real number system is provided by the axioms, definitions and theorems considered thus far. Specifically, the system includes

1. A set of elements called real numbers.
2. An equality relation between these numbers.
3. Operations of addition and multiplication on these numbers.
4. A set of axioms (R−1 through R−9) that assume: commutativity and associativity of addition and multiplication; the existence of unique additive and multiplicative identity elements; the existence of unique additive and multiplicative inverses; distributivity of multiplication over addition.

Any mathematical system based upon 1 through 4 as listed above is called a **field**, hence the name "field axioms." The real number system is one example of a field. Another field will be discussed later in this text. All fields share, as common features of their structures, those theorems that are direct consequences of the field axioms.

If an order relation that satisfies Axioms O−1 through O−3 is introduced as an additional property of a field, then that field is called an **ordered field**. The real number system is an ordered field. If the rational numbers are taken as the elements of a field, it can be shown that the rational number system is also an ordered field.

Exercise Set 0.3

(P)   I.   Prove Theorem 0.2. (Hint: See the suggestion following Theorem 0.2, page 11.)

II.   Fill in the blanks to complete the following proof of Theorem 0.4.
IF: $a, b \in R$,   THEN: $(-a) + (-b) = -(a+b)$.

1. $(-a) + (-b) = (-a) + (-b) + 0$ 1. Ax. __(a)__
2. $= (-a) + (-b) + \left((a+b) + [-(a+b)] \right)$ 2. Ax. __(b)__
3. $= [a + (-a)] + [b + (-b)] + [-(a+b)]$ 3. Ax. R−1, R−2
4. $= 0 + 0 + [-(a+b)]$ 4. Ax. __(c)__
5. $= -(a+b)$ 5. Ax. __(d)__

III.   Fill in the blanks to complete the following proof of Theorem 0.5.
IF: $a \in R$,   THEN: $a \cdot 0 = 0$

Proof By Axiom __(a)__, $0 + 0 = 0$. Multiplying each member of this equation by a (Theorem __(b)__), we have $a(0+0) = a \cdot 0$. But $a(0+0) = a \cdot 0 + a \cdot 0$ by Axiom __(c)__ and $a \cdot 0 = a \cdot 0 + 0$ by Axiom __(d)__. Hence, substituting appropriately, it follows that

$$a \cdot 0 + a \cdot 0 = a \cdot 0 + 0.$$

Applying Theorem __(e)__, it follows that $a \cdot 0 = 0$.

IV.   Construct a proof of Theorem 0.7−I that parallels the proof of Theorem 0.5 above. Begin with $1 + (-1) = 0$ and use $0 = a + (-a)$.

V. Prove Theorem 0.7–II.

VI. Prove Theorem 0.7–III.

VII. Use Theorem 0.2 to prove Theorem 0.8.

VIII. Use Theorem 0.8 to prove Theorem 0.9.

IX. Fill in the blanks to complete the following proof of Theorem 0.10.

IF: $a, b, c \in R$; $c \neq 0$, THEN: $\dfrac{a}{c} + \dfrac{b}{c} = \dfrac{a+b}{c}$.

1. $\dfrac{a}{c} + \dfrac{b}{c} = a \cdot \dfrac{1}{c} + b \cdot \dfrac{1}{c}$ 1. Def. __(a)__

2. $ = (a+b)\dfrac{1}{c}$ 2. Ax. __(b)__

3. $ = \dfrac{a+b}{c}$ 3. Def. __(c)__

X. Prove the following theorem (needed to Prove Theorem 0.11).

"Theorem A: If $b, d \in R$; $b \neq 0$, $d \neq 0$; then $\dfrac{1}{b} \cdot \dfrac{1}{d} = \dfrac{1}{bd}$."

$\left(\text{Hint: } bd \cdot \dfrac{1}{bd} = 1 \cdot 1 \quad \leftrightarrow \quad bd \cdot \dfrac{1}{bd} = b \cdot \dfrac{1}{b} \cdot d \cdot \dfrac{1}{d}. \right)$

XI. Use Theorem A of Exercise X to prove Theorem 0.11.

$\left(\text{Hint: } \dfrac{a}{b} \cdot \dfrac{c}{d} = a \cdot \dfrac{1}{b} \cdot c \cdot \dfrac{1}{d} \quad \leftrightarrow \quad \dfrac{a}{b} \cdot \dfrac{c}{d} = a \cdot c \cdot \dfrac{1}{b} \cdot \dfrac{1}{d}. \right)$

XII. Use Theorem 0.13 and Theorem 0.10 to prove Theorem 0.14.

XIII. Prove Part 2 of Theorem 0.15. (Hint: See the proof of Part 1.)

XIV. Construct a proof of Theorem 0.16–I on the basis of the following "skeleton" of an argument.

If $a > b$, then $b + p = a$ and $p > 0$. Hence, $bc + pc = ac$. Because $c > 0$ and $p > 0$, it follows that $pc > 0$ and we conclude that $ac > bc$.

XV. Construct a proof of Theorem 0.16–II similar to that of Exercise XIV. (Hint: If pc is negative, then $-pc$ is positive.)

0.4   GRAPHS; ABSOLUTE VALUE

The familiar **number line** is based upon the existence of a one-to-one correspondence between the set of real numbers and the points in a line. That is, to each point in a line there is associated exactly one real number called the **coordinate** of the point; to each real number there is associated exactly one point called the **graph** of the number. The point with coordinate zero is called the **origin**. Figure 0.4–1 on page 16 shows a number line on which the points with coordinates $-6, -2, \frac{3}{2}$ and $\sqrt{30}$ are marked. The resulting figure is called the **graph** of $\{-6, -2, \frac{3}{2}, \sqrt{30}\}$. The arrow-head pointing to the right indicates the direction in which coordinates are increasing. That is, if the graph of the number b is to the right of the graph of the number a, then $b > a$ or, equivalently, $a < b$.

Figure 0.4-1

Intervals In Figure 0.4–2, observe that $a < x$ and $x < b$. We say that x is **between** a and b, and write $a < x < b$ or $b > x > a$. The set of all real numbers between a and b is called an **interval** from a to b; the numbers a and b are called **endpoints** of the interval. Intervals are: **closed intervals** if both endpoints are included; **open intervals** if neither endpoint is included; **half-open** or **half-closed** if only one endpoint is included. Figure 0.4–3 shows how each such interval can be specified and graphed. Observe the use of an "open" circle to indicate that an endpoint is *not* included.

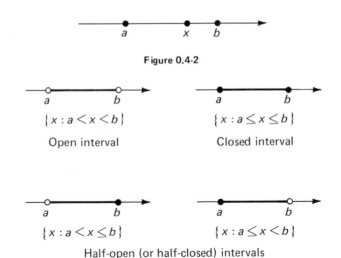

Figure 0.4-2

$\{x : a < x < b\}$

Open interval

$\{x : a \leq x \leq b\}$

Closed interval

$\{x : a < x \leq b\}$

$\{x : a \leq x < b\}$

Half-open (or half-closed) intervals

Figure 0.4-3

We sometimes refer to **unbounded intervals**, as illustrated in Figure 0.4–4.

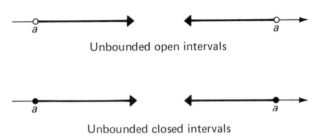

Unbounded open intervals

Unbounded closed intervals

Figure 0.4-4

For certain applications we shall be concerned with unions and inter-
sections of intervals.

EXAMPLE 1   Use set notation to specify each of the following graphs in two
ways: first, using the symbol **U**; second, using the connective **or**.

a.    b.

SOLUTION

    a. $\{x: x < -6\} \cup \{x: x \geq 3\}$; $\{x: x < -6$ or $x \geq -3\}$.

    b. $\{x: -4 \leq x \leq -1\} \cup \{x: x > 3\}$; $\{x: -4 \leq x \leq -1$ or $x > 3\}$.

EXAMPLE 2   Graph each intersection and specify the resulting interval in set
notation.

    a. $\{x: x < 2\} \cap \{x: x \geq -3\}$.

    b. $\{x: x > 2\} \cap \{x: x \geq -3\}$.

    c. $\{x: x > 2\} \cap \{x: x \leq -3\}$.

SOLUTION   For clarity, we first sketch each individual interval above the
number line and then sketch the actual graph of the intersection on the
number line.

a.

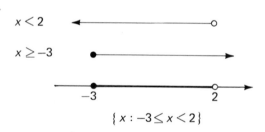

b.

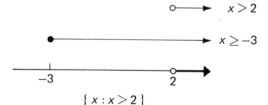

c.

Absolute Value   In Figure 0.4–5, observe that the distance from the origin to the point with coordinate −6 is six units. As you may remember, this distance can be denoted by the symbol |−6|, read "the absolute value of −6."

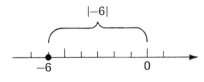

Figure 0.4-5

Definition 0.5   For each $r \in R$, the **absolute value** of r is:
$|r| = r$, if $r \geq 0$; $|r| = -r$ if $r < 0$.

One consequence of Definition 0.5 is a theorem with the simple geometric interpretation that the two points with coordinates r and $-r$ are equidistant from the origin.

Theorem 0.18   If $r \in R$, then $|r| = |-r|$.   **(P)**

As noted above, the expression $|r|$ may be interpreted as the distance between the origin and the point with coordinate r. Similarly, the expression $|a - b|$ may be interpreted as *the distance between the points with respective coordinates a and b*. The next theorem implies that we may use either $|a - b|$ or $|b - a|$ to compute the distance.

Theorem 0.19   If $a, b \in R$, then $|a - b| = |b - a|$.   **(P)**

The final two theorems of this section are concerned with the absolute value of a product or a quotient.

Theorem 0.20   If $a, b \in R$, then $|ab| = |a| \cdot |b|$.   **(P)**

Theorem 0.21   If $a, b \in R$ and $b \neq 0$, then $\left|\dfrac{a}{b}\right| = \dfrac{|a|}{|b|}$.   **(P)**

Exercise Set 0.4
Graph each of the following sets and specify whether the graph is an open interval, a closed interval, a half-open interval, an unbounded open interval or an unbounded closed interval.

1. $\{x: -2 < x \leq 3\}$
2. $\{x: -3 \leq x < 2\}$
3. $\{x: -2 \leq x \leq 3\}$
4. $\{x: -3 < x < 2\}$
5. $\{x: -2 < x\}$
6. $\{x: x \geq 3\}$
7. $\{x: x \leq 3\}$
8. $\{x: -5 < x\}$

*Use set notation to specify each of the following graphs in two ways: using the symbol **U**; using the connective **or**. (See Example 1.)*

9.

10.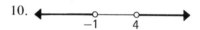

11.

−3 2 6

12.

−2 2 5

13.

1 5 8 11

14.

−6 −2 0 4

15.

−8 −5 −1 3

16.

−11 −9 −5 −3

Graph each intersection and specify the resulting interval in set notation. (See Example 2.)

17. $\{x: x \geq -5\} \cap \{x: x < 5\}$ 18. $\{x: x \leq 5\} \cap \{x: x > -5\}$

19. $\{x: x < -6\} \cap \{x: x \leq -1\}$ 20. $\{x: x \geq -6\} \cap \{x: x > -1\}$

21. $\{x: x \geq 4\} \cap \{x: x \leq 1\}$ 22. $\{x: x \leq -6\} \cap \{x: x > 0\}$

23. $\{x: x \geq 1\} \cap \{x: 2 \leq x \leq 6\}$ 24. $\{x: x < 4\} \cap \{x: -5 \leq x \leq -1\}$

Prove (cite a definition, theorem or axiom for each step): If $x < 0$, then

25. $|-6(-x)| = -6x$

26. $\left|\dfrac{-3}{x}\right| = \dfrac{-3}{x}$

Specify a replacement set for x such that

27. $\dfrac{|x|}{x} = 1$ 28. $\dfrac{|x|}{x} = -1$

(P)   I. Prove Theorem 0.18. (Hint: By Axiom O-1, there are three cases to consider: $r > 0$, $r = 0$ and $r < 0$. Apply Definition 0.5 in each case.)

II. Prove Theorem 0.19 (Hint: Use Theorem 0.18, substituting $a - b$ for r.)

To prove Theorems 0.20 and 0.21, consider four cases: i. $a \geq 0$ and $b \geq 0$; ii. $a \geq 0$ and $b < 0$; iii. $a < 0$ and $b \geq 0$; iv. $a < 0$ and $b < 0$. (Of course, for Theorem 0.21 we have that $b \neq 0$.)

III. Prove Theorem 0.20.

IV. Prove Theorem 0.21.

If x is any nonzero real number then, by Theorem 0.17, x^2 is a positive number. It follows that there are no *real* numbers x such that $x^2 = -1$. Historically, the desire to find numbers x such that $x^2 = -1$ led to the extension of the real number system to a "larger" system called the **complex number system** C. One way to construct such an extension is to begin by introducing a new number denoted by the symbol i, called the **imaginary unit**, with the property that

$$i \cdot i = i^2 = -1.$$

Next, we introduce symbols of the form $a + bi$ to represent members of the set C of complex numbers, as follows:

$$C = \{a + bi: a, b \in R\}.$$

We call a the *real part* and b the *imaginary part* of the complex number $a + bi$. For example, $6 + (-2)i$ is a complex number with real part 6 and imaginary part -2. Customarily, the number $6 + (-2)i$ is written in the form $6 - 2i$.

Equality between complex numbers and the sum and product of complex numbers is defined in terms of equality, addition and multiplication of real numbers. (Division is reviewed in Section 0.8.)

Definition 0.6   For real numbers a, b, c and d:

   I $a + bi = c + di$ if and only if $a = c$ and $b = d$.

  II $(a + bi) + (c + di) = (a + c) + (b + d)i$.

 III $(a + bi)(c + di) = (ac - bd) + (ad + bc)i$.

EXAMPLE 1

  a. $11 + yi = x - 2i$ if and only if $x = 11$ and $y = -2$.

  b. $(3 + 2i) + (4 - 3i) = (3 + 4) + [2 + (-3)]i = 7 - i$.

  c. $(3 + 2i)(4 - 3i) = [3 \cdot 4 - 2(-3)] + [3(-3) + 2 \cdot 4]i = 18 - i$.

Note that $i^2 = -1$ is a consequence of Definition 0.6–III, as follows:

$$i^2 = (0 + i)(0 + i) = (0 \cdot 0 - 1 \cdot 1) + (0 \cdot 1 + 0 \cdot 1)i = -1.$$

The Field of Complex Numbers   Definition 0.6 enables us to prove that the complex numbers satisfy field axioms R–1 through R–9. The next two examples illustrate two such proofs. Some other proofs are considered in the exercises.

EXAMPLE 2 Prove that addition is commutative in C.

SOLUTION
$$\begin{aligned}(a+bi)+(c+di)&=(a+c)+(b+d)i \quad \text{(Definition 0.6–II)}\\&=(c+a)+(d+b)i \quad \text{(Axiom R–1, commutativity)}\\&=(c+di)+(a+bi) \quad \text{(Definition 0.6–II)}\end{aligned}$$

EXAMPLE 3 Prove that $1+0i$ is the multiplicative identity element in C.

SOLUTION We must show that $(a+bi)(1+0i)=a+bi$. Applying Definition 0.6–III and Theorem 0.5, we have

$$(a+bi)(1+0i)=(a-0)+(0+b)i=a+bi.$$

Henceforth, we shall accept that *the complex numbers C constitute a field under addition and multiplication*. By the closure properties, it follows that sums and products of complex numbers are complex numbers. All of the field axioms, together with all those theorems in R that are not consequences of any order relations, are applicable in C. For example, it is usually simpler to compute the product of two complex numbers by the distributivity axiom (R–9) rather than by Definition 0.6–III. Thus,

$$\begin{aligned}(3+2i)(4-3i)&=3(4-3i)+2i(4-3i)\\&=12-9i+8i-6i^2\\&=12-i-6(-1)\\&=18-i.\end{aligned}$$

Note that i^2 in the second line is replaced by -1 in the third line, and we obtain the same result as in Example 1c.

Subsets of C Consider a one-to-one correspondence between the elements of the two sets

$$C_1=\{z\colon z=a+0i,\ a\in R\} \quad \text{and} \quad R.$$

In particular, if $z_1=a_1+0i$ corresponds to a_1 and $z_2=a_2+0i$ corresponds to a_2, then

$$z_1+z_2=(a_1+a_2)+0i \quad \text{and} \quad z_1z_2=a_1a_2+0i.$$

Observe that z_1+z_2 corresponds to a_1+a_2 and z_1z_2 corresponds to a_1a_2. That is, the sum or the product of any two elements in C_1 corresponds to the sum or the product, respectively, of the corresponding elements in R. Such a correspondence is technically known as an *isomorphism*. For our purposes, we adopt the somewhat simpler viewpoint that $a+0i$ and a are different names for the same number because they "behave" in the same manner under addition and multiplication. The significance of this viewpoint is that the set R of real numbers can now be regarded as a proper subset of the set C of complex numbers. Hence, every real number is a complex number, but not every complex number is a real number. To avoid ambiguities, we make the following classification. For $a, b \in R$: numbers of the form $a + bi$ with $b \neq 0$ are called **imaginary numbers**; numbers of the form bi with $b \neq 0$ are called **pure imaginary numbers**; numbers of the form $a + 0i$, or simply a, are real numbers.

EXAMPLE 4

a. $4 + 3i$ and $-7i$ are imaginary numbers.

b. $-7i$ and i are pure imaginary numbers.

c. $-7i$, $4 + 3i$, 14 and 0 are complex numbers.

In Section 0.8 we consider the concept of finding complex numbers x such that $x^2 = k$, where k is a *negative* number.

Exercise Set 0.5

Specify the real part and the imaginary part of each complex number.

1. $2 + 3i$ 2. $3 - 4i$
3. $-3 + 5i$ 4. $-6 - 2i$

Find the values of x and y for which each sentence is true.

5. $6 + xi = y - 2i$ 6. $x + 2i = y + yi$
7. $-x + yi = x - 3i$ 8. $x^2 - 6i = 4 + 2yi$

Perform the indicated operations and express the results as a complex number in $a + bi$ form.

9. $(2 - 3i) + (-4 + 6i)$ 10. $(-8 + 2i) + (6 - 5i)$
11. $(3 + 4i) - (-6 + 2i)$ 12. $(-5 - 2i) - (6 - 7i)$
13. $(2 + 3i)(2 - 3i)$ 14. $(3 - i)(2 + i)$
15. $(6 + 5i) \cdot 2i$ 16. $(3 - 4i)(-2i)$
17. $(-i)(-i)(-i)(-i)$ 18. $i \cdot i \cdot i \cdot i$
19. $3i \cdot 3i$ 20. $(-3i)(-3i)$
21. $(-2i)(-2i)(-2i)(-2i)$ 22. $2i \cdot 2i \cdot 2i \cdot 2i$

In Exercises 23 through 27, use Definition 0.6.

23. Prove: Multiplication in C is commutative.
24. Prove: $0 + 0i$ is the additive identity element in C.
25. Prove: $-a - bi$ is the additive inverse of $a + bi$.
26. Prove: The multiplicative inverse of $a + bi$ is

$$\frac{a}{a^2 + b^2} + \frac{-b}{a^2 + b^2}i, \qquad \text{where } a \text{ and } b \text{ are not both zero.}$$

27. Prove: If $a, b \in R$, then $(a + bi)(a - bi)$ is a real number.

0.6   RATIONAL EXPONENTS

Indicated products involving two or more of the same factors, such as $5 \cdot 5 \cdot 5 \cdot 5$ can be written in concise form by the use of *exponents*.

Definition 0.7   For each real number b and natural number n,

$$b^n = \overbrace{b \cdot b \cdot b \cdot \cdots \cdot b}^{n \text{ factors}}.$$

Thus, $5 \cdot 5 \cdot 5 \cdot 5$ may be written as 5^4. In the notation of Definition 0.7, the number b is called the **base**, n is called the **exponent** and b^n is called the nth **power of** b.

Definition 0.7 refers only to *natural* number exponents. The next definition extends the concept of exponents to include *integral* exponents.

Definition 0.8   For each nonzero real number b and natural number n,

$$b^0 = 1; \qquad b^{-n} = \frac{1}{b^n}.$$

Definition 0.8 asserts than any nonzero real number to the power zero is equal to 1; no meaning is given to 0^0. The definition also asserts that a negative integral exponent may be interpreted as indicating "take the reciprocal of."

EXAMPLE 1 $5^0 \cdot 4^{-2} = 1 \cdot \dfrac{1}{4^2} = \dfrac{1}{16}.$

Principal n^{th} Roots   If r is a real number such that $r^2 = b$, then r is called a *square root* of b and b is the *square* of r. If $r^3 = b$, then r is a *cube root* of b and b is the *cube* of r. In general, if $r^n = b$, where n is a natural number and $n > 1$, then r is an nth *root* of b and b is an nth *power* of r.

EXAMPLE 2

a.   5 and -5 are square roots of 25 because $5^2 = 25$ and $(-5)^2 = 25$.

b.   2 and -2 are fourth roots of 16 because $2^4 = 16$ and $(-2)^4 = 16$.

c.   3 is a cube root of 27 because $3^3 = 27$.

d. $2i$ and $-2i$ are square roots of -4 because $(2i)^2 = 4i^2 = -4$ and $(-2i)^2 = 4i^2 = -4$.

In Example 2: parts (a) and (b) suggest that a real number may have more than one real n^{th} root; part (c) suggests that a real number may have only one real n^{th} root; part (d) suggests that a real number may not have any real n^{th} roots. The actual situation with respect to n^{th} roots of real numbers may be summarized by the following five statements. For each $n \in N$, $n \geq 2$:

1. Each nonzero real number b has exactly n complex n^{th} roots—some, all, or none of which may be real numbers. That is, b has exactly two square roots, three cube roots, four fourth roots, . . . , n n^{th} roots.

2. Each real number b has exactly one real *odd** n^{th} root—any other odd n^{th} roots are imaginary numbers.

3. Each *positive* number b has exactly two real *even* n^{th} roots, each the additive inverse of the other—any other even n^{th} roots are imaginary numbers.

4. Each *negative* number b has no real even n^{th} roots—all of its even n^{th} roots are imaginary numbers.

5. For any $n \in N$, zero has exactly one n^{th} root, namely, zero.

Definition 0.9   For each $n \in N$ and $b \in R$, the **principal n^{th} root** of b, denoted by $b^{1/n}$, is:

I   the positive real n^{th} root of b if b is positive;

II   0 if $b = 0$;

III   the real n^{th} root of b if b is negative and n is odd.

From Statement 4 above, a negative number has no real *even* n^{th} roots. Hence, in the case that b is negative and n is even we say that $b^{1/n}$ *is not defined*, or that $b^{1/n}$ *is not a real number*. In all other cases, Definition 0.9 implies that $b^{1/n}$ names a unique real number. Henceforth, unless specified otherwise, when a symbol such as $b^{1/n}$ is used it is to be understood that the case "$b < 0$ and n even" is excluded.

EXAMPLE 3

a. $25^{1/2} = 5$; $81^{1/4} = 3$. By Definition 0.9, the principal n^{th} roots of positive numbers must be *positive*.

b. $-81^{1/4} = -(81)^{1/4} = -3$; $(-81)^{1/4}$ is not defined.

c. $(-27)^{1/3} = -3$.

Rational Exponents   We can now define rational exponents and assign a meaning to the symbol $b^{m/n}$.

Definition 0.10   For each $n \in N$, $m \in J$ and real number b such that $b^{1/n}$ is a real number,

$$b^{m/n} = (b^{1/n})^m = (b^m)^{1/n}.$$

---

*"Odd n^{th} roots" means that n is an odd number; "even n^{th} roots" means that n is an even number.

Essentially, Definition 0.10 enables us to view the number $b^{m/n}$ as either the mth power of the nth root of b, or as the nth root of the mth power of b, provided that the nth root of b is a real number.

EXAMPLE 4

a. $9^{3/2} = (9^{1/2})^3 = 3^3 = 27$; or $9^{3/2} = (9^3)^{1/2} = 729^{1/2} = 27$.

b. $(-8)^{2/3} = [(-8)^{1/3}]^2 = (-2)^2 = 4$; or $(-8)^{2/3} = [(-8)^2]^{1/3} = 64^{1/3} = 4$.

c. $8^{-2/3} = \dfrac{1}{8^{2/3}} = \dfrac{1}{(8^{1/3})^2} = \dfrac{1}{4}$; or $8^{-2/3} = \dfrac{1}{8^{2/3}} = \dfrac{1}{(8^2)^{1/3}} = \dfrac{1}{4}$.

d. Definition 0.10 does not apply to $(-9)^{3/2}$ because $(-9)^{1/2}$ is not a real number.

Properties of Rational Exponents   Operations on expressions that involve bases raised to rational number exponents (including integral and natural number exponents) are described in the next theorem.

Theorem 0.22 (Properties of Rational Exponents)   If p and q are rational numbers and b and c are nonzero real numbers such that b^p, b^q, c^p and c^q are real numbers, then

I $b^p \cdot b^q = b^{p+q}$

II $\dfrac{b^p}{b^q} = b^{p-q}$

III $(b^p)^q = b^{pq}$

IV $(bc)^p = b^p c^p$

V $\left(\dfrac{b}{c}\right)^p = \dfrac{b^p}{c^p}$.

EXAMPLE 5

a. $2^3 \cdot 2^4 = 2^{3+4} = 2^7$, or 128.

b. $(5^3)^{4/3} = 5^{3(4/3)} = 5^4$, or 625.

c. $\dfrac{16^{3/4}}{16^{1/2}} = 16^{3/4-(1/2)} = 16^{1/4}$, or 2.

EXAMPLE 6   Write the expression $\dfrac{xy^3 z^{-2/3}}{x^{1/2} y^2 z^{-1}}$ so that each variable occurs only once and all exponents are positive. Assume that replacement sets for x, y and z are such that Theorem 0.22 applies.

SOLUTION

$$\frac{xy^3 z^{-2/3}}{x^{1/2} y^2 z^{-1}} = x^{1-1/2} \cdot y^{3-2} \cdot z^{(-2/3)-(-1)} = x^{1/2} y z^{1/3}.$$

A sometimes troublesome case is that of an nth root of an nth power of a real number b, where n is an even natural number. For example, consider $(b^2)^{1/2}$. By Theorem 0.17, b^2 is a positive number and, by Definition 0.9, it follows that $(b^2)^{1/2}$ names the *positive* square root of b^2. Now, if $b > 0$, then the square root of b^2 is equal to b; if $b = 0$, then the square root of b^2 is equal to b (or zero); if $b < 0$ then $-b > 0$ and the square root of b^2 is equal to $-b$. That is,

$$(b^2)^{1/2} = b \text{ if } b \geq 0; \qquad (b^2)^{1/2} = -b \text{ if } b < 0.$$

Comparing these results with Definition 0.5 (absolute value), we see that for all $b \in R$,

$$(b^2)^{1/2} = |b|.$$

More generally, if n is an *even* natural number and $b \in R$, then

$$(b^n)^{1/n} = |b|.$$

EXAMPLE 7

   a. $[(-2)^4]^{1/4} = |-2| = 2.$

   b. $[16(x-3)^2]^{1/2} = 16^{1/2}[(x-3)^2]^{1/2} = 4|x-3|.$

Exercise Set 0.6

Simplify.

1. $4^2 \cdot 2^{-3}$ 2. $5^3 \cdot 3^{-3}$ 3. $(-4)^0(-3)^{-2}$ 4. $6^0(-5)^3$

5. $9^{-2} + 3^{-3}$ 6. $(-3)^2 - (-3)^{-2}$ 7. $25^0 - 5^{-2}$ 8. $3^{-3} + 3^{-2}$

9. $\dfrac{2^{-3}}{3^{-2}}$ 10. $\left(\dfrac{2}{5}\right)^{-2}$ 11. $\left(\dfrac{5}{2}\right)^{-3}$ 12. $\dfrac{4^{-2}}{4^0}$

Find the indicated real root(s) of each number, if any.

13. 25; square root   14. 121; square root

15. 27; cube root   16. -64; cube root

17. 81; fourth root   18. -16; fourth root

19. -32; fifth root   20. 243; fifth root

21. Show that $3i$ and $-3i$ are all square roots of -9.

22. Show that $4, -4, 4i$ and $-4i$ are all fourth roots of 256.

Specify how many real and imaginary square roots, cube roots and fourth roots there are for each of the following numbers.

23. -114 24. -206 25. 114   26. 206

Simplify.

27. $8^{1/3}$ 28. $16^{1/4}$ 29. $9^{1/2}$ 30. $-32^{1/5}$

31. $(-64)^{1/3}$ 32. $-81^{1/2}$ 33. $-64^{1/2}$ 34. $(-64)^{1/2}$

35. $16^{3/4}$ 36. $32^{2/5}$ 37. $243^{3/5}$ 38. $(-27)^{4/3}$

39. $25^{-3/2}$ 40. $49^{-1/2}$ 41. $(-64)^{-4/3}$ 42. $(-8)^{-2/3}$

Write each of the following expressions so that each variable occurs at most once and all exponents are positive. Assume that replacement sets for x and y are such that Theorem 0.22 applies.

43. $(x^2y^{-2})^3$
44. $(6x^{-1})^3$
45. $(9x^4y^2)^{1/2}$

46. $(64x^3y^{-9})^{2/3}$
47. $\left(\dfrac{81x^4}{y^2}\right)^{-1/2}$
48. $\left(\dfrac{27x^4y^{-2}}{64x^{-2}y}\right)^{-2/3}$

49. $\dfrac{(9x^2)^{2/3}}{(3x)^{1/3}}$
50. $\left(\dfrac{125x^3y}{8xy^2}\right)^{4/3}$

Simplify as in Example 7.

51. $(x^2)^{1/2}$
52. $[(-x)^4]^{1/4}$
53. $[(-9x)^2]^{1/2}$

54. $(16x^4)^{1/4}$
55. $[9(x+5)^2]^{1/2}$
56. $(x^2y^2)^{1/2}$

0.7   EXPRESSIONS INVOLVING RADICALS

An alternate notation for the *principal* n^{th} *root of a real number* involves the radical symbol $\sqrt{}$.

Definition 0.11   For each natural number n, $n \geq 2$, and $b^{1/n} \in R$,

$$\sqrt[n]{b} = b^{1/n}.$$

The number b is called the **radicand**, n is the **index** and $\sqrt{b}$ is called a **radical of order** n. Symbols such as $\sqrt{b}$, where no index appears, are understood to mean radicals of *order two* (square root).

Theorem 0.23   If $m, n \in N$, $n \geq 2$, and b, c, $\sqrt[n]{b}$ and $\sqrt[n]{c}$ are real numbers, then

  I $\sqrt[n]{b^n} = b$ (*n* odd),

  II $\sqrt[n]{b^n} = |b|$ (*n* even),

  III $\sqrt[n]{b^m} = (\sqrt[n]{b})^m$,

  IV $\sqrt[n]{b} \cdot \sqrt[n]{c} = \sqrt[n]{bc}$,

  V $\dfrac{\sqrt[n]{b}}{\sqrt[n]{c}} = \sqrt[n]{\dfrac{b}{c}}$. $(c \neq 0)$.

A radical expression is said to be in **simplest form** if it satisfies each of the following criteria:

1. A radicand does not include any factor with an exponent greater than or equal to the index.

2. A radicand does not include any fractions.

3. No radicals appear in denominators of fractions.

Theorem 0.23 is used to transform radical expressions into simplest form.

EXAMPLE 1   If $x > 0$ and $y > 0$, then each of the following radical expressions can be simplified as shown.

a. $\sqrt{16x^3y^2} = \sqrt{16x^2y^2}\,\sqrt{x} = 4xy\sqrt{x}.$

b. $\sqrt[3]{135x^2y^5} = \sqrt[3]{27y^3}\,\sqrt[3]{5x^2y^2} = 3y\sqrt[3]{5x^2y^2}.$

c. $3\sqrt[4]{\dfrac{x^2}{4y}} = 3\sqrt[4]{\dfrac{x^2 \cdot 4y^3}{4y \cdot 4y^3}} = \dfrac{3\sqrt[4]{4x^2y^3}}{\sqrt[4]{16y^4}} = \dfrac{3\sqrt[4]{4x^2y^3}}{2y}.$

d. $\dfrac{x^2y}{\sqrt[3]{x^2y}} = \dfrac{x^2y\,\sqrt[3]{xy^2}}{\sqrt[3]{x^2y}\,\sqrt[3]{xy^2}} = \dfrac{x^2y^3\sqrt[3]{xy^2}}{\sqrt[3]{x^3y^3}} = x\sqrt[3]{xy^2}.$

EXAMPLE 2   Let x and y be real numbers (not necessarily positive) such that each of the following radicals names a real number. Then:

a. $\sqrt{16x^3y^2} = \sqrt{16x^2y^2}\,\sqrt{x} = 4x|y|\,\sqrt{x}.$ Note: If $\sqrt{16x^3y^2}$ is to name a real number, then x^3 must be *positive*, which implies that $x > 0$.

b. $\sqrt[4]{32(x-3)^4} = \sqrt[4]{2^4 \cdot 2(x-3)^4} = 2|x-3|\,\sqrt[4]{2}.$

Sums and Differences   If two or more terms of a sum or difference have a radical expression as a common factor, then the distributivity axiom (Axiom R–9) can be applied to rewrite that sum or difference with fewer terms, or as a single term. For example

$$5\sqrt{3} - \sqrt{3} = [5 + (-1)]\,\sqrt{3} = 4\sqrt{3}.$$

Often, one or more of the terms may have to be simplified first.

EXAMPLE 3

a. $2\sqrt{18} - \sqrt{32} + 5\sqrt{2} = 2\sqrt{9}\,\sqrt{2} - \sqrt{16}\,\sqrt{2} + 5\sqrt{2}$
$$= 6\sqrt{2} - 4\sqrt{2} + 5\sqrt{2}$$
$$= 7\sqrt{2}.$$

b. $4\sqrt[3]{27x^4} + x\sqrt[3]{8x} = 4\sqrt[3]{27x^3}\,\sqrt[3]{x} + x\sqrt[3]{8}\,\sqrt[3]{x}$
$$= 12x\sqrt[3]{x} + 2x\sqrt[3]{x}$$
$$= 14x\sqrt[3]{x}.$$

Products and Quotients   The distributivity axiom, together with Theorem 0.23, can be used to express certain products as sums or differences in which radical expressions are in simplest form.

EXAMPLE 4

a. $\sqrt{5}(\sqrt{2} - 2\sqrt{15}) = \sqrt{5}\,\sqrt{2} - \sqrt{5} \cdot 2\sqrt{15}$
$$= \sqrt{10} - 2\sqrt{75}$$
$$= \sqrt{10} - 10\sqrt{3}.$$

b. $(x - \sqrt[3]{5})(x + 2\sqrt[3]{5}) = (x - \sqrt[3]{5})x + (x - \sqrt[3]{5}) \cdot 2\sqrt[3]{5}$
$$= x^2 - x\sqrt[3]{5} + 2x\sqrt[3]{5} - 2\sqrt[3]{25}$$
$$= x^2 + x\sqrt[3]{5} - 2\sqrt[3]{25}.$$

When a fraction that includes a radical in the denominator is rewritten in an equivalent fractional form in which the denominator does not include a radical, we say that the given denominator has been *rationalized*. A similar remark applies to rationalizing a numerator. Two expressions of the form $a + b$ and $a - b$ are sometimes called **conjugates**. For example, each of the pairs of expressions

$$a + \sqrt{b} \ \text{ and } \ a - \sqrt{b}; \qquad \sqrt{a} + \sqrt{b} \ \text{ and } \ \sqrt{a} - \sqrt{b};$$

is a pair of conjugates. The process of rationalizing a denominator (or numerator) of a fraction involves multiplication by a conjugate expression, as in the next example.

EXAMPLE 5

a. Rationalizing a denominator:
$$\frac{2 + \sqrt{3}}{3 - \sqrt{3}} = \frac{(2 + \sqrt{3})(3 + \sqrt{3})}{(3 - \sqrt{3})(3 + \sqrt{3})} = \frac{9 + 5\sqrt{3}}{6}.$$

b. Rationalizing a numerator:
$$\frac{x - \sqrt{7}}{2x + \sqrt{7}} = \frac{(x - \sqrt{7})(x + \sqrt{7})}{(2x + \sqrt{7})(x + \sqrt{7})} = \frac{x^2 - 7}{2x^2 + 3x\sqrt{7} + 7}.$$

Exercise Set 0.7
Find each root, if it exists.

1. $\sqrt{36}$
2. $\sqrt{49}$
3. $\sqrt[3]{27}$
4. $\sqrt[3]{64}$
5. $\sqrt{-64}$
6. $\sqrt{-16}$
7. $\sqrt[6]{64}$
8. $\sqrt[4]{16}$
9. $\sqrt[3]{-64}$
10. $\sqrt[3]{-27}$
11. $\sqrt{\dfrac{25}{121}}$
12. $\sqrt{\dfrac{81}{144}}$

Simplify. Assume that $x > 0$ and $y > 0$.

13. $\sqrt{32x^2y^4}$
14. $\sqrt{72x^3y}$
15. $\sqrt[3]{32x^2y^4}$
16. $\sqrt[3]{72x^3y}$
17. $\sqrt[4]{16x^6y^5}$
18. $\sqrt[4]{32x^6y^3}$
19. $\dfrac{2x}{\sqrt{18x}}$
20. $\dfrac{\sqrt{25x^3}}{\sqrt{125x^2}}$
21. $\dfrac{64y}{\sqrt[3]{4y^3}}$
22. $\sqrt[3]{\dfrac{5}{6x^2}}$
23. $\dfrac{\sqrt{xy}\,\sqrt{x^3y}}{\sqrt{xy^3}}$
24. $\dfrac{\sqrt[3]{xy}\,\sqrt[3]{x^3y}}{\sqrt[3]{xy^3}}$

Simplify. Assume that $x, y \in R$ and each radical expression names a real number.

25. $\sqrt{144x^2y^3}$
26. $\sqrt{50x^2y^2}$
27. $\sqrt{18(x+4)^2}$
28. $\sqrt{x^2(y-2)^2}$
29. $\sqrt[4]{16x^4}$
30. $\sqrt[4]{x^2(y-2)^4}$

Simplify.

31. $\sqrt{2x} + \sqrt{32x} - \sqrt{18x}$

32. $\sqrt[3]{54x^3} - 5x\sqrt[3]{16} + 7\sqrt[3]{2x^3}$

33. $\sqrt{2}(\sqrt{3} - \sqrt{6})$

34. $\sqrt{5}(\sqrt{2} + \sqrt{5})$

35. $(\sqrt{5} - \sqrt{3})(\sqrt{5} + 3\sqrt{3})$

36. $(7 - \sqrt{3})(7 + \sqrt{3})$

37. $(x + \sqrt{y})(x - \sqrt{y})$

38. $(\sqrt{a} + \sqrt{b})(\sqrt{a} - \sqrt{b})$

39. $(\sqrt[3]{3x} + 2)(\sqrt[3]{9x^2} - 2\sqrt[3]{3x} + 4)$

40. $(3 - \sqrt[3]{2x})(9 + 3\sqrt[3]{2x} + \sqrt[3]{4x^2})$

Rationalize the denominator.

41. $\dfrac{2x - \sqrt{5}}{x + \sqrt{5}}$

42. $\dfrac{\sqrt{x} + \sqrt{3}}{2\sqrt{x} - 5\sqrt{3}}$

Rationalize the numerator.

43. $\dfrac{3 - \sqrt{x}}{\sqrt{x} + 5}$

44. $\dfrac{x + \sqrt{7}}{x - \sqrt{7}}$

45. Show that each of the numbers $-1 + \sqrt{3}\,i$ and $-1 - \sqrt{3}\,i$ is a cube foot of 8.

0.8 FURTHER PROPERTIES OF COMPLEX NUMBERS

In Section 0.5 we introduced the imaginary unit i as a number such that $i^2 = -1$. It follows that i is a square root of -1, that is, $i = \sqrt{-1}$. The number i may be used to define square roots of negative numbers.

Definition 0.12   For each positive real number b, **the square root of $-b$** is given by

$$\sqrt{-b} = \sqrt{b}\,i.$$

EXAMPLE 1

a. $\sqrt{-16} = \sqrt{16}\,i = 4i$; $-\sqrt{-16} = -\sqrt{16}\,i = -4i$.

b. $2\sqrt{-5} = 2\sqrt{5}\,i$ or $2i\sqrt{5}$.*

Compare the following two chains of equations:

$$\sqrt{-4}\,\sqrt{-9} = \sqrt{36} = 6; \qquad \sqrt{-4}\,\sqrt{-9} = 2i \cdot 3i = -6.$$

Clearly, one of the chains must be incorrect. In fact, the lefthand chain is wrong because of a misapplication of Theorem 0.23, which requires that both $\sqrt{a}$ and $\sqrt{b}$ name *real* numbers in order for $\sqrt{a}\,\sqrt{b}$ to equal $\sqrt{ab}$. Hence, operations between expressions involving square roots of negative numbers must be handled with care. In particular, if $b > 0$, then expressions such as $\sqrt{-b}$ should be changed to $\sqrt{b}\,i$ (or $i\sqrt{b}$) form *before* any further operations are undertaken.

*We write i in front of a radical factor so that i does not appear to be under the radical.

EXAMPLE 2

a. $\sqrt{-25}+\sqrt{-49}=i\sqrt{25}+i\sqrt{49}=5i+7i=12i.$

b. $\sqrt{-2}(2+\sqrt{-3})=i\sqrt{2}(2+i\sqrt{3})=2i\sqrt{2}+i^2\sqrt{6}=-\sqrt{6}+2i\sqrt{2}.$

c. $\dfrac{\sqrt{-10}}{\sqrt{5}}=\dfrac{i\sqrt{10}}{\sqrt{5}}=i\sqrt{\dfrac{10}{5}}=i\sqrt{2}.$

Each of the numbers $a+bi$ and $a-bi$ is called the **complex conjugate** or, simply, the **conjugate** of the other. Thus, $3+2i$ and $3-2i$ are conjugates of each other. A method for finding the quotient of two imaginary numbers depends upon the fact that the product of a complex number and its conjugate is a real number, which can be established as follows:

$$(a+bi)(a-bi)=a^2-abi+abi-b^2i^2=a^2+b^2.$$

Because sums and products of real numbers are real numbers (closure properties), we have that a^2+b^2 is a real number.

A method for finding the $a+bi$ form of the quotient of two imaginary numbers involves multiplying by the conjugate of the divisor.

EXAMPLE 3

$$\frac{2+i}{3-2i}=\frac{(2+i)(3+2i)}{(3-2i)(3+2i)}=\frac{4+7i}{13}, \text{ or } \frac{4}{13}+\frac{7}{13}i.$$

If z is a complex number and $z\neq0+0i$, then the multiplicative inverse, or reciprocal, of z is given by $\dfrac{1}{z}$. To obtain the $a+bi$ form of the reciprocal of z, we proceed as in the next example.

EXAMPLE 4   Find the $a+bi$ form of the reciprocal of $1-4i$.

SOLUTION

$$\frac{1}{1-4i}=\frac{1(1+4i)}{(1-4i)(1+4i)}=\frac{1+4i}{17}, \qquad \text{or} \qquad \frac{1}{17}+\frac{4}{17}i.$$

You should verify that $(1-4i)\left(\dfrac{1}{17}+\dfrac{4}{17}i\right)=1.$

Recall, from Section 0.5, that the complex numbers constitute a field and that any real number theorem that is a consequence only of Axioms R-1 through R-9 is also applicable to complex numbers. In particular, Theorem 0.22, is so applicable. We consider only integral exponents.

EXAMPLE 5

a. $i^4\cdot i^3=i^7=(i^2)^3\cdot i=(-1)^3\cdot i,$ or $-i.$

b. $[(2i)^3]^2=(2i)^6=64i^6=-64.$

c. $\dfrac{1}{(2+i)^{-2}}=(2+i)^2=3+4i.$

Exercise Set 0.8

Express each radical expression in bi form.

1. $\sqrt{-36}$ 2. $\sqrt{-144}$ 3. $-\sqrt{-49}$ 4. $\sqrt{-50}$

Express each imaginary number in radical form.

5. $8i$ 6. $11i$ 7. $-16i$ 8. $9\sqrt{3}\,i$

Simplify and express each answer in $a + bi$ form.

9. $\sqrt{-9} + \sqrt{-16}$ 10. $\sqrt{-25} - \sqrt{-36}$

11. $\sqrt{-98} + \sqrt{-8}$ 12. $\sqrt{-12} - \sqrt{-27}$

13. $\sqrt{-2}(\sqrt{-8} - \sqrt{-32})$ 14. $\sqrt{-3}(\sqrt{-6} + \sqrt{-3})$

15. $(\sqrt{-5} + \sqrt{-2})(\sqrt{-5} + \sqrt{-2})$ 16. $(\sqrt{-8} - \sqrt{-6})(\sqrt{-8} + \sqrt{-6})$

17. $(\sqrt{-5} + 2\sqrt{-3})(2\sqrt{-5} - \sqrt{-3})$ 18. $(3 - \sqrt{-5})(2 + 3\sqrt{-5})$

Write the conjugate of each of the following complex numbers.

19. $3 + 4i$ 20. $3 - 4i$ 21. $-4i$ 22. 3

Find the $a + bi$ form of the reciprocal of each complex number.

23. i 24. $-i$ 25. $1 - 2i$ 26. $3 + i$

Compute and express each complex number in $a + bi$ form.

27. $\dfrac{2}{3 - i}$ 28. $\dfrac{-4}{2 + i}$ 29. $\dfrac{3 - i}{3 + i}$

30. $\dfrac{6 - 2i}{1 + i}$ 31. $\dfrac{3 + 4i}{5 - 6i}$ 32. $\dfrac{7 - 3i}{-2i}$

33. i^9 34. i^{-9} 35. $i^3 \cdot i^2$

36. $[(2i)^3]^{-2}$ 37. $(i^{-6})^{-3}$ 38. $(4 + i)^{-2}$

39. Simplify $\dfrac{1}{a + bi}$, $(a, b \neq 0)$, and thus obtain a general form of the multiplicative inverse of $a + bi$.

0.9   POLYNOMIAL EXPRESSIONS

The major emphasis in this book is on the set R of real numbers. However, for the sake of generality, in this review of polynomials we shall refer to the set C of complex numbers. As you read this section, keep in mind that C includes *real numbers* together with *imaginary numbers*.

    A **monomial** is either a complex number, or an expression that can be written as a product of a complex number and nonzero powers of variables. The numerical factor is called the **numerical coefficient** or, simply, the **coefficient**. Thus, $3x^2$ and $-4xy$ are monomials with coefficients 3 and -4, respectively. Occasionally, each factor of a monomial is referred to as the coefficient of the other factors. For example, in $-4xy$, $-4x$ is the coefficient of y, x is the coefficient of $-4y$, etc. The context will make it clear when this meaning is being used. The **degree of a monomial** with respect to one of its variables is given by the exponent on that variable; with respect to more than one variable the degree is given by the sum of the exponents on the specified variables. For instance, $3x^2$ is a monomial of degree two; $-4xy$ is of degree one with respect to either x or y, and of degree two with respect to both x and y.

An expression that is a monomial or a finite sum of monomials is called a **polynomial** (in one or more variables). Each of the monomials is a **term** of the polynomial. Terms that differ only in their numerical coefficients are called **like terms**. Polynomials of two and three terms of different degree are called **binomials** and **trinomials**, respectively. The **degree of a polynomial** with respect to one variable is equal to the degree of the term of highest degree in that variable; with respect to more than one variable the degree is equal to the degree of the term of highest degree in the specified variables. The polynomial consisting of just a number k is called a **constant polynomial** and is considered to be of degree zero, except that no degree is assigned to the *zero polynomial*.

EXAMPLE 1

a. $2x^3 - 11x^2$ is a binomial of degree three.

b. $x + xy - y$ is a trinomial of degree one with respect either to x or to y; of degree two with respect to both x and y.

Symbols such as $A(x)$ and $B(x, y)$, read "A of x" and "B of x and y," respectively, are often used to name polynomials. The letters within the parentheses name the variables that appear in the polynomials.

Each polynomial $A(x)$ can be written as a finite sum of monomials of the form ax^n, where each such monomial is a product of a coefficient a and n factors x. Hence, because sums and products of complex numbers are complex numbers (closure properties), it follows that for each $x \in C$, the monomial x^n names a complex number (real or imaginary). It follows that the polynomial $A(x)$ itself names a complex number for each $x \in C$. This line of reasoning is equally valid for polynomials in more than one variable. Symbols such as $A(-2)$ are used to name the value of the polynomial $A(x)$ when x is replaced by -2.

EXAMPLE 2

a. If $A(x) = 4x^2 - 2x + 1$, then $A(-2) = 4(-2)^2 - 2(-2) + 1 = 21$.

b. If $B(x, y) = 5x^2 - y^2$, then $B(1, -1) = 5(1)^2 - (-1)^2 = 4$.

Sums and Differences   From the preceding discussion we see that polynomials may be viewed as symbols representing complex numbers. Hence, we may apply properties such as commutativity, associativity and distributivity to rewrite polynomials into *equivalent expressions*—that is, expressions that name the same number for the same replacements of the variable. Polynomials are said to be in *simple form* when no two terms are like terms.

EXAMPLE 3   Given $A(x)=5x^2+9x+8$ and $B(x)=3x^2-4x$. Write each of the following as an equivalent polynomial in simple form.

a. $A(x)+B(x)$ b. $A(x)-B(x)$.

SOLUTION

a. $(5x^2+9x+8)+(3x^2-4x)=5x^2+9x+8+3x^2-4x$
$$=5x^2+3x^2+9x-4x+8$$
$$=(5+3)x^2+(9-4)x+8$$
$$=8x^2+5x+8.$$

b. $(5x^2+9x+8)-(3x^2-4x)=5x^2+9x+8-3x^2+4x$
$$=5x^2-3x^2+9x+4x+8$$
$$=(5-3)x^2+(9+4)x+8$$
$$=2x^2+13x+8.$$

Products   The field axioms, together with Theorem 0.22 as applied to positive integral exponents, enable us to write a product of polynomials as a single polynomial in simple form.

EXAMPLE 4

a. $x(4x^2-3x+6)=x\cdot 4x^2+x(-3x)+x\cdot 6$
$$=4x^3-3x^2+6x.$$

b. $(2x-5)(3x+7)=(2x-5)\cdot 3x+(2x-5)\cdot 7$
$$=6x^2-15x+14x-35$$
$$=6x^2-x-35.$$

Examples 3 and 4 emphasize the application of *distributivity*. As you probably recall from previous algebra courses, such products can often be obtained mentally, or at least with fewer details and fewer steps.

A type of problem similar to that of Example 2 is illustrated next.

EXAMPLE 5   Given $A(x)=7x^2-4x+2$.

a. $A(-x)=7(-x)^2-4(-x)+2=7x^2+4x+2.$

b. $A(x+1)=7(x+1)^2-4(x+1)+2$
$$=7(x^2+2x+1)-4x-4+2$$
$$=7x^2+10x+5.$$

c. $A(x+h)-A(x)=7(x+h)^2-4(x+h)+2-(7x^2-4x+2)$
$$=7(x^2+2hx+h^2)-4x-4h+2-7x^2+4x-2$$
$$=14hx+7h^2-4h.$$

Techniques similar to those illustrated above can be used to find products of expression other than polynomials.

EXAMPLE 6

   a. $x^{1/3}(x^{2/3}-2)=x^{1/3}\cdot x^{2/3}-x^{1/3}\cdot 2=x-2x^{1/3}$.

   b. $(a^{1/5}-2)(a^{1/5}+2)=a^{2/5}-2a^{1/5}+2a^{1/5}-4=a^{2/5}-4$.

Exercise Set 0.9

Specify the degree of each polynomial.

1. $2x+3x^2+5$ 2. $5x^3-7x$

3. 12 4. $12+3i$

Given $A(x)=x^2-5x+6$. *Find each of the following:*

5. $A(0)$ 6. $A(3)$ 7. $A(-3)$

8. $A(2+i)$ 9. $A(-x)$ 10. $A(x+h)-A(x)$

Given $B(x)=2x^2-xy+5y$. *Find each of the following:*

11. $B(-2,\ 3)$ 12. $B(-x,\ -y)$

13. $B(x-2,\ y+3)$ 14. $B(i,\ 2i)$

Given $C(x)=3x^2-8x+2$ *and* $D(x)=x^2+2x-5$. *Write each of the following as a simple form polynomial.*

15. $C(x)+D(x)$ 16. $C(x)-D(x)$

17. $D(x)-C(x)$ 18. $2[C(x)]+3[D(x)]$

Simplify.

19. $(x^2+3x)+(2-x^2)$ 20. $(x+x^2-2)-(3-x^2)$

21. $(3x-x^2)-(x-1-x^2)$ 22. $(x^2+2x-3)-(3-2x+x^2)$

23. $(x+3)(x-5)$ 24. $(x-3)(x+3)$

25. $(3x-2)(x+5)$ 26. $x(x+2)(x+4)$

27. $(3+x)(2-x)+(x-5)(x+7)$ 28. $[x-(2-3x)][2x-(3-x)]$

29. $(x+2)(x^2-2x+4)$ 30. $(x-3)(x^2+3x+9)$

31. $x^{1/2}(x^{1/3}+x^{1/4})$ 32. $x^{2/3}(x^{1/3}-x^{2/3})$

33. $(x^{1/2}+x^{1/4})(x^{1/2}-x^{1/4})$ 34. $(x^{2/3}+y)(x^{4/3}-x^{2/3}y+y^2)$

35. If $R(x)$ is a polynomial of degree one less than the degree of $x-2$, specify the degree of $R(x)$.

0.10   FACTORING

The process of writing a polynomial as a product of two or more polynomial factors is called **factoring**; the resulting indicated product is called a **factored form**, or a **factorization**, of the polynomial. For example, $(x-3)(x+7)$ is a factorization of $x^2+4x-21$, as you can verify by a multiplication.

Complete Factorization over J   A polynomial $A(x)$ can be classified according to the set of numbers—J, Q, H, etc.—from which its coefficients are obtained. If all of the coefficients of $A(x)$ are integers, we say that $A(x)$ is a polynomial over the set J of integers or, more concisely, $A(x)$ **is a polynomial over** J. For example, $6x^2 - 5x + 1$ is a polynomial over J; $6x^2 - \frac{5}{3}x + \frac{1}{3}$ is a polynomial over Q; $3x^2 - \sqrt{5}x - 4$ is a polynomial over R.

There may be more than one factorization of a given polynomial. For instance, each of the following products is a factorization of $x^2 - 9$:

$$(x+3)(x-3); \qquad -(x+3)(3-x); \qquad \tfrac{1}{3}(3x-9)(x+3).$$

However, if $A(x)$ is a polynomial *over* J, then $A(x)$ can be factored into exactly one *completely* factored form (except possibly for signs and the order of listing of the factors). $A(x)$ is in **completely factored form over** J if:

1. Each factor is a polynomial over J.
2. No polynomial factor can be further factored into polynomial factors over J.
3. Each monomial factor, if any, is in simple form.

With respect to Statement 2, if a monomial factor is a constant, we do *not* write it in factored form. Thus, if 6 is a factor, we write 6 rather than $2 \cdot 3$.

The following examples display frequently encountered types of factoring, the first of which is often referred to as *factoring out a common factor*.

EXAMPLE 1 Completely factor $24x^3 - 6x^2$ over J.

SOLUTION The polynomial $24x^3 - 6x^2$ may be written as

$$6x^2 \cdot 4x - 6x^2 \cdot 1, \qquad\qquad <1>$$

from which it is apparent that each term contains the *common factor* $6x^2$. Polynomial $<1>$ can now be written in completely factored form: $6x^2(4x - 1)$.

In any factoring problem, it is good practice to first check for a common factor.

The next example illustrates a trial and error technique for factoring polynomials of the form $ax^2 + bx + c$.

EXAMPLE 2 Completely factor $6x^2 - 5x + 1$ over J.

SOLUTION We seek two binomial factors whose product is the given polynomial. The product of the first two terms of the desired binomial factors must be $6x^2$, so we consider $6x$ and x, or $3x$ and $2x$. The product of the last two terms of the desired binomial factors must be 1, so we consider 1 and 1, or -1 and -1. Testing some possible combinations, we obtain:

$$(6x+1)(x+1)=6x^2+7x+1; \qquad (6x-1)(x-1)=6x^2-7x+1;$$
$$(3x+1)(2x+1)=6x^2+5x+1; \qquad (3x-1)(2x-1)=6x^2-5x+1;$$

from which it becomes apparent that the completely factored form of $6x^2 - 5x + 1$ is $(3x - 1)(2x - 1)$.

The method illustrated next is called *factoring by grouping*.

EXAMPLE 3   Completely factor $2x^2 + xy - 6x - 3y$ over J.

SOLUTION   We group the first and second terms, and the third and fourth terms, as follows:

$$2x^2 + xy - 6x - 3y = (2x^2 + xy) + (-6x - 3y)$$
$$= x(2x + y) - 3(2x + y)$$
$$= (x - 3)(2x + y).$$

You will find it instructive to reconsider Example 3 using different pairings of the terms.

Certain special form polynomials can be factored in a straightforward manner. These forms, and their complete factorizations, are:

$a^2 - b^2 = (a - b)(a + b)$ [Difference of two squares]

$a^3 + b^3 = (a + b)(a^2 - ab + b^2)$ [Sum of two cubes]

$a^3 - b^3 = (a - b)(a^2 + ab + b^2)$ [Difference of two cubes]

EXAMPLE 4

a. $x^4 - 16 = (x^2)^2 - 4^2 = (x^2 - 4)(x^2 + 4)$
$$= (x - 2)(x + 2)(x^2 + 4)$$

b. $27x^3 - 8 = (3x)^3 - 2^3 = (3x - 2)[(3x)^2 + 3x \cdot 2 + 2^2]$
$$= (3x - 2)(9x^2 + 6x + 4)$$

Factoring over Specified Sets   For some of our work we need to factor polynomials over sets of numbers other than J.

EXAMPLE 5   Factor $x^2 - 2$ over R.

SOLUTION   Viewing $x^2 - 2$ as the difference of two squares:

$$x^2 - 2 = x^2 - (\sqrt{2})^2 = (x - \sqrt{2})(x + \sqrt{2}).$$

Note that $x^2 - 2$ is not factorable over J.

EXAMPLE 6   Factor $x^3 - \frac{1}{8}$ over Q.

SOLUTION $$x^3 - \frac{1}{8} = x^3 - \left(\frac{1}{2}\right)^2 = \left(x - \frac{1}{2}\right)\left(x^2 + \frac{1}{2}x + \frac{1}{4}\right).$$

The imaginary unit $i = \sqrt{-1}$ provides a method for factoring the sum of two squares over the set C of complex numbers. Consider:

$$-(bi)^2 = -b^2 i^2 = -(-b^2) = b^2.$$

This result enables us to factor as follows:

$$a^2 + b^2 = a^2 - (bi)^2 = (a - bi)(a + bi),$$

where the final factorization is patterned on the difference of two squares.

EXAMPLE 7 $4x^2+9$ can be factored over C as follows:

$$4x^2+9=(2x)^2-(3i)^2=(2x-3i)(2x+3i).$$

Exercise Set 0.10

Completely factor over J.

1. $3x^2+6x^4$
2. $12x^3-16x$
3. $x(y-3)-6(y-3)$
4. $x^2y^2-3xy+xy^2$
5. $x^2-3x-28$
6. $2x^2-7x+3$
7. $2x^2-7xy-4y^2$
8. $x^2-16x+64$
9. $4x^2-9$
10. $25x^2-81y^4$
11. x^3+8
12. $125x^6-27$
13. x^3-2x^2+x
14. $72x^4-162x^2y^2$
15. $x^2y^4-64y^2$
16. $16x^3z-2zy^3$
17. x^3+4x^2-2x-8
18. $x^4-3x^3-6x+18$
19. bx^2+x+b^2x+b
20. x^3-x^2+xy-y
21. $16x^4-8x^2+1$
22. x^4-6x^2-27
23. $2x^4-29x^2-48$
24. x^4-y^4

Factor each polynomial completely over the specified set.

25. $x^2-\frac{3}{2}x+\frac{9}{16}$; Q
26. $4y^2-\frac{1}{9}$; Q
27. $y^3+\frac{8}{27}$; Q
28. $4x^2-\frac{4}{3}x+\frac{1}{9}$; Q
29. $9x^2-5$; R
30. x^3+6; R
31. x^2+64; C
32. $3y^2+25$; C

0.11   RATIONAL EXPRESSIONS

A fraction in which both numerator and denominator are polynomials is called a **rational expression**. If a variable appears in the denominator then, because division by zero is not defined, the replacement set for that variable is restricted to those numbers for which the denominator polynomial is *not* equal to zero. For example, the fraction

$$\frac{3x^2+4}{(x-2)(x+3)}$$

is a rational expression with the restrictions $x\neq 2,\ -3$. *For the remainder of this section it is to be understood that replacement sets for variables in denominators are such that no denominator is equal to zero.*

In Section 0.9 we stated that a polynomial names a complex number (real or imaginary) for each complex number replacement of the variable. Furthermore, from Section 0.5 we have that those real number theorems that are not consequences of order relations may be applied to complex numbers. Theorems 0.8 through 0.14 are particularly useful for operations on rational expressions.

A fraction is said to be in *lowest terms* when the numerator and the denominator do not include common factors. The Fundamental Theorem of Fractions (Theorem 0.9) can be used to reduce a fraction to lowest terms.

EXAMPLE 1

a. $\dfrac{x^2-4}{x^2-6x+8} = \dfrac{(x-2)(x+2)}{(x-2)(x-4)} = \dfrac{x+2}{x-4}$.

b. $\dfrac{(3-y)(2+y)}{2y^2-5y-3} = \dfrac{(3-y)(2+y)}{(y-3)(2y+1)} = \dfrac{-(y-3)(2+y)}{(y-3)(2y+1)} = \dfrac{-(2+y)}{2y+1}$.

Sums and Differences   A sum or difference of fractions with like denominators can readily be written as a single fraction in lowest terms. Thus:

$$\frac{x^2}{x-3} + \frac{x-12}{x-3} = \frac{x^2+x-12}{x-3} = \frac{(x-3)(x+4)}{x-3} = x+4;$$

$$\frac{x^2}{x-3} - \frac{x-12}{x-3} = \frac{x^2-(x-12)}{x-3} = \frac{x^2-x+12}{x-3}.$$

If the denominators are different, then we find a polynomial that will serve as a common denominator (called the *Least Common Denominator*, or L.C.D.) by the following method.

1. Completely factor each denominator, including numerical coefficients.

2. Form the product of all the different factors that appear, including each factor the greatest number of times that it appears in any one of the denominators.

For example, if $3 \cdot 5x^2$, 5^2x and $3(x-1)$ are denominators in factored form, then the L.C.D. is $3 \cdot 5^2x^2(x-1)$.

EXAMPLE 2   Write $\dfrac{2x-5}{x^2-5x+6} - \dfrac{x-2}{2x^2-11x+15}$ as a single fraction in lowest terms.

SOLUTION $\dfrac{2x-5}{x^2-5x+6} - \dfrac{x-2}{2x^2-11x+15} = \dfrac{2x-5}{(x-2)(x-3)} - \dfrac{x-2}{(x-3)(2x-5)}$.

From the factored denominators on the right we see that the L.C.D. is $(x-2)(x-3)(2x-5)$. Hence, using Theorem 0.9,

$$\frac{2x-5}{x^2-5x+6} - \frac{x-2}{2x^2-11x+15} = \frac{(2x-5)(2x-5)}{(x-2)(x-3)(2x-5)} - \frac{(x-2)(x-2)}{(x-3)(2x-5)(x-2)}$$

(*Continued on next page*)

$$\frac{2x-5}{x^2-5x+6} - \frac{x-2}{2x^2-11x+15} = \frac{(4x^2-20x+25)-(x^2-4x+4)}{(x-2)(x-3)(2x-5)}$$
$$= \frac{3x^2-16x+21}{(x-2)(x-3)(2x-5)}$$
$$= \frac{(3x-7)(x-3)}{(x-2)(x-3)(2x-5)}$$
$$= \frac{3x-7}{(x-2)(2x-5)}.$$

Products and Quotients   In finding a product or quotient of fractions, it is usually desirable to first completely factor each polynomial and then to apply the Fundamental Theorem of Fractions to reduce each product to lowest terms. From Theorem 0.12 we have that a quotient of two fractions may be found by multiplying the first fraction by the reciprocal of the divisor.

EXAMPLE 3

a.
$$\frac{x^2+2x-8}{x^2+5x+6} \cdot \frac{x^2+x-6}{x^2-4x+4} = \frac{(x+4)(x-2)(x-2)(x+3)}{(x+3)(x+2)(x-2)(x-2)}$$
$$= \frac{x+4}{x+2}.$$

b.
$$\frac{x^2-1}{2x+3} \div \frac{x+1}{4x^2+12x+9} = \frac{x^2-1}{2x+3} \cdot \frac{4x^2+12x+9}{x+1}$$
$$= \frac{(x-1)(x+1)(2x+3)(2x+3)}{(2x+3)(x+1)}$$
$$= (x-1)(2x+3)$$
$$= 2x^2+x-3.$$

EXAMPLE 4

$$\frac{\dfrac{2}{x}}{3-\dfrac{5}{x}} = \frac{\dfrac{2}{x}}{\dfrac{3x-5}{x}} = \frac{2}{x} \cdot \frac{x}{3x-5}$$
$$= \frac{2}{3x-5}.$$

Expressions such as the fraction considered in Example 4 are called *complex fractions* and are said to be *simplified* when they are written equivalently as a polynomial in simple form or as a rational expression in lowest terms.

EXAMPLE 5

$$\frac{x^{-1}-y^{-1}}{x^{-1}y^{-1}} = \frac{\dfrac{1}{x}-\dfrac{1}{y}}{\dfrac{1}{x} \cdot \dfrac{1}{y}} = \frac{\dfrac{y-x}{xy}}{\dfrac{1}{xy}} = \frac{(y-x) \cdot xy}{xy \cdot 1} = y-x.$$

Division of Polynomials　　If $A(x)$ and $B(x)$ are polynomials, then the rational expression $\dfrac{A(x)}{B(x)}$ may be viewed as the quotient of $A(x)$ and $B(x)$, provided that $B(x) \neq 0$. For example, if $x \neq 0$, then

$$\frac{6x^2 + 3x + 5}{3x} = \frac{6x^2}{3x} + \frac{3x}{3x} + \frac{5}{3x}$$

$$\frac{6x^2 + 3x + 5}{3x} = (2x + 1) + \frac{5}{3x}, \tag{<1>}$$

where $3x$ is called the **divisor** polynomial, $2x + 1$ is called the **quotient** polynomial and 5 is called the **remainder** polynomial. If each member of $\langle 1 \rangle$ is multiplied by $3x$, we obtain

$$6x^2 + 3x + 5 = 3x(2x + 1) + 5,$$

a form that is useful for many purposes.

Next, consider the long division process for dividing the polynomial $x^3 + 5x^2 - 7x + 2$ by $x^2 - x - 2$;

$$
\begin{array}{r}
x + 6 \qquad\qquad \text{(Quotient)} \\
x^2 - x - 2 \overline{\smash{\big)}\, x^3 + 5x^2 - 7x + 2} \\
\underline{x^3 -\ x^2 - 2x} \\
6x^2 - 5x + 2 \\
\underline{6x^2 - 6x - 12} \\
x + 14 \quad \text{(Remainder)}
\end{array}
$$

(Divisor)

This result can be given either in the form

$$x^3 + 5x^2 - 7x + 2 = (x^2 - x - 2)(x + 6) + (x + 14)$$

or in the form

$$\frac{x^3 + 5x^2 - 7x + 2}{x^2 - x - 2} = (x + 6) + \frac{x + 14}{x^2 - x - 2}.$$

The operation of division of polynomials, as illustrated above, presupposes that we can always find quotient and remainder polynomials to complete the process. This is indeed the case whenever the divisor is not equal to zero.

Theorem 0.24 (The Division Algorithm for Polynomials)　　If $A(x)$ and $B(x)$ are polynomials and $B(x) \neq 0$, then there exist unique polynomials $Q(x)$ and $R(x)$ such that

$$A(x) = B(x) \cdot Q(x) + R(x), \tag{<2>}$$

where the degree of $R(x)$ is less than the degree of $B(x)$, or where $R(x) = 0$.

Equation $\langle 2 \rangle$ of Theorem 0.24 may also be written in the form

$$\frac{A(x)}{B(x)} = Q(x) + \frac{R(x)}{B(x)}.$$

When dividing $A(x)$ by $B(x)$, if one or more of the powers of the variable in $A(x)$ is "missing," it is convenient to "fill in" with zeros.

EXAMPLE 6   Divide $8x^3 - 27$ by $2x - 3$.

SOLUTION

$$
\begin{array}{r}
4x^2 + 6x + 9 \\
2x - 3 \overline{\smash{)}8x^3 + 0 \cdot x^2 + 0 \cdot x - 27} \\
\underline{8x^3 - 12x^2} \\
12x^2 + 0 \cdot x \\
\underline{12x^2 - 18x} \\
18x - 27 \\
\underline{18x - 27} \\
0
\end{array}
$$

Exercise Set 0.11

In these exercises, assume that no denominator is equal to zero.

Reduce each rational expression to lowest terms.

1. $\dfrac{3xy^2}{6x^4y}$

2. $\dfrac{3x - 3y}{x^2 - 2xy + y^2}$

3. $\dfrac{(y - x)^2}{x^2 - y^2}$

4. $\dfrac{x^2 + 4xy + 4y^2}{x^2 - 4y^2}$

5. $\dfrac{x^2 - x - 6}{x^2 - 6x + 9}$

6. $\dfrac{x^2 - x}{x^3 - 2x^2 + x}$

Write each expression as a single fraction in lowest terms.

7. $\dfrac{2x^3}{2x + 1} + \dfrac{7x^2 + 3x}{2x + 1}$

8. $\dfrac{2x^3}{2x + 1} - \dfrac{5x^2 + 3x}{2x + 1}$

9. $\dfrac{3x + 1}{25x} - \dfrac{x - 2}{15x^2}$

10. $\dfrac{3x + 1}{25x} + \dfrac{x - 2}{15x^2}$

11. $\dfrac{5}{y - 6} + \dfrac{3}{6 - y}$

12. $\dfrac{x + 1}{x - 1} - \dfrac{x - 1}{x + 1}$

13. $\dfrac{x - 2}{x^2 + x} + \dfrac{3x}{x + 1}$

14. $\dfrac{6 - x}{4 - x^2} - \dfrac{x}{(2 + x)^2}$

15. $\dfrac{2x - 1}{x^2 - 2x - 15} - \dfrac{x + 3}{2x^2 - 11x + 5}$

16. $\dfrac{x^2 - 2x - 8}{2x^2 - 11x + 15} + \dfrac{x^2 - 3x}{2x^2 - 3x - 5}$

17. $\dfrac{x^2 + 5x + 6}{x^2 - x - 20} \cdot \dfrac{x^2 - 2x - 15}{x^2 - 3x - 10}$

18. $\dfrac{x^3 + 4x^2}{2x^3 - 8x} \cdot \dfrac{x^2 + 3x + 2}{x^2 + 5x}$

19. $\dfrac{4x^2 + 12x + 9}{x^2 - 7x + 6} \div \dfrac{2x^2 + x - 3}{x^2 - 2x + 1}$

20. $\dfrac{x^3 - 8}{x^2 + 8x - 20} \div \dfrac{x^2 + 2x + 4}{x^2 + 7x - 30}$

Simplify.

21. $\dfrac{\dfrac{1}{x} + \dfrac{1}{y}}{\dfrac{1}{x^2} - \dfrac{1}{y^2}}$

22. $\dfrac{x + 1 - \dfrac{3}{x - 1}}{x - \dfrac{x + 4}{x + 1}}$

23. $\left(\dfrac{x}{y^{-1}} + \dfrac{x^{-1}}{y} \right)^{-1}$

24. $(x^{-1} + y^{-1})^{-1}$

Write each rational expression $\dfrac{A(x)}{B(x)}$ in the form $Q(x) + \dfrac{R(x)}{B(x)}$, where $Q(x)$ is the quotient, $R(x)$ is the remainder and $B(x)$ is the divisor.

25. $\dfrac{6x^2 - 4x + 3}{3x + 1}$

26. $\dfrac{8x^3 - 1}{2x - 1}$

27. $\dfrac{2x^3 - 3x^2 + 8x + 10}{x^2 + x - 1}$

28. $\dfrac{3x^4 + 6x}{x^2 - 2x + 5}$

29. $\dfrac{x^6 - 1}{x - 1}$

30. $\dfrac{x^5 + 1}{x + 1}$

0.12   GRAPHING ORDERED PAIRS

In Section 0.4 we assumed the existence of a one-to-one correspondence between the set R of real numbers and the points in a line. In turn, that assumption led to the construction of a number line on which we may graph sets of real numbers. Number lines are also used to construct coordinate systems in a plane on which we may graph sets of ordered pairs of real numbers.

Ordered Pairs   Consider the pair of numbers 3 and 5. If we specify that 3 is to be designated as the *first* number and 5 as the *second* number of the pair, then we have an **ordered pair** of numbers which we denote by the symbol (3, 5). If 5 is first and 3 is second, then we write (5, 3), which is not the same as the ordered pair (3, 5). Each of the numbers of an ordered pair is called a **component**—*first component* and *second component*—of the ordered pair. Thus, 3 is the first component and 5 is the second component of (3, 5).

The Rectangular Coordinate System   A rectangular (or Cartesian) coordinate system in a plane is established by introducing two perpendicular number lines, called **coordinate axes**, whose origins coincide, as shown in Figure 0.12–1. The point of intersection of the two axes is

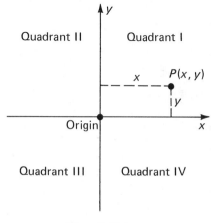

Figure 0.12-1

called the **origin** of the coordinate system. Each axis is labeled with an associated variable, most commonly x and y. The horizontal axis is called the **x-axis**; the vertical axis is called the **y-axis**. The parts of the x-axis to the right and to the left of the origin are sometimes referred to as the *positive x-axis* and the *negative x-axis*, respectively. Similarly, the parts of the y-axis above and below the origin are referred to as the *positive y-axis* and the *negative y-axis*, respectively.

We assume that there is a one-to-one correspondence between the set of all ordered pairs of real numbers $\{(x, y): x \in R,\ y \in R\}$ and the set of points in the plane. Each ordered pair (x, y) is associated with the point in the plane that is both $|x|$ units (left or right) from the y-axis *and* $|y|$ units (below or above) from the x-axis. The unique point located in this manner is called the **graph** of the ordered pair; the components of the ordered pair are called the **coordinates** of the point. The x-coordinate is sometimes called the *abscissa*; the y-coordinate is sometimes called the *ordinate*. Note that the origin is the graph of $(0, 0)$. A plane into which coordinate axes have been introduced may be referred to as a *coordinatized plane*.

It is customary to use "ordered pair" and "point" interchangeably. Thus, we may use "point $P(x, y)$" or just "$P(x, y)$" to mean "the point P with coordinates given by the ordered pair (x, y)." Although the symbol $P(x, y)$ may also be used to name an expression in two variables x and y, the context in which the symbol appears will clarify the intended meaning in each case. To "plot" or "graph" a point $P(x, y)$ means to locate the point with coordinates (x, y) in a coordinatized plane. If each ordered pair of a set of ordered pairs is graphed, the resulting set of points is called the *graph of the set*.

EXAMPLE 1   Figure $0.12-2$ shows the graphs of points $A(4, 3)$, $B(-5, 4)$, $C(-4, -1)$ and $D(2, -4)$. The resulting set of points is the graph of $\{(4, 3), (-5, 4), (-4, -1), (2, -4)\}$.

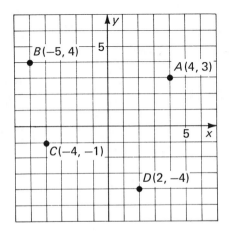

Figure 0.12-2

Exercise Set 0.12

Graph each of the following sets. (You may wish to use different scales on the x-axis and the y-axis.)

1. $\{(0, 0), (3, 3), (6, 6)\}$
2. $\{(0, 0), (-3, -3), (-6, -6)\}$
3. $\{(1, 5), (-1, 5), (1, -5), (-1, -5)\}$
4. $\{(5, 1), (5, -1), (-5, 1), (-5, -1)\}$
5. $\{(1, 10), (2, 20), (3, -30), (4, -40)\}$
6. $\{(-2, 12), (-1, 6), (0, 0), (4, 18)\}$

*Specify the quadrant or axis in which each point is located. If in an axis, specify whether it is in the **positive** or the **negative** part.*

7. $(-6, 11)$ 8. $(0, -11)$ 9. $(-6, 0)$ 10. $(6, 11)$

11. $(6, -11)$ 12. $(-6, -11)$ 13. $(6, 0)$ 14. $(0, 11)$

15. a. Graph: $\{(1, 0), (5, 0), (-2, 0), (-4, 0), (-6, 0)\}$.
    b. Describe a property that is a characteristic of the coordinates of all points in the x-axis.

16. a. Graph: $\{(0, 1), (0, 5), (0, -2), (0, -4), (0, -6)\}$.
    b. Describe a property that is a characteristic of the coordinates of all points in the y-axis.

Describe a property that is a characteristic of the coordinates of all points:

17. In Quadrant I   18. In Quadrant II

19. In Quadrant III   20. In Quadrant IV

21. To the right of the y-axis   22. To the left of the y-axis

23. Below the x-axis   24. Above the x-axis.

Given points $A_1(2, 1)$, $B_1(6, 8)$, $C_1(3, -2)$ and $D_1(-3, -4)$.

25. Locate point A_2 such that the pair of points A_1 and A_2 are on opposite sides of the x-axis and equidistant from the x-axis; specify the coordinates of A_2. Follow the same instructions for the pairs B_1 and B_2; C_1 and C_2; D_1 and D_2.

26. Follow the instructions of Exercise 25, but locate points A_3, B_3, C_3 and D_3 such that the pairs of points are on opposite sides of the y-axis and equidistant from the y-axis.

27. Specify a property that is a characteristic of the coordinates of pairs of points that are on opposite sides of, and also equidistant from, the x-axis. (Hint: See Exercise 25.)

28. Specify a property that is a characteristic of the coordinates of pairs of points that are on opposite sides of, and also equidistant from, the y-axis. (Hint: See Exercise 26.)

1

EQUATIONS AND INEQUALITIES

1.1   EQUIVALENT EQUATIONS; LINEAR EQUATIONS

If $A(x)$ and $B(x)$ are expressions in x, then the sentence

$$A(x) = B(x)$$

is an *equation in one variable x*. If x is replaced by a member of its replacement set, say b, and the resulting equation $A(b) = B(b)$ is a true statement, then b is said to *satisfy* the equation and b is a **solution** of the equation. For example, in the equation

$$x + 4 = 2x - 7,$$

if x is replaced by 11 we obtain

$$11 + 4 = 2 \cdot 11 - 7, \qquad \text{or} \qquad 15 = 15,$$

a true statement. Hence, 11 satisfies the equation and 11 is a solution of the equation.

Unless specified otherwise, in this text the replacement set for a variable is the set R of real numbers. Because division by zero is not defined, it is to be understood that if a variable appears in a denominator, then its replacement set does not include numbers for which the denominator is equal to zero. That is, if $A(x)$ names a denominator, then $A(x) \neq 0$. For example, given the expression

$$\frac{3x + 2}{x(x - 3)},$$

the replacement set for x does not include 0 or 3; we may write $x \neq 0, 3$ for emphasis.

The set of all solutions of an equation is called the **solution set** of the equation; to *solve an equation* means to *find its solution set*. Because any solution of an equation must be a member of the replacement set for the variable, it follows that the solution set of an equation is a subset of the replacement set for the variable. If the solution set is a *proper* subset of the replacement set, then the equation is called a **conditional equation**; if the solution set is equal to the replacement set, then the equation is called an **identity**.

Equivalent Equations It may be possible to determine the solution set of an equation by inspection. Thus, for $x+1=3$, it is clear that 2 is a solution and so its solution set is $\{2\}$. If the solution set cannot readily be determined by inspection, then it may be possible to transform the given equation into a second equation, the second into a third, etc., thus generating a sequence of equations each of which has the same solution set as the first and such that the solution set of the last equation can be determined by inspection. The equations in such a sequence are called *equivalent equations*.

Definition 1.1 Equations are **equivalent equations** if and only if they have equal solution sets.

EXAMPLE 1 The equations $x+4=2x-7$ and $x=11$ are equivalent equations because $\{11\}$ is the solution set of each equation.

Generating Equivalent Equations A general technique for solving equations is that of generating sequences of equivalent equations until an equation is obtained whose solution set is readily seen. The field axioms (see Section 0.2), together with the following theorem, furnish methods for transforming equations into equivalent equations.

Theorem 1.1 For all x such that $A(x)$, $B(x)$ and $C(x)$ denote real numbers:

I $A(x)=B(x)$ and $A(x)+C(x)=B(x)+C(x)$
are equivalent equations.

II   If $C(x)\neq 0$, then
$A(x)=B(x)$ and $A(x)\cdot C(x)=B(x)\cdot C(x)$
are equivalent equations.   **(P)**

EXAMPLE 2 Solve $(x+2)(x-3)=2+x^2-5x$.

SOLUTION
$$(x+2)(x-3)=2+x^2-5x$$
$$x^2-x-6=2+x^2-5x$$
$$x^2-x-6+(-x^2+5x+6)=2+x^2-5x+(-x^2+5x+6)$$
$$4x=8$$
$$\tfrac{1}{4}\cdot 4x=\tfrac{1}{4}\cdot 8$$
$$x=2.$$

The solution set of the last equation is $\{2\}$. But, by Theorem 1.1, we have a sequence of *equivalent* equations. Hence, the solution set of the given equation is also $\{2\}$.

The next example illustrates the need for the restriction $C(x) \neq 0$ in Theorem 1.1–II.

EXAMPLE 3   Solve $\dfrac{x}{x-5} = 3 + \dfrac{5}{x-5}$. $\langle 1 \rangle$

SOLUTION   To "remove" the denominators, we multiply each member of $\langle 1 \rangle$ by $x-5$.

$$\frac{x}{x-5}(x-5) = 3(x-5) + \frac{5}{x-5}(x-5) \qquad \langle 2 \rangle$$
$$x = 3x - 15 + 5 \qquad \langle 3 \rangle$$
$$x = 5.$$

The only solution of the last equation is 5. But, if x is equal to 5, then $x-5$ is equal to zero, and in equation $\langle 2 \rangle$ we have (unwittingly) multiplied by zero to obtain equation $\langle 3 \rangle$. In fact, 5 is not even in the replacement set for x because it leads to a zero denominator in equation $\langle 1 \rangle$, as you can verify. Thus, 5 cannot be a solution of the given equation. We conclude that there are no solutions. That is, the solution set of the given equation is the empty set $\emptyset$.

In general, whenever each member of an equation is multiplied by an expression involving the variable, it is a sound practice to check each "apparent" solution to be sure that it satisfies the given equation.

EXAMPLE 4   Solve $\dfrac{2x}{x-2} - \dfrac{1}{x+1} = 2$. $\langle 4 \rangle$

SOLUTION   To "remove" denominators, we first multiply by the L.C.D. (the lowest common denominator) of the denominators $x-2$ and $x+1$, namely $(x-2)(x+1)$.

$$\frac{2x}{x-2}(x-2)(x+1) - \frac{1}{x+1}(x-2)(x+1) = 2(x-2)(x+1)$$
$$2x(x+1) - (x-2) = 2(x^2 - x - 2)$$
$$2x^2 + 2x - x + 2 = 2x^2 - 2x - 4$$
$$x + 2 = -2x - 4$$
$$x = -2.$$

Each member of equation $\langle 4 \rangle$ was multiplied by an expression involving the variable. Hence, we check to see that -2 satisfies the given equation before we decide that it actually is a solution. Substituting -2 for x in $\langle 4 \rangle$, we have

$$\frac{2(-2)}{-2-2} - \frac{1}{-2+1} = 2, \quad \text{or} \quad 2 = 2.$$

Hence, -2 is a solution and the solution set is $\{-2\}$.

Linear Equations   An equation in the form

$$ax + b = 0, \qquad a \neq 0 \qquad \qquad \langle 5 \rangle$$

where one member is a polynomial of degree one in one variable and the other member is zero, is called a **linear equation**, or a **first degree equation**, in one variable. We sometimes say that the equation is *linear in x.* Any equation that can be transformed through a sequence of equivalent equations into the form of equation $\langle 5 \rangle$ is equivalent to a linear equation. An important property of any linear equation in one variable (or any equation equivalent to it) is that it has exactly one solution. To see this, suppose that a and b are real numbers, $a \neq 0$. Then, by Theorem 1.1, the following is a sequence of equivalent equations*:

$$ax + b = 0 \quad \leftrightarrow \quad ax = -b \quad \leftrightarrow \quad x = \frac{-b}{a}.$$

But $\dfrac{-b}{a}$ is the unique solution of the last equation. It follows that $\dfrac{-b}{a}$ is the unique solution of the linear equation $ax + b = 0$.

Equations with More Than One Variable   Given an equation in which two or more symbols denoting variables and/or constants are included, we often need to solve for a specified symbol in terms of the other symbols. By this we mean that the given equation is to be transformed into an equivalent equation in which the specified symbol appears by itself as one member of the equation and that same symbol does *not* appear in the other member. One technique is to treat all the symbols that are different from the one to be solved for as if they are constants, and apply Theorem 1.1 as needed.

EXAMPLE 5   Solve for n:

$$\ell = a + (n-1)d$$
$$\ell = a + nd - d$$
$$\ell + (-a+d) = a + nd - d + (-a+d)$$
$$\ell - a + d = nd$$
$$\frac{1}{d}(\ell - a + d) = \frac{1}{d} \cdot nd$$
$$n = \frac{\ell - a + d}{d}.$$

Exercise Set 1.1

(A)   *Solve each equation.*

1. $3x + 4 = -5$
2. $3 - 2x = 15$
3. $(x-2)(x+2) + 4 = (x+1)^2$
4. $(x-3)(x+2) = (x-2)(x+3)$
5. $\dfrac{3}{x} + 4 = 5$
6. $5 - \dfrac{2}{x} = 7$

*We sometimes use the symbol $\leftrightarrow$ to mean "is equivalent to."

7. $\dfrac{x}{x-3}+3=\dfrac{3}{x-3}$

8. $\dfrac{2}{x-2}-1=\dfrac{x}{x-2}$

9. $\dfrac{2x}{3x+5}+\dfrac{1}{x+1}=\dfrac{2}{3}$

10. $\dfrac{3x}{x+3}-\dfrac{1}{x-2}=3$

Find a value of k so that the two equations are equivalent.

11. $3x+8=2x$; $\dfrac{x+3}{k}=2-x$

12. $kx+4=x-9$; $2x+5=0$

Solve each equation for the indicated variable in terms of the others.

13. $d=rt$, for t

14. $E=IR$, for R

15. $x-r=3x+t$, for x

16. $2y+6b=b-3y$, for y

17. $ay=cy+d$, for y

18. $\dfrac{p}{q}z=2z-q$, for z

19. $\dfrac{1}{x}=\dfrac{1}{a}+\dfrac{1}{b}$, for x

20. $\dfrac{a}{y}=\dfrac{1}{b}+\dfrac{1}{c}$, for y

21. $\dfrac{y-3}{x+5}=-4$, for y

22. $\dfrac{y-b}{x}=m$, for y

23. Given $S=\tfrac{1}{2}n(a+s_n)$ and $s_n=a+(n-1)d$, express S in terms of a, n and d.

24. Given $A=\pi r^2$ and $C=2\pi r$, express A in terms of C and π.

Solve each "word" problem by solving a related linear equation.

25. Find two numbers whose sum is 94 and such that 3 times the smaller number is 6 more than the larger number.

26. Find three consecutive odd integers whose sum is 117.

27. A company sold 80 chairs, consisting of rockers and folding chairs, for which $1590 was paid. The bookkeeper then discovered that the bill for the chairs was lost. If rockers cost $21 each and folding chairs cost $19 each, how many of each kind of chair were sold?

28. How many pounds of peanuts at 60¢ per pound must be mixed with walnuts at $1.10 per pound to obtain 100 pounds of mix to be sold at $1.00 per pound?

29. Two trains leave the same terminal at the same time but in opposite directions. After 7 hours, they are 826 miles apart. If the rate of the faster train is 2 miles per hour less than three times the rate of the slower train, find the rate of each train. (Hint: distance = rate × time.)

30. A boat travels 21 miles per hour. A Coast Guard boat leaves from the same dock 2 hours later, traveling 35 miles per hour in the same direction. After how many hours will the second boat overtake the first? How far from the dock will this happen?

(P)   I.   Proof of Theorem 1.1–I

Hypothesis: $A(x)$, $B(x)$, and $C(x)$ denote real numbers.

Conclusion: $A(x)=B(x)$ and $A(x)+C(x)=B(x)+C(x)$ are equivalent equations.

Proof   Let b be any solution of the equation

$$A(x) = B(x). \qquad \langle 1 \rangle$$

That is, $A(b) = B(b)$.

By hypothesis, we have that $A(b)$, $B(b)$, and $C(b)$ denote real numbers. Hence, by Theorem 0.1, it follows that

$$A(b) + C(b) = B(b) + C(b),$$

and therefore that b is a solution of the equation

$$A(x) + C(x) = B(x) + C(x). \qquad \langle 2 \rangle$$

Conversely, let b be any solution of equation $\langle 2 \rangle$. That is,

$$A(b) + C(b) = B(b) + C(b). \qquad \langle 3 \rangle$$

Adding $-C(b)$ to each member of $\langle 3 \rangle$, we obtain

$$A(b) + C(b) + [-C(b)] = B(b) + C(b) + [-C(b)]$$
$$A(b) = B(b),$$

and therefore b is a solution of the equation

$$A(x) = B(x). \quad \square$$

II. Prove Theorem 1.1–II (Hint: Use the preceding proof as a guide.)

1.2   QUADRATIC EQUATIONS

An equation of the form

$$ax^2 + bx + c = 0, \qquad a \neq 0 \qquad \langle 1 \rangle$$

is a **quadratic equation**, or a **second degree equation**, in one variable x. We sometimes say that $\langle 1 \rangle$ is *quadratic in x*. Any equation that can be transformed through a sequence of equivalent equations into the form of equation $\langle 1 \rangle$ is equivalent to a quadratic equation. In $\langle 1 \rangle$, ax^2 is called the *second degree*, or *quadratic*, *term*, bx is called the *linear term*, and c is called the *constant term*. Quadratic equations in the form of equation $\langle 1 \rangle$ are said to be in *standard form*. For example, the quadratic equation $2x = 4 - x^2$ appears in standard form as $x^2 + 2x - 4 = 0$.

Solving by Factoring   If $P(x) = 0$ is a quadratic equation in standard form and $P(x)$ is factorable into a product of two linear factors (two polynomials of degree one), then the equation can be solved by the **method of factoring**, a consequence of Theorem 0.6.

EXAMPLE 1   Solve by the method of factoring: $2x^2 + 3x - 5 = 0$.

SOLUTION $2x^2 + 3x - 5 = 0$

$$(2x + 5)(x - 1) = 0$$

(Continued on next page)

Setting each factor equal to zero, we have

$$2x+5=0 \qquad \text{or} \qquad x-1=0$$
$$x=-\tfrac{5}{2} \qquad \text{or} \qquad x=1,$$

and the solution set is $\{-\tfrac{5}{2}, 1\}$.

Solving by the Square Root Method   Equations in the form

$$(ax+b)^2=k,$$

where $a \neq 0$ and $k \geq 0$, imply that the number $ax+b$ must be either the positive or the negative square root of k. Hence, such equations can be solved by solving each of the two related linear equations.

$$ax+b=\sqrt{k} \qquad \text{or} \qquad ax+b=-\sqrt{k},$$

as illustrated next. This technique is referred to as the **square root method**.

EXAMPLE 2   Solve by the square root method: $(2x+7)^2=5$.

SOLUTION $2x+7=\sqrt{5} \qquad \text{or} \qquad 2x+7=-\sqrt{5}$

$$x=\frac{-7+\sqrt{5}}{2} \qquad \text{or} \qquad x=\frac{-7-\sqrt{5}}{2}.$$

Hence, the solution set is $\left\{\dfrac{-7+\sqrt{5}}{2}, \dfrac{-7-\sqrt{5}}{2}\right\}$ or, more concisely, $\left\{\dfrac{-7\pm\sqrt{5}}{2}\right\}$.

Solving by Completing the Square   The square root method may not be immediately applicable to a given quadratic equation. However, by the process of **completing the square** it is always possible to transform the given equation into an equivalent equation which can be solved by the square root method. The next example illustrates this process and reviews the technique of completing the square. (See also Exercise 1.2-39.)

EXAMPLE 3   Solve by completing the square: $2x^2=12x-5$.

SOLUTION   In standard form, we have the equation

$$2x^2-12x+5=0. \tag{2}$$

First, we require that *the coefficient of the second degree term shall be 1.* Hence, we multiply each member of $\langle 2 \rangle$ by $\tfrac{1}{2}$ to obtain

$$x^2-6x+\tfrac{5}{2}=0$$

$$x^2-6x \qquad =-\tfrac{5}{2}. \tag{3}$$

Second, we *add the square of one-half the coefficient of the linear term to each member* of equation $\langle 3 \rangle$. That is, we add $[\frac{1}{2}(-6)]^2$, or 9, to each member:

$$x^2 - 6x + 9 = -\tfrac{5}{2} + 9. \qquad \langle 4 \rangle$$

The left member of $\langle 4 \rangle$ can now be factored into the product of two like factors, $(x-3)(x-3)$, and we write

$$(x-3)^2 = \tfrac{13}{2},$$

which equation can be solved by the square root method to obtain the

solution set $\left\{ 3 \pm \sqrt{\dfrac{13}{2}} \right\}$, or $\left\{ 3 \pm \dfrac{\sqrt{26}}{2} \right\}$.

Solving by Formula   The following theorem is a direct consequence of the method of completing the square.

Theorem 1.2   If a, b and c are complex numbers*, $a \neq 0$ and $x \in C$, then the solution set of $ax^2 + bx + c = 0$ is

$$\left\{ \dfrac{-b + \sqrt{b^2 - 4ac}}{2a}, \dfrac{-b - \sqrt{b^2 - 4ac}}{2a} \right\}. \qquad \textbf{(P)}$$

The conclusion of Theorem 1.2 is most often remembered as the familiar **quadratic formula**:

$$x = \dfrac{-b \pm \sqrt{b^2 - 4ac}}{2a}.$$

In Theorem 1.2, the replacement set for x is the set C of complex numbers (which includes the real numbers) because we shall be concerned with quadratic equations with imaginary solutions. However, we shall rarely consider equations in which a, b and c are imaginary numbers.

EXAMPLE 4   Solve each equation by the quadratic formula.

a. $2x^2 + 1 - 5x = 0$ b. $3x^2 - 2x + 4 = 0$

SOLUTION

a. In standard form we have the equation $2x^2 - 5x + 1 = 0$, from which $a = 2$; $b = -5$; $c = 1$. By the quadratic formula:

$$x = \dfrac{-(-5) \pm \sqrt{(-5)^2 - 4 \cdot 2 \cdot 1}}{2 \cdot 2},$$

the solution set is $\left\{ \dfrac{5 \pm \sqrt{17}}{4} \right\}$.

(Continued on next page)

*Complex numbers are reviewed in Sections 0.5 and 0.8.

b. From the given equation we have $a=3$; $b=-2$; $c=4$. Hence,

$$x=\frac{2\pm\sqrt{-44}}{6} \leftrightarrow x=\frac{1\pm\sqrt{11}\,i}{3},$$

and the solution set is $\left\{\frac{1}{3}\pm\frac{\sqrt{11}}{3}i\right\}$.

Quadratic Equations in Two Variables   The methods of this section are applicable to quadratic equations in more than one variable.

EXAMPLE 5   Solve for y: $9x^2+16y^2=144$.

SOLUTION $9x^2+16y^2=144 \leftrightarrow 16y^2=144-9x^2$

$$y^2=\frac{144-9x^2}{16}$$

$$y=\frac{\pm\sqrt{144-9x^2}}{4}$$

$$y=\pm\tfrac{3}{4}\sqrt{16-x^2}.$$

EXAMPLE 6   Solve for x: $3x^2+xy-y^2=0$.

SOLUTION   We view the equation as a *quadratic equation in* x, and apply the quadratic formula with $a=3$, $b=y$ and $c=-y^2$:

$$x=\frac{-y\pm\sqrt{y^2-4(3)(-y^2)}}{6} \leftrightarrow =\frac{-y\pm\sqrt{13y^2}}{6}$$

from which

$$x=\frac{-y\pm|y|\sqrt{13}}{6}.$$

Now, if $y\geq0$, then $|y|=y$ and so $\pm|y|=\pm y$. If $y<0$, then $|y|=-y$ and so $\pm|y|=\mp y$ which does not really differ from $\pm y$. Hence, for all y:

$$x=\frac{-y\pm y\sqrt{13}}{6} \leftrightarrow x=\frac{-1\pm\sqrt{13}}{6}y.$$

Exercise Set 1.2
The replacement set for x and y is the set C of complex numbers.

(A)   *Solve by factoring. (See Example 1.)*

1. $x^2+x-12=0$

2. $x^2-6x-16=0$

3. $6+x(2x-13)=0$

4. $1=6x^2-x$

5. $\dfrac{x}{x+1}+\dfrac{4}{3}=\dfrac{x}{x-1}$

6. $\dfrac{x}{x+3}+\dfrac{2x}{3x+1}=0$

Solve by the square root method. (See Example 2.)

7. $x^2=16$

8. $x^2=-16$

9. $(x-8)^2=-9$

10. $(x-8)^2=9$

11. $\left(x-\tfrac{3}{4}\right)^2=\tfrac{11}{16}$

12. $\left(x+\tfrac{4}{5}\right)^2=\tfrac{17}{25}$

Solve: a.   *By completing the square. (See Example 3.)*
              b.   *By the quadratic formula. (See Example 4.)*

13. $x^2 - 6x = 7$ 14. $x^2 + 5x + 6 = 0$

15. $3x^2 - 2x - 5 = 0$ 16. $6x^2 - x - 2 = 0$

17. $x^2 - 5x + 3 = 0$ 18. $x^2 + 3x - 6 = 0$

19. $2x^2 - 3x + 4 = 0$ 20. $3x^2 - x + 7 = 0$

Solve each equation for y. (See Example 5.)

21. $9x^2 - 16y^2 = 144$ 22. $4y^2 - 25x^2 = 100$

23. $25y^2 + 4x^2 = 100$ 24. $9x^2 + 4y^2 = 36$

25. $\dfrac{x^2}{a^2} + \dfrac{y^2}{b^2} = 1, \; a > 0, \; b > 0$ 26. $\dfrac{x^2}{a^2} - \dfrac{y^2}{b^2} = 1, \; a > 0, \; b > 0$

Solve each equation for x or y, as indicated. (See Example 6.)

27. $y^2 + 3xy + x^2 = 0; \; y$ 28. $y^2 + 3xy + x^2 = 0; \; x$

29. $4x^2 - 5xy + y^2 = 0; \; x$ 30. $4x^2 - 5xy + y^2 = 0; \; y$

(B)   *Solve by factoring*

31. $(x - 2)(2x^2 + 9x + 4) = 0$ 32. $(x + 1)(6x^2 - 7x + 2) = 0$

33. $3x^3 + x^2 - 2x = 0$ 34. $x^4 + x^3 - 42x^2 = 0$

35. $x^4 - 13x^2 + 36 = 0$ 36. $4x^4 - 5x^2 + 1 = 0$

Solve by completing the square.

37. $x^2 + 2rx + s = 0$ 38. $x^2 + px + q = 0$

39. Write the product $(x + q)^2$ as a trinomial. From your result, describe how to obtain the constant term (the term that does not include x) from the coefficient of the linear term in x.

(C)   *The following BASIC computer program can be used to compute the solutions of a quadratic equation with real coefficients. Use this program to solve the equations of Exercises 13–20.*

```
 5   INPUT A,B,C
10   LET D=B↑2-4*A*C
15   IF D<0 THEN 40
20   LET P=(-B+SQR(D))/(2*A)
25   LET Q =(-B-SQR(D))/(2*A)
30   PRINT A;B;C;P;Q
35   GO TO 5
40   LET M=-B/(2*A)
45   LET N=SQR(-D)/(2*A)
50   PRINT A;B;C;M;"+";N;"I";M;"-";N;"I"
55   GO TO 5
60   END
```

(P)   I.   Prove Theorem 1.2. (Hint: Complete the square in x.)

1.3   MORE ON QUADRATICS

Some of the consequences of the quadratic formula (Theorem 1.2) are explored in this section.

Number and Nature of Solutions   In general, Theorem 1.2 indicates that a quadratic equation has two solutions. However, note that if $b^2 - 4ac$ is equal to zero, then

$$\frac{-b \pm \sqrt{b^2 - 4ac}}{2a} = \frac{-b \pm 0}{2a} = \frac{-b}{2a},$$

and there is only one solution. We view this as the case where the two solutions are equal and say that there is one *solution of multiplicity two*, or a *double solution*. As an example, you may verify by factoring that the equation $x^2 - 10x + 25 = 0$ has 5 as a solution of multiplicity two.

The number $b^2 - 4ac$ that appears under the radical in the quadratic formula is called the **discriminant**. If a, b and c are real numbers, $a \neq 0$, then a considerable amount of information concerning the nature of the solutions of the quadratic equation

$$ax^2 + bx + c = 0 \qquad\qquad <1>$$

can be obtained by examining the discriminant. For instance, if $b^2 - 4ac$ is a negative number, then $\sqrt{b^2 - 4ac}$ is an imaginary number and equation $<1>$ has imaginary solutions. Table 1 summarizes the relation between the discriminant and the solutions of a quadratic equation. For all a, b, $c \in R$:

If	Then
$b^2 - 4ac > 0$	There are exactly two, real, unequal solutions.
$b^2 - 4ac = 0$	There is exactly one real solution of multiplicity two.
$b^2 - 4ac < 0$	There are exactly two imaginary solutions, each the conjugate of the other.

Table 1

Furthermore, if a, b and c are *rational* numbers and $b^2 - 4ac$ is the square of a rational number, then the solutions of equation $<1>$ are rational numbers. If $b^2 - 4ac$ is not such a square, then the solutions are irrational. In the following example you will find it instructive to solve each of the four quadratic equations and verify the conclusions that are reached.

EXAMPLE 1   Determine the nature of the solutions of each quadratic equation.

a. $x^2 + 6x + 8 = 0$ b. $x^2 + x - 5 = 0$

c. $x^2 + 2x + 1 = 0$ d. $x^2 + 3x + 4 = 0$

SOLUTION

a. For $x^2 + 6x + 8 = 0$: $b^2 - 4ac = 6^2 - 4 \cdot 1 \cdot 8 = 4$.
    Because $4 > 0$ and 4 is the square of the rational number 2, we conclude that the equation has two real, unequal, rational solutions.

b. For $x^2 + x - 5 = 0$: $b^2 - 4ac = 1^2 - 4 \cdot 1 \cdot (-5) = 21$.
    Because $21 > 0$ but 21 is not the square of a rational number, the equation has two real, unequal, irrational solutions.

c. For $x^2 + 2x + 1 = 0$: $b^2 - 4ac = 2^2 - 4 \cdot 1 \cdot 1 = 0$.
The equation has one real solution of multiplicity two.

d. For $x^2 + 3x + 4 = 0$: $b^2 - 4ac = 3^2 - 4 \cdot 1 \cdot 4 = -7$.
Because $-7 < 0$, the equation has two conjugate imaginary solutions.

Writing Quadratic Equations   In Section 1.2 we solved quadratic equations by the factoring method. Here we consider a "reverse" problem –given a solution set, write a quadratic equation having that solution set.

Suppose that $\{r_1, r_2\}$ is the solution set of a quadratic equation. The equations $x - r_1 = 0$ and $x - r_2 = 0$ have r_1 and r_2 as their respective solutions. Hence, the factored-form equation

$$(x - r_1)(x - r_2) = 0 \qquad \langle 2 \rangle$$

has $\{r_1, r_2\}$ as its solution set. By performing the indicated multiplication and simplifying as needed, equation $\langle 2 \rangle$ can be transformed into a quadratic equation in standard form.

EXAMPLE 2   Write a quadratic equation in standard form with integral coefficients that has as its solution set

a. $\{2, -\frac{1}{2}\}$ b. $\{3\}$ c. $\{i, -i\}$.

SOLUTION   In each case we start with a factored form equation.

a. $(x - 2)[x - (-\frac{1}{2})] = 0 \quad \leftrightarrow \quad x^2 - \frac{3}{2}x - 1 = 0$
$$2x^2 - 3x - 2 = 0.$$

b. Because a quadratic equation is required, we view 3 as a solution of multiplicity two. Hence:

$$(x - 3)(x - 3) = 0 \quad \leftrightarrow \quad x^2 - 6x + 9 = 0.$$

c. $(x - i)[x - (-i)] = 0 \quad \leftrightarrow \quad (x - i)(x + i) = 0$
$$x^2 + 1 = 0.$$

Factoring Quadratic Trinomials   In addition to solving quadratic equations, the quadratic formula can be used to factor quadratic trinomials in the form $ax^2 + bx + c$. We shall consider only the case where $a, b, c \in R$ and $ax^2 + bx + c$ is to be factored over the set R of real numbers (see Section 0.10).

EXAMPLE 3   Factor $3x^2 - 6x - 4$ over R.

SOLUTION   We apply the quadratic formula to the equation $3x^2 - 6x - 4 = 0$

(check the details) to obtain the solution set $\left\{\dfrac{3 \pm \sqrt{21}}{3}\right\}$. Following the form of the left member of equation $\langle 2 \rangle$, we have that

(*Continued on next page*)

$$3x^2 - 6x - 4 = \left(x - \frac{3 + \sqrt{21}}{3}\right)\left(x - \frac{3 - \sqrt{21}}{3}\right).$$

As a "fringe" benefit, the technique of Example 3 furnishes a test for factorization of a quadratic trinomial over the set Q of rational numbers (including the set J of integers). If the discriminant $b^2 - 4ac$ is the square of a rational number, then $ax^2 + bx + c$ is factorable over Q.

EXAMPLE 4   For the trinomial $32x^2 + 20x - 75$, we have that

$$b^2 - 4ac = 10,000 = (100)^2.$$

Hence, the given trinomial is factorable over Q. You will find it instructive to factor the trinomial, as in Example 3.

Exercise Set 1.3
Solution sets may include imaginary numbers.

(A)   *Rewrite each equation as a standard form quadratic equation and determine the nature of the solutions. (See Example 1.) Verify your answers by solving each equation.*

1. $17 + 3x^2 = 5x$
2. $4x(x + 1) = 5$
3. $2(x - 1) = (x - 3)(x - 2)$
4. $4x(x + 1) + 3 = x - 1$
5. $\dfrac{x}{x - 2} - \dfrac{2}{x + 1} = \dfrac{5}{2}$
6. $\dfrac{4x}{x + 1} + \dfrac{17}{x^2 - 1} = \dfrac{8}{x - 1}$
7. $\dfrac{12}{x} - 9 = \dfrac{4}{x^2}$
8. $\dfrac{2x + 1}{5} + \dfrac{1}{7x} + \dfrac{2}{7} = 0$

Write a quadratic equation in standard form with integral coefficients that has the given solution set. (See Example 2.)

9. $\{5, \frac{1}{3}\}$
10. $\{-3, \frac{2}{5}\}$
11. $\{-\frac{5}{4}\}$
12. $\{-6\}$
13. $\{2i, -2i\}$
14. $\{2 - i, 2 + i\}$

Use the quadratic formula to factor each trinomial over R. (See Example 3.)

15. $x^2 - 2x - 1$
16. $x^2 - 6x + 6$
17. $x^2 + 6x + 4$
18. $x^2 + 4x - 3$
19. $x^2 + 6x + 12$
20. $2x^2 - 5x + 4$

Test each trinomial for factorability over Q. (See Example 4.) Factor over Q, if possible.

21. $x^2 + 10x - 144$
22. $2x^2 - 9x - 143$
23. $2x^2 + 13x - 210$
24. $4x^2 - 17x - 120$
25. $2x^2 - 19x + 91$
26. $2x^2 - 13x + 210$

For each equation, determine k so that the equation has a solution of multiplicity two.

27. $2x^2 - kx + 8 = 0$
28. $kx^2 + 2x - 7 = 0$

29. A box with no lid is to be formed from a square piece of metal by cutting 2 inch squares from each corner and folding up the sides. If

the volume of the box is to be 98 inches, find the length of each side of the original piece of metal.

30. A circle has a radius of 8 inches. By how much should the radius be increased in order to increase the area by 80π square inches?

31. The distance s (in feet) that an object falls is given by $s = v_0 t + 16t^2$, where v_0 is the initial velocity in ft/sec and t is the time in seconds. How long will it take an object to fall 160 feet if: a. It is thrown downwards with an initial velocity of 48 ft/sec? b. It is dropped?

32. By increasing his usual average speed by 10 miles per hour, a trucker finds that he can decrease his time on a 240 mile route by two hours. Find the usual average speed of the truck.

(B)   33. If $\{r_1, r_2\}$ is the solution set of $ax^2 + bx + c = 0$, $a \neq 0$, show that
$$r_1 + r_2 = -\frac{b}{a} \text{ and } r_1 r_2 = \frac{c}{a}.$$

34. Prove: If $m + ni$ is a solution of $ax^2 + bx + c = 0$, $a \neq 0$, then $m - ni$ is the other solution.

(C)   *The following BASIC computer program can be used to examine the discriminant and determine the nature of the solutions of a quadratic equation with integer coefficients. Use this program to determine the nature of the solutions of the equations in Exercises 1–8.*

```
 5   PRINT "ENTER A, B AND C IN THIS FORM: A,B,C"
10   INPUT A,B,C
15   LET D=B↑2-4*A*C
20   PRINT "THE DISCRIMINANT IS" D
25   IF D=0 THEN 70
26   IF D<0 THEN 60
35   IF SQR(D)-INT(SQR(D))=0 THEN 50
40   PRINT "THE SOLUTIONS ARE REAL, UNEQUAL, AND IRRATIONAL"
45   GO TO 5
50   PRINT "THE SOLUTIONS ARE REAL, UNEQUAL, AND RATIONAL"
55   GO TO 5
60   PRINT "THE SOLUTIONS ARE COMPLEX CONJUGATES"
65   GO TO 5
70   PRINT "THERE IS ONE REAL SOLUTION OF MULTIPLICITY TWO"
75   GO TO 5
80   END
```

1.4   MORE EQUATION-SOLVING METHODS

In this section we consider some equation-solving techniques that do *not* necessarily transform a given equation into an equivalent equation.

Equations Containing Radicals   Let us investigate the results of raising each member of an equation to the same natural number power. In particular, let us consider the effect of raising each member of the two equations $x = 3$ and $x = -3$ to the fourth power. Thus:

$$x = 3 \qquad\qquad x = -3$$
$$x^4 = 3^4 \qquad\qquad x^4 = (-3)^4$$
$$x^4 = 81 \qquad\qquad x^4 = 81,$$

and observe that each time the same equation, $x^4 = 81$, has been generated. Now, in the set R of real numbers, the solution set of $x^4 = 81$ is $\{-3, 3\}$, the solution set of $x = 3$ is $\{3\}$ and the solution set of $x = -3$ is $\{-3\}$. Because their solution sets are not equal, neither $x = 3$ nor $x = -3$ is equivalent to the equation $x^4 = 81$, and we see that raising each member of an equation to the same natural number power does not necessarily generate an equivalent equation. However, note that $\{3\}$ and $\{-3\}$ are *subsets* of $\{-3, 3\}$. The preceding discussion suggests the next theorem.

Theorem 1.3   For each natural number n and for all x such that $A(x)$ and $B(x)$ name real numbers, the solution set of the equation $A(x) = B(x)$ is a subset of the solution set of the equation $[A(x)]^n = [B(x)]^n$.

An immediate implication of Theorem 1.3 is that if each member of a given equation is raised to the same natural number power, each solution of the resulting equation must be checked to determine whether or not it satisfies the given equation.

Theorem 1.3 is particularly useful when solving equations involving radical expressions, where we "remove" radical expressions by squaring both members, cubing both members, etc.

EXAMPLE 1   Solve $\sqrt{x} = x - 2$. <1>

SOLUTION   Squaring each member of the given equation, we have

$$x = x^2 - 4x + 4$$
$$x^2 - 5x + 4 = 0$$
$$(x - 1)(x - 4) = 0. \qquad <2>$$

By inspection, 1 and 4 are solutions of equation <2>, and we check to determine whether or not 1 and 4 are solutions to equation <1>. Replacing x by 1 in equation <1>, we have

$$\sqrt{1} = 1 - 2, \qquad \text{or} \qquad \sqrt{1} = -1,$$

a false statement. Hence, 1 is not a solution. Replacing x by 4, we have

$$\sqrt{4} = 4 - 2, \qquad \text{or} \qquad \sqrt{4} = 2,$$

a true statement. Hence, 4 is a solution and, by Theorem 1.3, the solution set of the given equation is $\{4\}$.

An equation containing more than one term involving radicals may require more than one application of Theorem 1.3.

EXAMPLE 2
$$\sqrt{x + 3} + \sqrt{x + 15} = 6$$
$$\sqrt{x + 3} = 6 - \sqrt{x + 15}$$
$$x + 3 = 36 - 12\sqrt{x + 15} + x + 15$$
$$-48 = -12\sqrt{x + 15}$$
$$\sqrt{x + 15} = 4$$
$$x + 15 = 16$$
$$x = 1.$$

Plainly, 1 is the only solution to the last equation. Replacing x by 1 in the given equation, we have

$$\sqrt{1+3}+\sqrt{1+15}=16 \quad \leftrightarrow \quad 2+4=6,$$

a true statement. Hence, the solution set of the given equation is $\{1\}$.

Equations Quadratic in Form   Some equations that are not quadratic can be rewritten as equations in the form

$$a[A(x)]^2+b[A(x)]+c=0, \tag{<3>}$$

where $A(x)$ names an expression in the variable x. A useful notational device for writing equations such as <3> in a simpler form is that of substituting another variable for $A(x)$. Thus, we may say "Let $y=A(x)$", in which case equation <3> takes the form $ay^2+by+c=0$.

EXAMPLE 3   Solve $x^{1/2}+x^{1/4}-12=0$.

SOLUTION   The given equation may be written as

$$[x^{1/4}]^2+[x^{1/4}]-12=0$$

or, letting $y=x^{1/4}$, as

$$y^2+y-12=0.$$

By the method of factoring, we have

$$(y+4)(y-3)=0$$

from which

$$y=-4 \quad \text{or} \quad y=3.$$

Replacing y by $x^{1/4}$, we obtain the two equations

$$x^{1/4}=-4 \quad \text{or} \quad x^{1/4}=3. \tag{<4>}$$

Raising each member of both equations <4> to the fourth power, we obtain

$$x=(-4)^4=256 \quad \text{or} \quad x=3^4=81.$$

Substituting 256 for x in the original equation, we find

$$(256)^{1/2}+(256)^{1/4}-12=0, \quad \text{or} \quad 16+4-12=0,$$

a false statement. Hence, 256 is not a solution. Substituting 81 for x in the original equation, we find

$$(81)^{1/2}+(81)^{1/4}-12=0, \quad \text{or} \quad 9+3-12=0,$$

a true statement. Hence, 81 is a solution and the solution set is $\{81\}$.

EXAMPLE 4   Solve $x^4 - 3x^2 - 4 = 0$, where $x \in C$.

SOLUTION   Viewing the equation as $[x^2]^2 - 3[x^2] - 4 = 0$ and letting $y = x^2$, we have

$$y^2 - 3y - 4 = 0$$

from which $y = 4$ or $y = -1$. Replacing y by x^2, we obtain the two equations

$$x^2 = 4 \quad \text{or} \quad x^2 = -1$$

from which

$$x = \pm 2 \quad \text{or} \quad x = \pm\sqrt{-1} = \pm i.$$

The solution set of the given equation is $\{-2, 2, -i, i\}$.

Exercise Set 1.4

(A)   *Solve each equation. (See Examples 1 and 2.)*

1. $\sqrt{x} = 12 - x$
2. $\sqrt{2x + 4} = x$
3. $\sqrt{x + 7} = x + 1$
4. $\sqrt{7 - x} = 5 + x$
5. $\sqrt{x + 4} = \sqrt{x - 3} + 1$
6. $\sqrt{x + 10} + \sqrt{-x} = 4$
7. $\sqrt{5 + x} + \sqrt{x + 5} = 0$
8. $\sqrt{x + 7} + \sqrt{x + 4} + 3 = 0$
9. $4 + \sqrt[3]{5 - 3x} = 0$
10. $\sqrt[3]{4x + 7} = 3$

Solve by substituting a new variable. (See Example 3.)

11. $x^{1/2} - x^{1/4} - 6 = 0$
12. $x^{1/2} + x^{1/4} - 2 = 0$
13. $x^{2/3} - 3x^{1/3} - 28 = 0$
14. $x^{2/3} - 5x^{1/3} + 6 = 0$
15. $x^{1/3} - 3x^{1/6} + 2 = 0$
16. $2x^{1/3} + 3x^{1/6} - 2 = 0$

Solve by substituting a new variable, $x \in C$. (See Example 4.)

17. $x^4 - 5x^2 - 36 = 0$
18. $x^4 + 8x^2 - 48 = 0$
19. $x^6 + 7x^3 + 10 = 0$
20. $x^6 + 5x^3 - 14 = 0$

(B)   *Solve each equation, $x \in R$.*

21. $(x + 2) + 3(x + 2)^{1/2} - 4 = 0$
22. $(x^2 - 6x)^2 - 2(x^2 - 6x) - 35 = 0$

Solve each equation for z. Specify the replacement set for x for which z will be real; for which z will be imaginary.

23. $z^4 + 5xz^2 + 6x^2 = 0$
24. $x^2z^4 + 3xz^2 + 2 = 0$

1.5   EQUIVALENT INEQUALITIES

If $A(x)$ and $B(x)$ are expressions in x, then

$$A(x) > B(x), \quad A(x) < B(x), \quad A(x) \geq B(x), \quad A(x) \leq B(x),$$

are *inequalities* in one variable x. Any member of the replacement set of *x* for which the given inequality is a true statement is a **solution** of that inequality. The set of all such solutions is the **solution set** of the inequality. To *solve an inequality* means to "find its solution set." If the solution set of an inequality is a proper subset of the replacement set of the variable, the inequality is a **conditional inequality**. If the solution set is equal to the replacement set, then the inequality is an **absolute inequality**.

EXAMPLE 1   If $x \in R$, then

a. $x > 2$ is a conditional inequality because its solution set is $\{x: x > 2\}$, the set of all real numbers greater than 2, and $\{x: x > 2\} \subset R$.

b. $x^2 \geq 0$ is an absolute inequality because the square of any real number is either a positive number or zero. Hence, the solution set of $x^2 \geq 0$ is equal to R.

Equivalent Inequalities   The solution set of Example 1a is readily determined by inspection. If the solution set of an inequality cannot be so determined, then it may be possible to generate a sequence of inequalities, each of which has the same solution set as the first, and such that the last inequality can be solved by inspection. The inequalities of such a sequence are *equivalent inequalities*.

Definition 1.2   Inequalities are **equivalent inequalities** if and only if they have equal solution sets.

Generating Equivalent Inequalities   The following two theorems which are generalizations of Theorems 0.15 and 0.16, together with the field axioms, provide methods for transforming given inequalities into equivalent inequalities.

Theorem 1.4   For all x such that $A(x)$, $B(x)$, and $C(x)$ denote real numbers,

$$A(x) > B(x) \qquad \text{and} \qquad A(x) + C(x) > B(x) + C(x)$$

are equivalent inequalities.   **(P)**

Theorem 1.5   For all x such that $A(x)$, $B(x)$, and $C(x)$ denote real numbers:

I   If $C(x) > 0$, then

$$A(x) > B(x) \qquad \text{and} \qquad A(x) \cdot C(x) > B(x) \cdot C(x)$$

are equivalent inequalities;

II   If $C(x) < 0$, then

$$A(x) > B(x) \qquad \text{and} \qquad A(x) \cdot C(x) < B(x) \cdot C(x)$$

are equivalent inequalities.   **(P)**

Theorems 1.4 and 1.5, appropriately reworded, remain valid with respect to the $<$, $\geq$, and $\leq$ relations. For example, the conclusion of Theorem 1.5–II can be stated as

II   If $C(x) < 0$, then

$$A(x) \leq B(x) \qquad \text{and} \qquad A(x) \cdot C(x) \geq B(x) \cdot C(x)$$

are equivalent inequalities.

In the next example, note particularly the "reversal" of sign from the third to the fourth inequalities, as specified by Theorem 1.5–II.

EXAMPLE 2   Solve $4 + x > 5x - 2$.

SOLUTION
$$4 + x > 5x - 2$$
$$4 + x + (-4 - 5x) > 5x - 2 + (-4 - 5x)$$
$$-4x > -6$$
$$-\tfrac{1}{4}(-4x) < -\tfrac{1}{4}(-6)$$
$$x < \tfrac{3}{2}$$

The solution set is $\{x: x < \tfrac{3}{2}\}$. The graph of this solution set is the open half-line shown in Figure 1.5–1. Note the use of an "open" circle to indicate a point that is *not* a point of the graph.

Figure 1.5-1

Theorems 1.4 and 1.5 are also applicable to solving "chains" of inequalities.

EXAMPLE 3   Solve $-1 \leq \dfrac{x-2}{4} < \dfrac{2}{3}$.

SOLUTION   We first multiply each member by the L.C.D., 12.

$$12(-1) \leq 12 \cdot \frac{x-2}{4} < 12 \cdot \tfrac{2}{3}$$
$$-12 \leq 3x - 6 < 8$$
$$-12 + 6 \leq 3x - 6 + 6 < 8 + 6$$
$$-6 \leq 3x < 14$$
$$\tfrac{1}{3}(-6) \leq \tfrac{1}{3} \cdot 3x < \tfrac{1}{3} \cdot 14$$
$$-2 \leq x < \tfrac{14}{3}$$

The solution set is $\{x: -2 \le x < \frac{14}{3}\}$. The graph of this solution set is the half-open interval shown in Figure 1.5-2.

Figure 1.5-2

If each member of an inequality is to be multiplied by a polynomial $P(x)$, then Theorem 1.5 indicates that we must distinguish between subsets of the replacement set of x for which $P(x) > 0$ and for which $P(x) < 0$.

EXAMPLE 4   Solve $\dfrac{x-4}{x+2} < 3$.   ⟨1⟩

SOLUTION   Note first that the replacement set of x is $\{x: x \in R,\ x \ne -2\}$. Now, one method of solving ⟨1⟩ involves multiplying each member of the inequality by $x+2$. But $x+2$ names a *positive* number if $x > -2$, and $x+2$ names a *negative* number if $x < -2$. Hence, we consider two cases. In Case 1, we take $x+2 > 0$ and seek a solution set that is a subset of $\{x: x > -2\}$. In Case 2, we take $x+2 < 0$ and seek a solution set that is a subset of $\{x: x < -2\}$. The union of these two subsets is the solution set of inequality ⟨1⟩.

Case 1 $x \in \{x: x > -2\}$

$$\frac{x-4}{x+2} < 3$$
$$\frac{x-4}{x+2}(x+2) < 3(x+3)$$
$$x-4 < 3x+6$$
$$-5 < x$$

Case 2 $x \in \{x: x < -2\}$

$$\frac{x-4}{x+2} < 3$$
$$\frac{x-4}{x+2}(x+2) > 3(x+2)$$
$$x-4 > 3x+6$$
$$-5 > x$$

From Case 1 we see that all numbers that are greater than -2 *and* also greater than -5 are solutions. But all numbers greater than -2 are certainly greater than -5. Hence, the solution set for Case 1 is $\{x: x > -2\}$. From Case 2 we see that all numbers that are less than -5 *and* also less than -2 are solutions. But all numbers less than -5 are also less than -2. Hence, the solution set for Case 2 is $\{x: x < -5\}$. The solution set of inequality ⟨1⟩ is

$$\{x: x > -2\} \cup \{x: x < -5\},$$

which can also be given as

$$\{x: x > -2 \text{ or } x < -5\}.$$

The graph of this solution set is the pair of open half-lines shown in Figure 1.5-3.

Figure 1.5-3

Quadratic Inequalities   Some inequalities that involve factorable polynomials can often be solved as in the next example.

EXAMPLE 5   Solve $x^2 - x - 6 \leq 0$. $\langle 2 \rangle$

SOLUTION   Factoring the left member of $\langle 2 \rangle$, we have

$$(x - 3)(x + 2) \leq 0,\qquad\qquad \langle 3 \rangle$$

and we first note that 3 and -2 are in the solution set of inequality $\langle 3 \rangle$. Next, the product in the left member of $\langle 3 \rangle$ will name a negative number for all values of x for which the factors $(x - 3)$ and $(x + 2)$ are numbers that are *opposite* in sign. Thus, we consider two cases.

Case 1 $(x - 3) > 0$ and $(x + 2) < 0$ or, equivalently,
           $x > 3$ and $x < -2$. $\langle 4 \rangle$

From $\langle 4 \rangle$, we see that all numbers that are greater than 3 and less than -2 are solutions. But there are no such numbers. Hence, the solution set for Case 1 is $\emptyset$.

Case 2 $(x - 3) < 0$ and $(x + 2) > 0$ or, equivalently,
           $x < 3$ and $x > -2$. $\langle 5 \rangle$

From $\langle 5 \rangle$, we see that all numbers less than 3 and greater than -2 are solutions. Hence, the solution set for Case 2 is $\{x: -2 < x < 3\}$.

Combining both cases, the solution set of inequality $\langle 2 \rangle$ is $\{3, -2\}$ $\cup \emptyset \cup \{x: -2 < x < 3\}$ or, more simply, $\{x: -2 \leq x \leq 3\}$. The graph of this solution set is the closed interval shown in Figure 1.5-4.

Figure 1.5-4

Solution by Sign Graphs   Examples 2 through 5 illustrate *algebraic* approaches to solving inequalities. Another approach, involving *sign graphs*, considers intervals in a line and is particularly useful for solving inequalities that involve products and quotients in one member and zero in the other member. Consider the expression $x - 1$, and note that

1. If $x = 1$, then $x - 1 = 0$;
2. If $x > 1$, then $x - 1 > 0$;
3. If $x < 1$, then $x - 1 < 0$.

These three facts are shown in Figure 1.5-5 by a diagram called a **sign graph**. The next examples illustrate how sign graphs can be used to solve inequalities containing products or quotients.

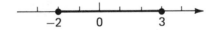

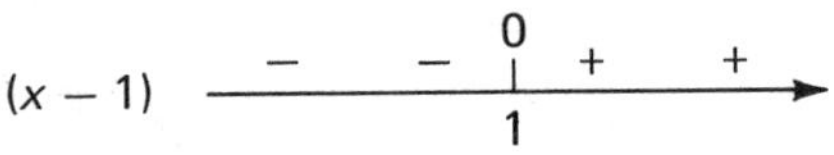

Figure 1.5-5

EXAMPLE 6   Use sign graphs to solve $x^2 + x - 2 \geq 0$.

SOLUTION   In factored form, we have

$$(x + 2)(x - 1) \geq 0. \qquad <6>$$

First, from $(x + 2)(x - 1) = 0$ we see that -2 and 1 are solutions. Next, the indicated product in $<6>$ will name a positive number for all values of x for which $x + 2$ and $x - 1$ are *both positive*, or *both negative*. From the sign graphs of $x + 2$ and $x - 1$ in Figure 1.5–6(a), we construct the sign graph of $(x + 2)(x - 1)$. As indicated by the figure, the solution set of the given inequality is

$$\{-2, 1\} \cup \{x: x < -2\} \cup \{x: x > 1\}, \qquad \text{or} \qquad \{x: x \leq -2 \text{ or } x \geq 1\}.$$

The graph of the solution set is given in Figure 1.5–6(b).

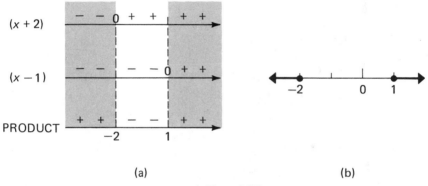

(a)   (b)

Figure 1.5-6

In sign graphs such as that in Figure 1.5–6(a), note that it is not essential to locate points "to scale," because we are concerned only with *changes in sign* on the different intervals. However, when graphing the solution set we do locate points to scale, as usual.

The diagram in Figure 1.5–6(a) can be used to solve other inequalities related to the given inequality of Example 6. For example, consider

$$(x + 2)(x - 1) < 0. \qquad <7>$$

By locating intervals in the sign graphs of $x + 2$ and $x - 1$ where the two factors are *opposite* in sign, we see that the solution set of $<7>$ is $\{x: -2 < x < 1\}$. You will find it instructive to use Figure 1.5–6(a) to solve $x^2 + x - 2 \leq 0$ and $x^2 + x - 2 > 0$.

EXAMPLE 7   Use sign graphs to solve

$$\frac{2}{x + 3} > \frac{1}{x - 4}, \qquad (x \neq -3, 4). \qquad <8>$$

SOLUTION   We begin by writing $<8>$ in the form $\dfrac{A(x)}{B(x)} > 0$:

(Continued on next page)

$$\frac{2}{x+3} - \frac{1}{x-4} > 0$$
$$\frac{x-11}{(x+3)(x-4)} > 0.$$

Next, we construct the sign graphs shown in Figure 1.5-7(a), where the symbol ♦ indicates that the quotient is not defined for that particular replacement value of x. By observing intervals over which the signs of the numerator and the denominator are *alike*, we see that the solution set of inequality $\langle 8 \rangle$ is

$$\{x: -3 < x < 4\} \cup \{x: x > 11\}, \quad \text{or} \quad \{x: -3 < x < 4 \text{ or } x > 11\}.$$

The graph of the solution set is shown in Figure 1.5-7(b).

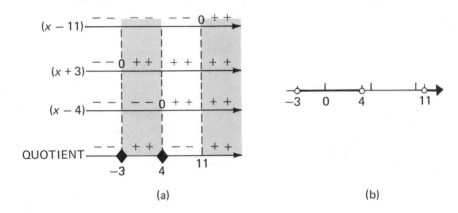

(a)   (b)

Figure 1.5-7

You will find it instructive to use Figure 1.5-7(a) to solve:

$$\frac{2}{x+3} \geq \frac{1}{x-4}; \quad \frac{2}{x+3} < \frac{1}{x-4}; \quad \frac{2}{x+3} \leq \frac{1}{x-4}.$$

Exercise Set 1.5

(A)   *Solve and graph each inequality. (See Examples 2 and 3.)*

1. $5 + x < 3x - 1$
2. $3 - 2x \leq 4 - x$
3. $6(x-1) \geq 9 + 4x$
4. $3(x-1) > 2(2+x) + 3x$
5. $-6 \leq \frac{1}{2}(x+3) \leq 6$
6. $-9 \leq 3(x-4) \leq 9$
7. $-2 < \frac{3x-5}{-4} < 3$
8. $-5 \leq \frac{3-x}{2} < 2$

9. Determine k so that $3x^2 + 2x - (k+1) = 0$ has no real solutions.
10. Determine k so that $2x^2 - 5x + k + 2 = 0$ has real and unequal solutions.

Solve and graph each inequality in two ways:
a.   algebraically (see Examples 4 and 5);
b.   by sign graphs (see Example 6).

11. $(x-4)(x+1) \leq 0$
12. $(x-5)(x-2) < 0$

13. $(2x+1)(x-3) > 0$

14. $(3x-2)(x+3) > 0$

15. $6x^2 + x - 2 \geq 0$

16. $15 - 2x - x^2 \leq 0$

17. $\dfrac{5-x}{x+1} < 0$

18. $\dfrac{x+4}{x-2} \geq 0$

19. $\dfrac{x-3}{x+1} \leq 2$

20. $\dfrac{2x-1}{x+2} < 1$

Solve and graph each inequality by sign graphs.

21. $\dfrac{3}{x-5} < \dfrac{1}{x+5}$

22. $\dfrac{-2}{x+5} > \dfrac{3}{x-4}$

(B) 23. $\dfrac{x}{x+2} \leq \dfrac{2}{x-1}$

24. $\dfrac{3x}{x-3} \geq \dfrac{-8}{x+2}$

25. $(x^2-5x)(3-x) < 0$

26. $(x^2+3x+2)(x+3) > 0$

27. $x^2 - 4x - 4 \geq 0$

28. $x^2 + 6x - 4 \leq 0$

(P)   I. Prove Theorem 1.4. (Hint: See Theorem 0.15.)

II. Prove Theorem 1.5. (Hint: See Exercises 0.3–XIV, XV.)

1.6   SENTENCES INVOLVING $|ax+b|$

In order to solve open sentences (equations or inequalities) that involve absolute value expressions such as $|ax+b|$, a much used technique is that of finding equivalent open sentences in which the absolute value expressions have been replaced by expressions that do not involve absolute value. One method for accomplishing this is to "separate" the set R of real numbers into two disjoint replacement sets for the variable x. To see how this comes about, we appeal to Definition 0.5 which states that

$$|r| = r \text{ if } r \geq 0; \qquad |r| = -r \text{ if } r < 0.$$

Now, if a and b are real numbers, then $ax+b$ names a real number for each $x \in R$ and, again by Definition 0.5, we have that

$$|ax+b| = ax+b \text{ if } ax+b \geq 0, \quad \left(\text{equivalently: if } x \geq -\frac{b}{a}\right);$$

$$|ax+b| = -(ax+b) \text{ if } ax+b < 0, \quad \left(\text{equivalently: if } x < -\frac{b}{a}\right).$$

For example, to solve the equation

$$|4x-3| = 9, \qquad\qquad <1>$$

where $x \in R$, we first note that

$$|4x-3| \text{ may be replaced by } 4x-3 \text{ if } x \geq \tfrac{3}{4},$$

$$|4x-3| \text{ may be replaced by } -(4x-3) \text{ if } x < \tfrac{3}{4}.$$

Next, we separate the set R of real numbers into the two disjoint replacement sets R_1 and R_2, where

$$R_1 = \{x\colon x \geq \tfrac{3}{4}\} \quad \text{and} \quad R_2 = \{x\colon x < \tfrac{3}{4}\}. \qquad \langle 2 \rangle$$

Finally, using the replacement sets listed in $\langle 2 \rangle$, from equation $\langle 1 \rangle$ we obtain the two associated equations

$$4x - 3 = 9, \quad x \in R_1; \qquad -(4x-3) = 9, \quad x \in R_2;$$

each of which can be readily solved. In actual practice we do not always need to specify the two disjoint replacement sets. Instead, when solving open sentences in forms such as:

$$|ax+b| = k; \qquad |ax+b| \geq k; \qquad |ax+b| > k;$$

$(k \geq 0)$, we generally proceed as in the following examples.

EXAMPLE 1   Solve $|4x-3| = 9$.

SOLUTION   We write and solve the two associated equations

$$\begin{array}{ccc} 4x-3=9 & \text{or} & -(4x-3)=9 \\ x=3 & & x=-\tfrac{3}{2}. \end{array}$$

The solution set is $\{-\tfrac{3}{2}, 3\}$.

EXAMPLE 2   Solve and graph the solution set:

    a. $|4x-3| > 9$;   b. $|4x-3| \geq 9$.

SOLUTION

    a. We write and solve the two associated inequalities

$$\begin{array}{ccc} 4x-3>9 & \text{or} & -(4x-3)>9 \\ 4x>12 & & 4x-3<-9 \\ x>3 & & x<-\tfrac{3}{2}. \end{array}$$

The solution set is $\{x\colon x < -\tfrac{3}{2} \text{ or } x > 3\}$. Figure 1.6-1 shows the two unbounded intervals that are the graph of the solution set. Note the "open" circles indicating that the points corresponding to $-\tfrac{3}{2}$ and 3 *are not* points of the graph.

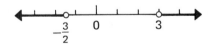

Figure 1.6-1

    b. The procedure here is identical to part (a), except that the symbols $>$ and $<$ are replaced by $\geq$ and $\leq$ respectively. The solution set is $\{x\colon x \leq -\tfrac{3}{2} \text{ or } x \geq 3\}$; the graph is shown in Figure 1.6-2. Note the "filled in" circles indicating that the points corresponding to $-\tfrac{3}{2}$ and 3 *are* points of the graph.

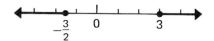

Figure 1.6-2

For inequalities in the form $|ax+b| < k$, $(k \geq 0)$, we modify slightly the procedure illustrated above. First, consider the two inequalities obtained by replacing $|ax+b|$ by $ax+b$ and by $-(ax+b)$:

$$ax+b < k \quad \text{or} \quad -(ax+b) < k. \qquad \langle 3 \rangle$$

Next, multiplying each member of the rightmost inequality of $\langle 3 \rangle$ by -1, by Theorem 1.5 we have

$$ax+b < k \quad \text{or} \quad ax+b > -k.$$

This pair of inequalities can be written as the "chain" of inequalities

$$-k < ax+b < k,$$

and solved as in Section 1.5.

EXAMPLE 3 Solve, and graph the solution set of: $|4x-3| < 9$.

SOLUTION From the two associated inequalities

$$4x-3 < 9 \quad \text{or} \quad -(4x-3) < 9$$

we obtain and solve the chain of inequalities

$$-9 < 4x-3 < 9$$
$$-6 < 4x < 12$$
$$-\tfrac{3}{2} < x < 3.$$

The solution set is $\{x: -\tfrac{3}{2} < x < 3\}$; Figure 1.6-3 shows the open interval that is the graph of the solution set. You will find it instructive to solve and graph the solution set of $|4x-3| \leq 9$ by following the above procedure after reading Example 2b again.

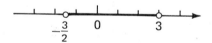

Figure 1.6-3

Consider the absolute value inequality

$$|x| \leq 4. \qquad \langle 4 \rangle$$

Applying the technique of Example 3, we have

$$x \leq 4 \quad \text{or} \quad -x \leq 4,$$

from which $-4 \leq x \leq 4$. Hence, the solution set of $\langle 4 \rangle$ is the closed interval $\{x: -4 \leq x \leq 4\}$. This result illustrates how we may use absolute value notation to specify closed or open intervals. In general, if $k > 0$, then either of the inequalities

$$|x| \leq k \qquad \text{or} \qquad -k \leq x \leq k$$

specifies a *closed interval*; either of the inequalities

$$|x| < k \qquad \text{or} \qquad -k < x < k$$

specifies an *open interval*. (See Figure 1.6-4.)

$$|x| \leq k \qquad\qquad\qquad |x| < k$$

Figure 1.6-4

Exercise Set 1.6

(A) *Solve each equation. (See Example 1.)*

1. $|x - 3| = 5$
2. $|x| = 5$
3. $|7x + 4| = 18$
4. $|7x + 4| = -18$
5. $\left|3 - \dfrac{x}{2}\right| = 5$
6. $\left|\dfrac{3 - x}{2}\right| = 5$

Solve and graph each inequality. (See Examples 2 and 3.)

7. $|x| < 5$
8. $|x| \geq 6$
9. $|x + 1| > 3$
10. $|x - 2| \leq 3$
11. $|x + 1| < 5$
12. $|x - 3| > 5$
13. $|x - 7| \geq 0$
14. $|x - 7| < 0$
15. $|2x + 3| \leq 4$
16. $|5 - 2x| \leq 3$
17. $|2x + 3| \geq 4$
18. $|7 - 2x| < 9$
19. $|3 - 2x| > 7$
20. $|3 - 4x| \geq 15$
21. $|4 - 3x| \leq 22$
22. $\left|4 - \dfrac{x}{2}\right| < 5$
23. $\left|\dfrac{x + 1}{3}\right| \geq 4$
24. $\left|\dfrac{x}{2} + \dfrac{1}{3}\right| \leq 3$
25. $\left|\dfrac{x - 3}{2}\right| > 4$
26. $\left|\dfrac{2 - x}{3}\right| < 2$
27. $\left|\dfrac{x - 3}{2}\right| \leq \dfrac{1}{4}$
28. $\left|\dfrac{x}{2} + \dfrac{1}{3}\right| > 3$
29. $\left|\dfrac{x + 1}{3}\right| < 4$
30. $\left|\dfrac{2 - x}{3}\right| \geq 2$

(B) *Equations in the form $|A(x)| = |B(x)|$ can be solved by considering the union of the solution sets of the two associated equations: $A(x) = B(x)$; $A(x) = -B(x)$. Solve each equation.*

31. $|5x - 6| = |x + 2|$
32. $|x - 2| = |2x + 4|$
33. $|x + 5| = 3|3x - 1|$
34. $|x + 1| = 2|6 - 5x|$

35. Show that $|x - a| < k$ specifies the open interval $a - k < x < a + k$ ($k \geq 0$). Specify the coordinate of the midpoint of the interval.
36. Determine how the graph of $|x - a| < k$ differs from the graph of $0 < |x - a| < k$.

Chapter 1 Self-test

[**1.1**] *Solve each equation.*

1. $(x-3)(x-2)=(x-9)(x+2)$

2. $\dfrac{3}{x}+\dfrac{x-2}{x+4}=1$

3. Solve for y: $ax+by+c=0$.

4. Tickets to a play sell for \$3, \$4 and \$5. A theater sold 270 tickets for a total of \$1020. If the number of \$3 tickets sold was twice the number of \$5 tickets sold, how many tickets were sold at each price?

[**1.2**] *Solve each equation.*

5. $x(x-7)+10=28$

6. $2x^2+8x-35=0$

7. Solve for y: $4x^2-25y^2=100$.

[**1.3**] *Determine the nature of the solutions of each equation.*

8. $10x^2+49x-33=0$

9. $10x^2-9x+33=0$

10. Towns A and B are 800 miles apart. Mr. Smith left town A at 8:00 A.M. to meet Ms. Jones, who left town B at 10:00 A.M. They met at a point half-way between the two towns. If Ms. Jones travels 10 mph faster than Mr. Smith, what time did they meet?

[**1.4**] *Solve each equation.*

11. $\sqrt{x-5}+\sqrt{x+7}=6$

12. $6x^{2/3}-13x^{1/3}+6=0$

Solve and graph each inequality.

[**1.5**] 13. $-3<\dfrac{x-4}{3}\le 2$

14. $\dfrac{2}{3x-1}>\dfrac{1}{x}$

[**1.6**] 15. $|2x-3|\le 7$

16. $|4x-1|>11$

RELATIONS AND FUNCTIONS

2.1   RELATIONS AND FUNCTIONS

You are already familiar with the ordinary usage of the word *relation*, as in relations between members of a family, relations between departments in a large company, etc. You are also familiar with mathematical usages of relation from your knowledge of relations between numbers, such as "is equal to" and "is less than." In this section we consider, in some detail, the mathematical concept of relation, and a special kind of relation called a *function*.

Cartesian Products   Let us begin by considering a "source" of ordered pairs of real numbers. Given sets A and B, the set of all possible pairs of real numbers with first components selected from A and second components selected from B is called the **Cartesian product** set of A and B, denoted by $A \times B$. Symbolically,

$$A \times B = \{(a, b): a \in A \text{ and } b \in B\}.$$

EXAMPLE 1   If $A = \{1, 2\}$ and $B = \{3, 4\}$, then

$A \times B = \{(1, 3), (1, 4), (2, 3), (2, 4)\};$

$B \times A = \{(3, 1), (3, 2), (4, 1), (4, 2)\};$

$A \times A = \{(1, 1), (1, 2), (2, 1), (2, 2)\}.$

Note that $A \times B$ is not equal to $B \times A$. This is a consequence of the fact that if the components of an ordered pair are reversed, the resulting ordered pair is not necessarily the same as the original ordered pair.

In this text, the Cartesian product set most often referred to is $R \times R$, constructed from the set R of real numbers.

Relations   If $A = \{1, 2\}$ and $B = \{1, 2, 3\}$, then

$$A \times B = \{(1, 1), (1, 2), (1, 3), (2, 1), (2, 2), (2, 3)\}.$$

Consider the subset G of $A \times B$, where

$$G = \{(1, 2), (1, 3), (2, 3)\},\qquad\qquad <1>$$

and observe that the components in each ordered pair of G are related in that *each second component is greater than its associated first component*. We say that G is the *is greater than* relation in $A \times B$. In fact, any subset of $A \times B$ is called a *relation*.

Definition 2.1   If X and Y are sets, not necessarily distinct, then a **relation** F in $X \times Y$ is any subset of

$$\{(x, y): x \in X \text{ and } y \in Y\}.$$

The **domain** of F is the set of all first components x such that $(x, y) \in F$; the **range** of F is the set of all second components y such that $(x, y) \in F$.

EXAMPLE 2   For the relation G as specified in $<1>$, the domain is $\{1, 2\}$ and the range is $\{2, 3\}$.

The variable used to represent first components (the elements in the domain) is often referred to as the *independent variable*; in which case the variable that represents second components (the elements in the range) is referred to as the *dependent variable*. Of course, variables other than x and y are used as independent and dependent variables, and letters other than F are used to name relations.

Specifying Relations   A relation is sometimes specified by verbal descriptions (is equal to; is greater than) or by a complete listing of its ordered pairs, as in relation G above. More frequently, relations are specified by means of an open sentence (or sentences), together with the domain over which the relation is defined. We shall return to this concept after a brief review of a second "source" of ordered pairs – solution sets of open sentences in two variables.

Consider the ordered pairs $(1, 3)$, $(2, 4)$, $(3, 5)$ and the equation in two variables

$$y = x + 2.\qquad\qquad <2>$$

If the first components 1, 2, and 3 of each ordered pair are successively substituted for x, together with the corresponding second components 3, 4, and 5 for y, we obtain the true statements

$$3 = 1 + 2;\qquad 4 = 2 + 2;\qquad 5 = 3 + 2.$$

From these results we conclude that each of the three ordered pairs *satisfies* equation $<2>$ and is therefore a *solution* of the equation. The set of all solutions of $<2>$ is the *solution set* of the equation. More generally, given an open sentence – equation or inequality – in two variables, each ordered pair of numbers that satisfies the sentence is a **solution**; the set of all solutions is the **solution set** of the sentence. Hence, the solution set of an open sentence in two variables is a subset of $R \times R$ and, by Definition 2.1, the solution set is therefore a relation in $R \times R$. The open sentence (or sentences) is said to "define the relation."

EXAMPLE 3   Consider the relation $H = \{(x, y): y = x + 2, -3 \leq x \leq 3\}$. The equation $y = x + 2$ defines relation H; the domain of H is $\{x: -3 \leq x \leq 3\}$, which is the closed interval that includes all the real numbers from -3 to 3, inclusive. Relation H may also be specified somewhat more simply by stating only the defining equation and the domain:

$$y = x + 2, \qquad -3 \leq x \leq 3.$$

Unless specified otherwise, relations in this text are assumed to be subsets of $R \times R$. That is, domains and ranges of relations are understood to be sets of real numbers. Furthermore, the *domain* of a relation is understood to be *the set of all real numbers for which a real number exists in the range.*

EXAMPLE 4   Specify the domain of the relation in $R \times R$ defined by

a. $y = \dfrac{2x}{x(x-4)};$ b. $y = \sqrt{x(x-4)}.$

SOLUTION

a.   The expression $\dfrac{2x}{x(x-4)}$ names a real number for every replacement of x for which the denominator is not equal to zero. (Division by zero is not defined.) Hence, the domain is $\{x: x \neq 0, 4\}$.

b.   The expression $\sqrt{x(x-4)}$ names a real number for every replacement of x for which $x(x-4) \geq 0$. As you can verify, this inequality has the solution set $\{x: x \leq 0 \text{ or } x \geq 4\}$. This set is the domain.

Functions   Of particular importance in mathematics are those relations in which no first component is paired with more than one second component. Such relations are called *functions.*

Definition 2.2   A **function** is a relation in which each element of the domain is paired with one and only one element of the range.

EXAMPLE 5   Consider the relations

$$f = \{(x, y): y = x + 4\} \qquad \text{and} \qquad F = \{(x, y): y > x + 4\}.$$

The relation f pairs each real number x with the sum $x + 4$. Because the sum of two real numbers is unique, it follows that each x is paired with one and only one real number, and so f is a function. The relation F pairs each real number x with all those real numbers that are greater than $x + 4$. Hence, each real number x is paired with more than one real number and so F is not a function.

A function in $R \times R$, that is, a function with sets of real numbers for its domain and range, is called a *real-valued function of a real variable*.

EXAMPLE 6 Find the domain and the range of the function G defined by $y = \sqrt{9 - x^2}$ so that G is a real-valued function of a real variable.

SOLUTION In order for $\sqrt{9 - x^2}$ to name a real number, we require that

$$9 - x^2 \geq 0.$$

The solution set of this inequality is $\{x: -3 \leq x \leq 3\}$. This set is the domain of G. To find the range, we first solve the defining equation for x to obtain

$$x = \pm\sqrt{9 - y^2}.$$

In order for $\sqrt{9 - y^2}$ to name a real number, we require that

$$9 - y^2 \geq 0.$$

This inequality has the solution set $\{y: -3 \leq y \leq 3\}$. However, because $\sqrt{9 - x^2}$ in the defining equation names a *non-negative* number, we see that y must be greater than or equal to zero. Hence, the range of G is $\{y: 0 \leq y \leq 3\}$.

Functional Notation A much used notation to name elements in the range of a function involves symbols such as $f(x)$, $g(x)$, $H(t)$, etc. Thus, if (x, y) is an ordered pair in a function f, then the dependent variable y is often denoted by $f(x)$, read "f of x" or "the value of f at x." For example, $f(-3)$ denotes the number in the range of f that is paired with the number -3 in the domain of f. Symbols y and $f(x)$ are used interchangeably. For example,

$$f = \{(x, y): y = x + 4\}, \qquad \{(x, f(x)): f(x) = x + 4\}$$

are alternate notations for the same function f. Still another notation is $f = \{(x, x + 4)\}$.

EXAMPLE 7 Given the function defined by $g(x) = x^2 + 4x - 1$:

a. $g(2) = 2^2 + 4(2) - 1 = 11$.

b. $g(-x) = (-x)^2 + 4(-x) - 1 = x^2 - 4x - 1$.

c. $g(x + a) = (x + a)^2 + 4(x + a) - 1 = x^2 + 2ax + a^2 + 4x + 4a - 1$.

Graphing Relations The components of an ordered pair of real numbers can be viewed as the coordinates of a point in a plane. Hence, we often use "point (x, y)" to mean "the point with coordinates given by the ordered pair (x, y)." With this agreement on language, we say that *the graph of a relation F is the set of all points (x, y) such that $(x, y) \in F$.* If a relation is defined by an equation or an inequality, then each ordered pair in the relation is a solution of the equation or inequality. It follows that the graph of a relation defined by an open sentence is the graph of the solution set of the sentence. As an important consequence of this concept, we have that: *a point is in the graph of a relation if and only if the co-ordinates of the point satisfy the open sentence that defines the relation.*

EXAMPLE 8   Graph the relation L defined by

$$y = 2x + 3, \qquad x \in \{-2, -1, 0, 2\}.$$

Show that (2, 5) is *not* a point in the graph of L.

SOLUTION   Successively substituting $-2, -1, 0, 1$, and 2 for x in the defining equation, we obtain $-1, 1, 3, 5$, and 7 as the corresponding values for y. Hence,

$$L = \{(-2, -1), (-1, 1), (0, 3), (1, 5), (2, 7)\}.$$

The graph of L is shown in Figure 2.1–1. Substituting 2 for x and 5 for y in the defining equation $y = 2x + 3$, we have

$$5 = 2 \cdot 2 + 3 \qquad \text{or} \qquad 5 = 7,$$

a false statement. Hence, (2, 5) is not in the graph of L.

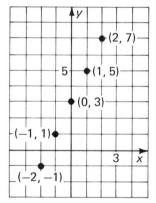

Figure 2.1-1

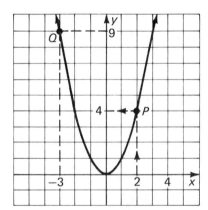

Figure 2.1-2

Mappings   The function concept may be geometrically interpreted as a **mapping** from one set to another. Figure 2.1–2 shows the graph of the function f defined by $y = x^2$. (In Section 2.4 we consider how such graphs are obtained.) By following the dashed line from 2 in the x-axis to point P in the graph and then over to the y-axis as shown, we find that function f pairs 2 and 4 and we say that "f maps 2 to 4." That is, $(2, 4) \in f$, and (2, 4) are the coordinates of P. Similarly, we observe that f maps -3 to 9 and $(-3, 9)$ are the coordinates of Q. More generally, we may say that "f maps x to x^2," and (x, x^2) are the coordinates of any point in the graph of f. The statement "f maps the set R of real numbers to the set of nonnegative numbers" asserts that R is the domain of f and that $\{y: y \geq 0\}$ is the range of f.

EXAMPLE 9   From the graph of function G shown in Figure 2.1–3, determine the number to replace each question mark in:

a. G maps -3 to ?   and G maps 0 to ?;

b. G maps ? to 5   and G maps ? to -1.

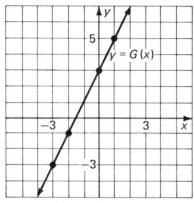

Figure 2.1-3

SOLUTION

a.   Reading successively from -3 and 0 in the x-axis to the graph, and then over to the y-axis, we find that:

G maps -3 to -3 and G maps 0 to 3.

b.   Reversing the procedure, we read from the y-axis to the graph, and then to the x-axis to find that:

G maps 1 to 5   and G maps -2 to -1.

The Vertical Line Test   In order for a relation to be a function, each number in the domain must be paired with one and only one number in the range. Now, if we inspect the graph of a relation, and mentally picture a vertical line "moving" across the graph from left to right, we can see whether any point in the x-axis maps to one and only one point in the y-axis or to more than one point in the y-axis. If, for any x, the vertical line intersects the graph more than once, then the graph is *not* the graph of a function. This procedure is referred to as the **vertical line test**.

EXAMPLE 10   By the vertical line test, the graph in Figure 2.1–4(a) on page 80 is that of a function; the graph in (b) is not that of a function.

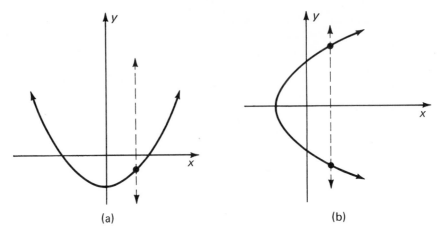

Figure 2.1-4

Exercise Set 2.1

(A)   *Given $A = \{1, 2, 3\}$ and $B = \{3, 4, 5\}$. Specify:*
   1. $A \times B$ 2. $B \times A$ 3. $A \times A$ 4. $B \times B$
   5. Specify: the "is less than or equal to" relation in $A \times A$; its domain; its range. (Hint: See Definition 2.1 and the discussion immediately preceding it.)
   6. Specify: the "is greater than" relation in $B \times B$; its domain; its range.

Specify the domain of the relation in $R \times R$ defined by each equation. (See Example 4.)

7. $y = \dfrac{1}{x^2 + 2x - 15}$

8. $y = \dfrac{1}{x^3 - 3x^2 - 4x}$

9. $y = \sqrt{x^2 + 2x - 15}$

10. $y = \sqrt{x^3 - 3x^2 - 4x}$

11. $y = \dfrac{1}{\sqrt{x^2 + 2x - 15}}$

12. $y = \dfrac{1}{\sqrt{x^3 - 3x^2 - 4x}}$

Find the domain and the range of the function defined by each equation. (See Example 6.)

13. $y = x^2$ 14. $y = x^2 - 6$ 15. $y = \sqrt{x - 1}$

16. $y = \dfrac{5x}{x - 1}$ 17. $y = \sqrt{x^2 - 16}$ 18. $y = \dfrac{3}{x - 2}$

Given $f(x) = x^2 - 3x + 2$, evaluate each expression. (See Example 7.)

19. $f(-3)$ 20. $f(5)$ 21. $f(0)$ 22. $f(x^2)$

23. $f(-x)$ 24. $f(x + 3)$ 25. $f(x - a)$ 26. $f(x + h)$

Given $f(x) = 6x - 12$ and $g(x) = \frac{1}{6}x + 2$. Evaluate:

27. $f[g(x)]$ 28. $g[f(x)]$

*Graph the relation defined by each equation over the domain $\{-3, -2,$
$-1, 0, 1, 2, 3\}$. Show that $(2, 3)$ is not a point in any of the graphs.*

29. $y = x$ 30. $y = -x$ 31. $y = 3 - x^2$

32. $y = x^2 + 3$ 33. $y = |x|$ 34. $y = |x - 2|$

*From the adjoining graph of a function F, determine the number (or
numbers) to replace each question mark. (See Example 9.)*

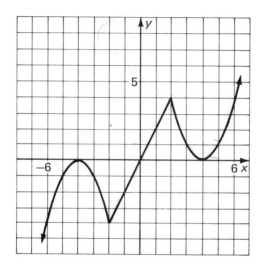

35. *F* maps -4 to ?   36. *F* maps 4 to ?   37. *F* maps ? to 4

38. *F* maps ? to -4 39. *F* maps 5 to ?   40. *F* maps -5 to ?

41. *F* maps ? to -2 42. *F* maps ? to 2

*Use the vertical line test to decide whether each of the following graphs
is, or is not, the graph of a function.*

43.

(a)

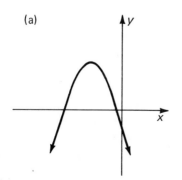

(b)

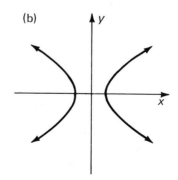

44. (a)

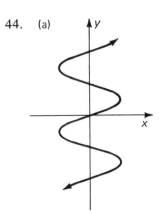

(b)

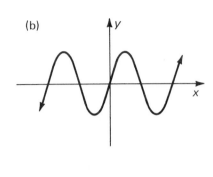

2.2   LINEAR FUNCTIONS

A *polynomial function* is a function defined by an equation of the form

$$y = P(x),$$

where $P(x)$ is a polynomial in x. In this section we consider polynomial functions where $P(x)$ is a polynomial of degree less than or equal to one.

Linear Functions   A **first degree equation** or **linear equation** in two variables x and y is any equation that can be written equivalently in the form

$$Ax + By + C = 0. \qquad \langle 1 \rangle$$

where A and B are not both zero. Equation $\langle 1 \rangle$ is called a **standard form** linear equation. Functions defined by linear equations are called **linear functions**. In the case that $B \neq 0$, we can reason that $\langle 1 \rangle$ defines a linear function with domain R as follows. First, we solve $\langle 1 \rangle$ for y:

$$y = -\frac{A}{B}x - \frac{C}{B}.$$

Next (for convenience) let us rename the constants $-\dfrac{A}{B}$ and $-\dfrac{C}{B}$ by a and b respectively, and write

$$y = ax + b. \qquad \langle 2 \rangle$$

Now, by the uniqueness of sums and products of real numbers, it follows that for each real number x the expression $ax + b$ denotes one and only one real number. Hence, equation $\langle 2 \rangle$, and equivalent equation $\langle 1 \rangle$, define linear functions.

Graphing Linear Functions   It can be shown that the graph of a linear function is a nonvertical line and that each nonvertical line is the graph of a linear function.

Customarily, we say "the graph of the equation" to mean "the graph of the set of ordered pairs that is the solution set of the relation defined by the equation." Furthermore, in those cases when the graph is readily identifiable, for example if it is a line, we may refer to the equation as "the line." Thus, we may say "the line $y = 3x + 5$."

Recall from your earlier study of algebra that we can construct the graph of a linear function by computing the coordinates of two points, together with a third point as a "check" point. Most often, we choose the points of intersection of the line and the axes. If the graph intersects the x-axis at $(a, 0)$, then a is called the **x-intercept**; if the graph intersects the y-axis at $(0, b)$, then b is called the **y-intercept**.

EXAMPLE 1   Sketch the graph of $2x + y = 6$; specify the x- and y-intercepts, if any.

SOLUTION   To find intercepts, we substitute 0 for x and solve for y; then we substitute 0 for y and solve for x. Thus,

$$2 \cdot 0 + y = 6 \quad \leftrightarrow \quad y = 6; \quad \text{the } y\text{-intercept is } 6;$$
$$2x + 0 = 6 \quad \leftrightarrow \quad x = 3; \quad \text{the } x\text{-intercept is } 3.$$

For a third point, we (arbitrarily) choose -1 for x:

$$2(-1) + y = 6 \quad \leftrightarrow \quad y = 8.$$

Figure 2.2-1 shows points $(0, 6)$, $(3, 0)$, $(-1, 8)$ and the line that is the required graph.

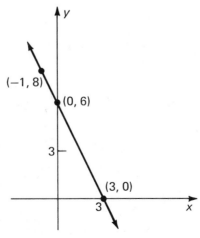

Figure 2.2-1

Two special cases with respect to $Ax+By+C=0$ need to be considered. First, if $B\neq0$ and $A=0$, we have

$$0\cdot x+By+C=0 \quad\leftrightarrow\quad y=k, \qquad\qquad <3>$$

where k is another name for the constant $-\dfrac{C}{B}$. For each real number k, the right-hand equation in $<3>$ defines a linear function called a **constant function**, because every real number in the domain is paired with the same real number k in the range. The graph of a constant function defined by $y=k$ is a horizontal line that is $|k|$ units above or below the x-axis, as in Figure 2.2-2(a). Next, if $B=0$ and $A\neq0$, then we have

$$Ax+0\cdot y+C=0 \quad\leftrightarrow\quad x=h,$$

where h is another name for the constant $-\dfrac{C}{A}$. The equation $x=h$ defines a relation that is *not* a function, because the only element in the domain is h, and h is paired with every real number in the range. The graph of a relation defined by $x=h$ is a vertical line $|h|$ units to the right or to the left of the y-axis, as in Figure 2.2-2(b).

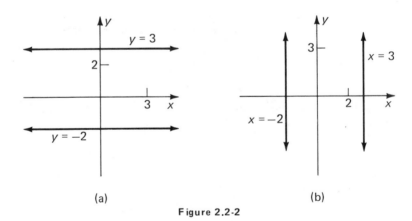

(a)　　　　　　　　　　　　　　(b)

Figure 2.2-2

Distance and Slope Two characteristic properties of a line involve the distance between any two of its points, and its "inclination" with respect to the x-axis.

In Figure 2.2-3, consider $\overline{P_1P_2}$ (read "segment P_1P_2") with endpoints $P_1(x_1, y_1)$ and $P_2(x_2, y_2)$. If we introduce $\overline{P_3P_2}$ and $\overline{P_1P_3}$, parallel to the x-axis and the y-axis, respectively, we see that the coordinates of P_3 are (x_1, y_2). Now, the distance between P_3 and P_2 is given by $|x_2-x_1|$ and the distance between P_3 and P_1 is given by $|y_2-y_1|$ as discussed on page 18. Thus, we can now compute the distance between two points if they are the endpoints of a segment *parallel* to either of the coordinate axes.

The distance between the endpoints of a segment not parallel to a co-ordinate axis can be found with the help of the *Pythagorean Theorem*, which asserts that the square of the length of the hypotenuse of a right triangle is equal to the sum of the squares of the lengths of the other two sides of the triangle. In Figure 2.2–3, consider right $\triangle P_1P_2P_3$. If d is the length of $\overline{P_1P_2}$, then

$$d^2 = |x_2 - x_1|^2 + |y_2 - y_1|^2,$$

from which

$$d = \sqrt{|x_2 - x_1|^2 + |y_2 - y_1|^2}.$$

But, $|x_2 - x_1|^2 = (x_2 - x_1)^2$ and $|y_2 - y_1|^2 = (y_2 - y_1)^2$, (see Exercise 2.2–43). Hence,

$$d = \sqrt{(x_2 - x_1)^2 + (y_2 - y_1)^2}. \qquad \langle 4 \rangle$$

Equation $\langle 4 \rangle$ is commonly referred to as the **Distance Formula**. By Theorem 0.18, $|a - b| = |b - a|$. Hence, when applying the distance formula, either of the two points may be regarded as the first point.

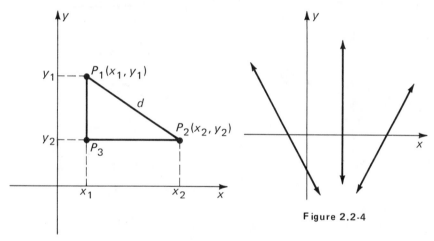

Figure 2.2-3

Figure 2.2-4

EXAMPLE 2   To find the distance between (2, 7) and (8, 4) by the distance formula, we have

$$d = \sqrt{(8-2)^2 + (4-7)^2} = \sqrt{45}, \qquad \text{or} \qquad 3\sqrt{5}.$$

Figure 2.2–4 shows three lines each of which displays a different "inclination" with respect to the x-axis. The fact that any two points determine a line enables us to assign a numerical measure to the inclination of a line.

Definition 2.3   For any two points $P_1(x_1, y_1)$ and $P_2(x_2, y_2)$ in a line, with $x_1 \neq x_2$, the **slope** m of the line is given by

$$m = \frac{y_2 - y_1}{x_2 - x_1}.$$

Note that, by Theorem 0.13,

$$\frac{y_2 - y_1}{x_2 - x_1} = \frac{-(y_2 - y_1)}{-(x_2 - x_1)} = \frac{y_1 - y_2}{x_1 - x_2},$$

from which it follows that either of the two points P_1 or P_2 may be viewed as the first point when computing a slope.

If P_1 and P_2 are any two points in the vertical line $x = h$ as in Figure 2.2-5(a), then the x-coordinates of P_1 and P_2 are the same, namely h. Then, $x_1 = x_2$, from which $x_2 - x_1 = 0$ and Definition 2.3 does not apply. We say that *vertical lines have no slope*, or that the slope is *not defined*. If Q_1 and Q_2 are any two points in the horizontal line $y = k$, as in Figure 2.2-5(b), then the y-coordinates of Q_1 and Q_2 are the same, namely k. Then, $y_1 = y_2$, from which $y_2 - y_1 = 0$ and, by Definition 2.3, the slope of the line is zero. Thus, *horizontal lines have zero slope*.

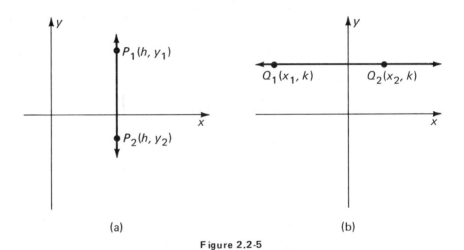

(a)   (b)

Figure 2.2-5

EXAMPLE 3   Find the slope of the line determined by each pair of points:

a. $(5, 2), (-3, 6)$;   b. $(5, 2), (-3, 2)$;   c. $(5, 2), (5, 6)$.

SOLUTION

a. $m = \dfrac{6 - 2}{-3 - 5} = \dfrac{4}{-8} = -\dfrac{1}{2}$.

b. $m = \dfrac{2 - 2}{-3 - 5} = \dfrac{0}{-8} = 0$.

c. $x_2 - x_1 = 5 - 5 = 0$; the line has no slope.

Forms of Linear Equations   A most significant property of any (non-vertical) line is that its slope m is the same no matter which two points are selected to compute m. This property can be used to obtain various useful forms of the equation of a line.

Consider the line that includes point $P_1(x_1, y_1)$ and that has slope m, and let $Q(x, y)$ be any other point in the line. Then, from the definition of slope,

$$m = \frac{y - y_1}{x - x_1},$$

we obtain the **point-slope form**

$$y - y_1 = m(x - x_1).$$

EXAMPLE 4 If a line includes point $(1, 3)$ and has slope -2, then the point-slope form of the equation of that line is

$$y - 3 = -2(x - 1).$$

If we know the y-intercept b of a line, then the definition of slope leads to a particularly useful form of equation. Consider a line that includes point $(0, b)$ and has slope m. If $Q(x, y)$ is another point in the line, then

$$m = \frac{y - b}{x - 0}.$$

Solving for y, we obtain the **slope-intercept form**

$$y = mx + b. \qquad \langle 5 \rangle$$

EXAMPLE 5 Given the line $2x - 3y = 14$, obtain the slope-intercept form equation of the line. Specify the slope and the y-intercept of the line.

SOLUTION Solving for y, we have the slope-intercept form

$$y = \tfrac{2}{3}x - \tfrac{14}{3}.$$

By comparison with equation $\langle 5 \rangle$, we see that the slope of the line (the coefficient of x) is $\tfrac{2}{3}$; the y-intercept is the *constant* term, $-\tfrac{14}{3}$.

Consider the line determined by $(a, 0)$ and $(0, b)$, where $a \neq 0$ and $b \neq 0$. The slope of the line is readily computed as $-\dfrac{b}{a}$, and so the slope-intercept form equation is

$$y = -\frac{b}{a}x + b,$$

from which we have

$$bx + ay = ab.$$

Multiplying each member of the latter equation by $\dfrac{1}{ab}$, we obtain

$$\frac{x}{a} + \frac{y}{b} = 1, \qquad \langle 6 \rangle$$

which we call the **double-intercept form** because a and b are the x- and y-intercepts, respectively.

EXAMPLE 6   Consider the line determined by the two points $(5, 0)$ and $(0, -4)$. Then, $a = 5$ and $b = -4$, and:

a.   The double-intercept form equation is: $\dfrac{x}{5} + \dfrac{y}{-4} = 1$.

b.   A standard form equation is: $4x - 5y - 20 = 0$.

c.   The slope-intercept form equation is: $y = \dfrac{4}{5}x - 4$.

Exercise Set 2.2

(A)   *Graph each line. Specify x-intercepts and y-intercepts, if any. (See Example 1.)*

1. $5x + y + 10 = 0$	2. $3x + 2y - 6 = 0$	3. $3x - 2y + 8 = 0$
4. $4x - 3y - 9 = 0$	5. $y = -6$	6. $x = -6$

In each of Exercises 7–10, graph two lines on the same axes and specify the point of intersection of the lines.

7. $x - y = 6$	8. $x - y = 6$
$\quad\ x = 5$	$\quad\ y = 5$
9. $3x - 5y = -15$	10. $4x + 3y = 9$
$\quad\ 2x + \ \ y = 16$	$\quad\ x - \ \ y = -3$

For each pair of points: Find the distance between the points. Find the slope of the line determined by the points. (See Examples 2 and 3.)

11. $(-3, 4), (2, -5)$	12. $(-4, 3), (5, -2)$	13. $(8, 0), (-3, 0)$
14. $(5, -2), (7, -2)$	15. $(0, 8), (0, -3)$	16. $(-2, 5), (-2, 7)$

Find the perimeter of the triangle with the given vertices.

17. $(0, 0), (0, 5), (3, 4)$	18. $(2, 4), (6, 2), (3, 6)$

Write each of the following equations in slope-intercept form (if possible). Specify the slope and the y-intercept of each line. (See Example 5.)

19. $3x + 5y - 6 = 0$	20. $-2x + y + 4 = 0$	21. $x - y + 2 = 0$
22. $3x - y + 5 = 0$	23. $5x - 6 = 0$	24. $y + 3 = 0$

Write the double-intercept form equation of the line with x-intercept a and y-intercept b. (See Equation $\langle 6 \rangle$.)

25. $a = 2, b = 5$	26. $a = \frac{1}{2}, b = -2$	27. $a = 3b$	28. $b = 5a$

Rewrite each equation in double-intercept form. Specify both intercepts.

29. $3x + 2y - 6 = 0$	30. $3x - 2y + 6 = 0$
31. $3x - 2y - 6 = 0$	32. $3x + 2y + 6 = 0$

Theorem: If m_1 and m_2 are the respective slopes of two nonvertical lines, then the two lines are:

I   Parallel if and only if $m_1 = m_2$;

II   Perpendicular if and only if $m_1 m_2 = -1$.

Each pair of lines are either: a. parallel; b. perpendicular; c. neither parallel nor perpendicular. Use the preceding theorem to determine which condition holds for each pair.

33. $2x + 3y + 5 = 0$
    $4x + 6y - 3 = 0$

34. $15x - 3y + 10 = 0$
    $5x - y + 8 = 0$

35. $x + y - 1 = 0$
    $x - y + 1 = 0$

36. $x - 3y + 2 = 0$
    $3x + y + 5 = 0$

37. $3x - y = 5$
    $x - 3y = -1$

38. $x + 2y = 1$
    $x + 1 = 2y$

(B) 39. A taxicab company charges 50¢ for the first quarter mile and 10¢ for each additional quarter mile of a trip. Express the fare F for a trip as a linear function of x, where x is the number of quarter miles after the first quarter mile. Find the total cost, and also the cost per mile, of a 3 mile trip; a 4 mile trip; a 5 mile trip.

40. A hardware company charges $6 each for the first five boxes of widgets, and $5 for each box after the first five, with a minimum purchase of 5 boxes. Express the cost C of boxes of widgets as a linear function of x, where x is the number of boxes after the first five. Find the total cost, and also the cost per box, of widgets if 10 boxes are ordered; 25 boxes are ordered; 100 boxes are ordered.

41. It costs $300 to make 200 toys and $580 to make 400 of the same toys. If the cost y is a linear function of the number x of toys, then $y = mx + b$. Determine m; determine b; find the cost of making 245 toys.

42. The monthly sales y of a company is a linear function of the amount x spent for advertising. If $2000 in advertising brings monthly sales of $52,000 and $8000 brings monthly sales of $61,000, how much in advertising will achieve monthly sales of $361,000?

43. Prove that $|a - b|^2 = (a - b)^2$. (Hint: Consider $a - b \geq 0$; $a - b < 0$.)

2.3   SOME SPECIAL FUNCTIONS

In this section we consider the graphs of some special functions that are defined by more than one equation. You will find that the ability to sketch graphs of linear functions will be helpful, even though the functions to be considered are not linear.

Absolute Value Functions   Consider the *absolute value* function f defined by $y = |x|$ or, synonymously, by

$$f(x) = |x|, \qquad x \in R. \qquad \langle 1 \rangle$$

By Definition 0.5 (absolute value), equation ⟨1⟩ is equivalent to the two-part definition of f:

$$\begin{bmatrix} f(x)= & x, & x \geq 0; \\ f(x)=-x, & x < 0. \end{bmatrix} \qquad \langle 2\rangle$$

Note that equations ⟨2⟩ define two linear functions, each over a different domain. The graph of function f is the union of the graphs of the two linear functions over their respective domains. Figure 2.3–1(a) shows the graph of $y=x$ over the domain $\{x: x \geq 0\}$; Figure (b) shows the graph of $y=-x$ over the domain $\{x: x < 0\}$; Figure (c) shows the graph of $f(x)= |x|$. By inspection of the graph it is apparent that the range of f is $\{y: y \geq 0\}$.

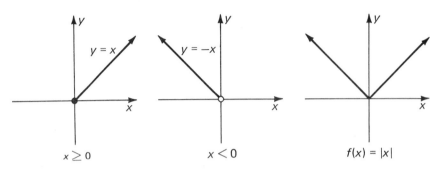

Figure 2.3-1

EXAMPLE 1   Graph the function F defined by $y= |x+1|$.

SOLUTION   By Definition 0.5, we have that

$|x+1|=x+1$ if $x+1 \geq 0$ or, equivalently, if $x \geq -1$;
$|x+1|=-(x+1)$ if $x+1 < 0$ or, equivalently, if $x < -1$;

from which we obtain the two-part definition of F:

$$\begin{bmatrix} F(x)=x+1, & x \geq -1; \\ F(x)=-(x+1), & x < -1. \end{bmatrix}$$

The graph of F is shown in Figure 2.3–2.

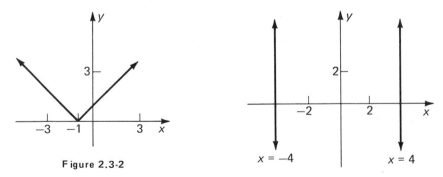

Figure 2.3-2

Figure 2.3-3

EXAMPLE 2   Graph $f = \{(x, y): |x| = 4\}$.

SOLUTION   By Definition 0.5, $|x| = x$ when $x \geq 0$, and $|x| = -x$ when $x < 0$. Hence, the required graph consists of the two vertical lines $x = 4$ and $x = -4$ or, equivalently,

$$x = 4 \qquad \text{and} \qquad x = -4,$$

as shown in Figure 2.3–3.

Step Functions   During certain hours of the night, telephone calls between two distant cities cost 20¢ per minute or fraction thereof. If y denotes the cost of a call in cents and x denotes the duration of a call in minutes, then we have a cost function defined by:

$$\left[\begin{array}{ll} y = 20, & 0 < x \leq 1; \\ y = 40, & 1 < x \leq 2; \\ y = 60, & 2 < x \leq 3; \\ \vdots & \vdots \end{array} \right.$$

Figure 2.3–4 shows the graph of this cost function, suggesting why such functions are called *step functions*. From the graph, observe that the domain is R^+, the set of positive real numbers; the range is $\{20, 40, 60, \ldots\}$.

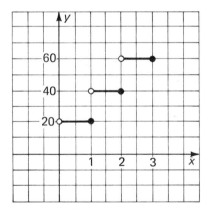

Figure 2.3-4

The symbol $[x]$, read "bracket x," may be used to mean *the greatest integer less than or equal to x*. Thus, $[-3] = -3$; $[-1.4] = -2$; $[0] = 0$; $[3.2] = 3$; etc. The function defined by $y = [x]$ is called the *greatest integer function*, or, sometimes, the *bracket function*.

EXAMPLE 3   Sketch the bracket function defined by $y = [x]$.

SOLUTION   We consider unit intervals along the x-axis. For example:

$$
\begin{array}{llll}
\text{If} & -2 \leq x < 1, & \text{then} & y = -2; \\
\text{If} & -1 \leq x < 0, & \text{then} & y = -1; \\
\text{If} & 0 \leq x < 1, & \text{then} & y = 0; \\
\text{If} & 1 \leq x < 2, & \text{then} & y = 1; \quad \text{etc.}
\end{array}
$$

The graph is shown in Figure 2.3–5. Observe that the domain of $y = [x]$ is the set R of real numbers; the range is the set J of integers.

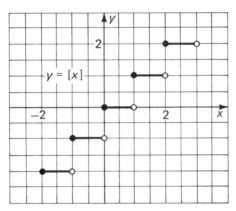

Figure 2.3-5

As a last example we introduce a function that has applications in higher mathematics. It is called the *signum function*, symbolized by $f(x) = \text{sgn } x$, and defined by

$$
\begin{array}{lll}
f(x) = 1, & x > 0, \\
f(x) = 0, & x = 0, \\
f(x) = -1, & x < 0.
\end{array}
$$

The graph of $f(x) = \text{sgn } x$ is shown in Figure 2.3–6. Observe that the domain is R; the range is $\{-1, 0, 1\}$.

EXAMPLE 4   Graph $G(x) = \text{sgn } (x + 1)$.

SOLUTION   Applying the above definition of the signum function, we have that $\text{sgn } (x + 1) = 1$ when $x + 1 > 0$ or, equivalently, when $x > -1$; $\text{sgn } (x + 1) = 0$ when $x = -1$; $\text{sgn } (x + 1) = -1$ when $x < -1$. Hence, we have the three-part definition

$$
\begin{array}{lll}
G(x) = 1, & x > -1, \\
G(x) = 0, & x = -1, \\
G(x) = -1, & x < -1.
\end{array}
$$

The graph is shown in Figure 2.3–7.

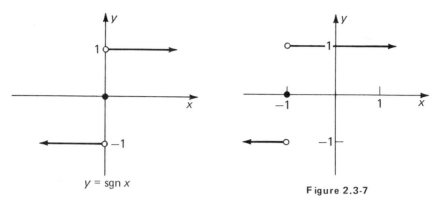

$y = \text{sgn } x$

Figure 2.3-6

Figure 2.3-7

Exercise Set 2.3

(A) *Graph each function over the indicated domain.*

1. $f(x)=x-3, \quad x \geq 3$
2. $f(x)=-x+3, \quad x<3$
3. $g(x)=-2x-5, \quad x<-\frac{5}{2}$
4. $g(x)=2x+5, \quad x \geq -\frac{5}{2}$
5. $\begin{bmatrix} F(x)=x+2, & x \geq -2 \\ F(x)=-x-2, & x<-2 \end{bmatrix}$
6. $\begin{bmatrix} G(x)=x-3, & x \geq 3 \\ G(x)=-x+3, & x<3 \end{bmatrix}$

Graph each function. (See Example 1.)

7. $y=|x+2|$
8. $y=|x-3|$
9. $y=-|x+2|$
10. $y=-|x-3|$
11. $y=|x|-2$
12. $y=|x|+2$

Graph each function. (See Example 2.)

13. $\{(x, y): |x|=3\}$
14. $\{(x, y): |x+2|=5\}$
15. $\{(x, y): |x-1|=2\}$
16. $\{(x, y): |2x-1|=4\}$
17. $\{(x, y): |y|=3\}$
18. $\{(x, y): |y+1|=2\}$

19. Air mail (at this writing) is 11 cents per ounce or fraction thereof. Sketch a graph of the cost function for air mail letters from 0 ounces to 5 ounces, inclusive.

20. A power tool is rented for $8 for the first hour (or fraction thereof). The rental is reduced by $1 per hour for each hour (or fraction thereof) up to a maximum of 5 hours. Sketch a graph of the cost-per-hour rental function for renting the tool from 0 to 5 hours, inclusive.

Graph each function. (See Examples 3 and 4.)

21. $y=[x+2]$
22. $y=[x-1]$
23. $y=[x]+2$
24. $y=[x]-1$
25. $y=\text{sgn } (x-2)$
26. $y=\text{sgn } (x+4)$
27. $y=2+\text{sgn } x$
28. $y=3-\text{sgn } x$

(B)   *Graph f(x) and g(x) on separate sets of axes.*

29. $\begin{bmatrix} f(x) = \dfrac{|x|}{x}, & x \neq 0 \\ f(x) = 0, & x = 0 \end{bmatrix}$ and $g(x) = \text{sgn } x.$

30. $\begin{bmatrix} f(x) = \dfrac{|x-2|}{x-2}, & x \neq 2 \\ f(x) = 0, & x = 2 \end{bmatrix}$ and $g(x) = \text{sgn } (x-2).$

31. $f(x) = |x|$ and $g(x) = \sqrt{x^2}.$

32. $f(x) = |x-2|$ and $g(x) = \sqrt{(x-2)^2}.$

33. In each of Exercises 29 to 32, what conclusion about f and g is suggested by the graphs?

2.4   QUADRATIC FUNCTIONS

A **second degree equation** or **quadratic equation** in two variables x and y is an equation that can be written equivalently in the form

$$Ax^2 + Bxy + Cy^2 + Dx + Ey + F = 0,$$

where A, B and C are not all zero. In this text we will rarely be concerned with equations that include the Bxy term. Hence, we focus our attention on the equation

$$Ax^2 + Cy^2 + Dx + Ey + F = 0, \qquad \qquad \langle 1 \rangle$$

where A and C are not both zero. In the case that $A \neq 0$, $C = 0$ and $E \neq 0$, (other cases are considered in Chapter 3), equation $\langle 1 \rangle$ can be solved for y to obtain an equation in the form

$$y = ax^2 + bx + c, \qquad \qquad \langle 2 \rangle$$

where a, b and c are real numbers and $a \neq 0$. For example,

$$3x^2 + 6x - y + 4 = 0 \quad \leftrightarrow \quad y = 3x^2 + 6x + 4.$$

The quadratic polynomial $ax^2 + bx + c$ pairs one and only one real number with each real number x. Hence, equation $\langle 2 \rangle$ defines a function; such functions are called **quadratic functions**.

Graphing Quadratic Functions   It can be shown that the graph of a quadratic function is a **parabola**. Figure 2.4–1 shows two typical parabolas. We say that the parabola in Figure 2.4–1(a) "opens upwards;" the "lowest" point in the graph is called the **vertex**. The parabola in Figure 2.4–1(b) "opens downwards;" the "highest" point is the **vertex**. In each case the vertical line through the vertex is called the **axis of symmetry** or, simply the **axis**, of the parabola, because the graph on each side of the axis is a "mirror image" of the graph on the opposite side. Observe that, by the vertical line test, each of the parabolas is the graph of a function.

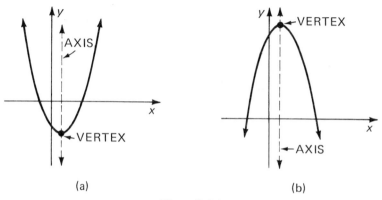

(a) (b)

Figure 2.4-1

Given a linear function, we need only a few points to efficiently sketch the graph, because we know that the graph is a line. Given a quadratic function, we know that the graph is a parabola but, in order to efficiently sketch the graph, it is desirable to obtain information other than just a few points. In particular, we consider: x- and y-intercepts*; whether the parabola opens up or down; the coordinates of the vertex; how to make use of the axis of symmetry. In order to obtain this information, we apply the technique of completing the square in x to the right member of

$$y = ax^2 + bx + c$$

(see Exercise 2.4-37), and transform it into an equivalent equation in the form

$$y = a(x - h)^2 + k, \qquad a \neq 0. \qquad \langle 3 \rangle$$

Equation $\langle 3 \rangle$ is called the **standard-form** equation of a parabola with a vertical axis. As an example,

$$\begin{aligned} y = 3x^2 + 6x + 4 \quad &\leftrightarrow \quad y - 4 = 3(x^2 + 2x \qquad) \\ & y - 4 + 3 = 3(x^2 + 2x + 1) \\ & y - 1 = 3(x + 1)^2 \\ & y = 3(x + 1)^2 + 1. \qquad \langle 4 \rangle \end{aligned}$$

From equation $\langle 4 \rangle$ we see that $a = 3$, $h = -1$ and $k = 1$. If $b = 0$, that is, if there is no linear term in x, then of course there is no need to complete the square in x.

The significance of standard-form equation $\langle 3 \rangle$ stems from the information to be derived from the numbers h, k and a. First, because the square of any real number is nonnegative, note that $(x - h)^2 \geq 0$. Next, consider the following cases:

1. $x = h$: In this case, $a(x - h)^2 = a(h - h)^2 = 0$ and

$$y = a(h - h)^2 + k \leftrightarrow y = k. \qquad \langle 5 \rangle$$

---

*Although the intercepts can always be found, for the purpose of sketching a graph it may not always be worth the effort involved in computing *irrational* x-intercepts.

2. $x \neq h$ and $a > 0$: In this case $a(x-h)^2 > 0$ and

$$a(x-h)^2 + k > k \leftrightarrow y > k. \qquad <6>$$

3. $x \neq h$ and $a < 0$: In this case $a(x-h)^2 < 0$ and

$$a(x-h)^2 + k < k \leftrightarrow y < k. \qquad <7>$$

Now, from results $<5>$ and $<6>$ we conclude that if a is positive, then y is always greater than or equal to k and so (h, k) is the lowest point of the parabola. It follows that *for $a > 0$ the parabola opens upwards and (h, k) is its vertex.* Similarly, reasoning from results $<5>$ and $<7>$, we conclude that *for $a < 0$ the parabola opens downwards and (h, k) is its vertex.* In each case the line $x = h$ is the *axis of symmetry*. The following example illustrates how this information may be used.

EXAMPLE 1   Sketch the graph of the function defined by $y = x^2 - 8x + 12$.

SOLUTION

1. Substituting 0 for x in the given equation, we find that 12 is a y-intercept. Substituting 0 for y, we obtain the equation

$$0 = x^2 - 8x + 12$$

with solution set $\{2, 6\}$. Hence, 2 and 6 are x-intercepts and we now have three points of the graph: (0, 12), (2, 0) and (6, 0).

2. Completing the square in x, we have

$$y = x^2 - 8x + 12 \quad \leftrightarrow \quad y - 12 = x^2 - 8x$$
$$y - 12 = x^2 - 8x + 16 - 16$$
$$y = (x - 4)^2 - 4,$$

from which we see that $a = 1$, $h = 4$ and $k = -4$. Hence, the graph is a parabola opening upwards, with vertex $(4, -4)$ and axis $x = 4$.

3. Point (0, 12) is four units to the *left* of the axis of symmetry $x = 4$. Hence, point (8, 12) which is four units to the *right* of the axis is also a point of the graph. From the given equation we can compute a few more (arbitrarily chosen) points. For example, you may verify that $(3, -3)$ and (1, 5) are points of the graph. Because $(3, -3)$ is one unit to the *left* of the axis, we also have point $(5, -3)$ one unit to the *right* of the axis. Similarly, paired with (1, 5) is (7, 5). Figure 2.4–2 shows all of the points obtained above, together with the axis of symmetry and the graph of the given function.

If the graph of a quadratic function is a parabola opening upwards, then the vertex is the lowest point and the y-coordinate of the vertex is the least, or *minimum*, function value. If the parabola opens downwards, then the vertex is its highest point and its y-coordinate is the greatest, or *maximum*, function value.

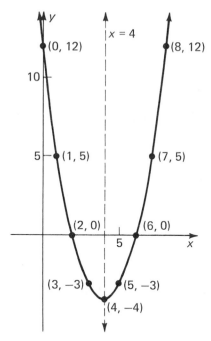

Figure 2.4-2

EXAMPLE 2   What is the greatest rectangular area that can be enclosed by 100 feet of fencing? Specify the dimensions of the rectangle of maximum area.

SOLUTION   Let x denote the length of one side. Then $50-x$ denotes the length of an adjacent side of the rectangle. The area A is given by

$$A = x(50-x) \quad \leftrightarrow \quad A = -x^2 + 50x.$$

The equation on the right defines a quadratic function whose graph is a parabola opening downwards. Hence, the maximum area A is given by the y-coordinate of the vertex, which can be found by completing the square.

$$y = -x^2 + 50x \quad \leftrightarrow \quad y = -(x^2 - 50x \qquad)$$
$$y - 625 = -(x^2 - 50x + 625)$$
$$y = -(x-25)^2 + 625.$$

From the last equation we find the vertex is $(25, 625)$. It follows that the maximum area is 625 square feet. The dimensions of the rectangle are given by $x = 25$ and $50 - x = 25$. That is, the rectangle of maximum area is a square, 25 feet by 25 feet.

Exercise Set 2.4

(A)   *Graph each quadratic function; specify the coordinates of the vertex and the equation of the axis. (See Example 1.)*

1. $y = x^2 + 3$

2. $y = x^2 - 3$

3. $y = -x^2 - 3$

4. $y = -x^2 + 3$

5. $y = 2(x + 3)^2$

6. $y = 2(x - 3)^2$

7. $y = -2(x - 3)^2 + 3$

8. $y = -2(x + 3)^2 - 3$

9. $y = x^2 - 2x$

10. $y = x^2 - 3x$

11. $y = x^2 + 4x + 8$

12. $y = -x^2 - 4x + 8$

13. $y = -x^2 + 4x - 8$

14. $y = -x^2 - 4x - 8$

15. $y = 4x^2 - 12x + 12$

16. $y = 4x^2 + 12x + 12$

17. $y = -6x^2 - 18x - 7$

18. $y = 6x^2 + 18x + 7$

19. $y = \frac{2}{3}x^2 - \frac{4}{3}x + 2$

20. $y = -\frac{1}{2}x^2 - 6x - 5$

Sketch both graphs on the same set of axes; specify the points of intersection of the graphs.

21. $y = -x^2 + 8$
    $y = x^2$

22. $y = -x^2 - 6x - 5$
    $y = 2x + 7$

23. $y = x^2 - 6x + 13$
    $y = x + 3$

24. $y = -x^2 + 6x - 1$
    $y = x^2 - 6x + 9$

On the same set of axes sketch the graph of each of the following functions for $t = 1, 2, 3$.

25. $y = tx^2$

26. $y = -tx^2$

27. $y = (x - t)^2$

28. $y = (x + t)^2$

29. $y = -x^2 + t$

30. $y = x^2 - t$

Example 2 suggests a method for Exercises 31–36.

31. What is the maximum rectangular area that can be enclosed by 100 feet of fencing if only three sides need to be fenced? Specify the dimensions of the rectangle.

32. Find two positive numbers whose sum is 40 and whose product is a maximum.

33. Find a number such that the difference between the number and the square of the number is a maximum.

34. If a ball is thrown vertically upwards with a velocity of 9.8 m/sec, its height s (in meters) after t seconds is given by $s = 9.8t - 4.9t^2$. Find the maximum height to which the ball will rise, and the time needed to reach that height.

(B) 35. A 40-seat restaurant makes a daily profit of $6 per seat. If the number of seats is increased to more than 40, the daily profit on each seat decreases by 5¢ for each seat in excess of 40. If x seats are added on, express the daily profit on each seat. Express the total daily profit. Find x so that the restaurant will earn a maximum daily profit.

36. A tour bus company charges $9 per passenger for a minimum of 32 passengers. The price is reduced 25¢ per ticket for each passenger in excess of 32. How many tickets should the company sell to provide a maximum amount of money in fares?

37. Show that the equation $y = ax^2 + bx + c$ can be written in the form
    $y = a(x - h)^2 + k$, where $h = -\dfrac{b}{2a}$ and $k = \dfrac{4ac - b^2}{4a}$.

2.5 VARIATION

Certain frequently encountered functions have traditionally been referred to by special names. These kinds of functions, together with some associated language, are considered in this section.

Direct Variation Any function defined by an equation of the form

$$y = kx^n \qquad\qquad \langle 1 \rangle$$

where n is a positive number and $k \in R$, is called a **direct variation**. We say that y *varies directly as* x^n. The constant k is called the **constant of variation**. Note that equation $\langle 1 \rangle$ defines a linear function when $n = 1$ and a quadratic function when $n = 2$.

EXAMPLE 1

a. The function defined by $y = 2x$ is a direct variation. We say that y varies directly as x; 2 is the constant of variation.

b. The area A of a circle varies directly as the square of the length of a radius. This is expressed by the familiar $A = \pi r^2$, where π is the constant of variation.

EXAMPLE 2 If y varies directly as the cube of x, write an equation expressing this variation. If $y = 54$ when $x = 3$, find y when $x = 4$.

SOLUTION The function is defined by an equation of the form $y = kx^3$. To find the constant of variation k, we substitute 54 for y and 3 for x and obtain

$$54 = k(3)^3, \quad \text{or} \quad k = 2.$$

Hence, $y = 2x^3$. Substituting 4 for x, we have

$$y = 2(4)^3, \quad \text{or} \quad y = 128.$$

Inverse Variation Any function defined by an equation in either of the forms

$$x^n y = k, \quad \text{or} \quad y = \frac{k}{x^n},$$

where n is a positive number, $x \neq 0$, and $k \in R$, is called an **inverse variation**. We say that y *varies inversely as* x^n.

EXAMPLE 3

a. The function defined by $y = \dfrac{10}{x^2}$ is an inverse variation. We say that y varies inversely as the square of x; 10 is the constant of variation.

b. If the electromotive force E applied across a wire is kept constant, then the current I varies inversely as the resistance R of the wire. This can be expressed by the equation $IR = E$, where E is the constant of variation.

EXAMPLE 4   If y varies inversely as the square root of x, write an equation expressing this variation. If $y = 50$ when $x = 4$, find y when $x = 25$.

SOLUTION   The function is defined by an equation of the form $\sqrt{x}\, y = k$ or $y = \dfrac{k}{\sqrt{x}}$, $x \neq 0$. To find the constant of variation k, we substitute 50 for y and 4 for x, and obtain

$$(\sqrt{4})(50) = k \quad \text{or} \quad k = 100.$$

Hence, $\sqrt{x}\, y = 100$. Substituting 25 for x, we have $\sqrt{25}\, y = 100$, or $y = 20$.

Joint Variation   The concept of variation applies to more than one variable, as well as to combinations of variations. If y varies directly as the product of powers of two (or more) variables, say x and z, we write

$$y = kx^n z^m$$

where n and m are positive numbers and $k \in R$. We say that y *varies jointly as x^n and z^m*; k is the constant of variation. For example, the volume V of a right circular cylinder varies jointly as the square of the length r of a base radius and the height h of the cylinder. Hence,

$$V = \pi r^2 h,$$

where π is the constant of variation.

EXAMPLE 5   If z varies jointly as x and the square of y, and inversely as t, write an equation expressing this combination of variations. If $z = 6$ when $x = 6$, $y = 2$ and $t = 8$, find z when $x = 2$, $y = 4$ and $t = 16$.

SOLUTION   The combination of variations is expressed by the equation

$$z = \frac{kxy^2}{t}, \quad t \neq 0.$$

To find k, we substitute the given values of x, y and t:

$$6 = \frac{k \cdot 6 \cdot 2^2}{8}, \quad \text{or} \quad k = 2.$$

Hence, for $x = 2$, $y = 4$ and $t = 16$:

$$z = \frac{2 \cdot 2 \cdot 4^2}{16}, \quad \text{or} \quad z = 4.$$

Exercise Set 2.5

(A) *Write an equation expressing each of the variations described. Use k as the constant of variation. (See Examples 1 and 3.)*

1. The perimeter P of a square varies directly as the length of a side.
2. The area A of a square varies directly as the square of the length s of a side.
3. The time t for a planet to revolve once around the sun varies directly as the $\frac{3}{2}$ power of the (average) distance d from the sun.
4. The number s of meters fallen by a freely falling body in a vacuum varies directly as the square of the time t in seconds.
5. The force F between two electrically charged particles varies inversely as the square of the distance d between them.
6. The density D of a given mass varies inversely as its volume V.
7. The temperature T of a gas varies jointly as the pressure P and the volume V.
8. The voltage drop E in an electrical circuit varies jointly as the current I and the resistance R.
9. The gravitational attraction force f between two objects varies jointly as their masses m_1 and m_2 and inversely as the square of the distance r between them.
10. The strength S of a rectangular cross-section beam varies jointly as the breadth b and the square of the depth d, and inversely as the length L.

For Exercises 11–16, refer to Examples 2, 4 and 5.

11. If y varies directly as the cube of x and $y=64$ when $x=2$, find y when $x=3$.
12. If y varies directly as the square root of x and $y=55$ when $x=25$, find x when $y=121$.
13. If s varies inversely as the cube root of t and $s=2$ when $t=8$, find s when $t=27$.
14. If z varies inversely as the fourth power of T and $z=4$ when $T=4$, find T when $z=1$.
15. If z varies directly as the square of y and inversely as the square root of x, and if $z=4$ when $x=4$ and $y=1$, find y when $x=9$ and $z=1$.
16. If z varies jointly as the cube of x and the square of y, and if $z=32$ when $x=2$ and $y=2$, find z when $x=\frac{1}{2}$ and $y=4$.
17. The exposure time for making an enlargement from a negative varies directly as the area of the enlargement. If it takes 10 seconds to make a 5 inch by 7 inch enlargement, how long will it take to make an enlargement that is (a) $2\frac{1}{2}$ inches by $3\frac{1}{2}$ inches? (b) 9 inches by 14 inches?
18. The simple interest i on a loan varies jointly as the amount P in dollars and the time t in years. If $i=\$67.50$ when $P=\$500$ and $t=3$ years, find i when $P=\$875$ and $t=2\frac{2}{3}$ years.

19. The volume of a gas is 100 cubic centimeters when the pressure is 4 pounds per square inch and the temperature is 500° Kelvin. (See Exercise 7.) Find the volume when the pressure is increased to 8psi and the temperature is dropped to 400° K.

20. The force between two electrically charged particles is 10 dynes when they are 0.1 cm. apart (See Exercise 5). Find the force when they are 1 cm. apart.

Chapter 2 Self-Test

[**2.1**] 1. Given $f(x)=3x+5$ and $g(x)=\dfrac{x-5}{3}$, evaluate: $f(3)$; $g(14)$; $f[g(x)]$; $g[f(x)]$.

2. Graph the relation defined by the equation $y=2-x^2$ over the domain $\{-2, -1, 0, 1, 2\}$.

3. Find the domain and range of the function defined by $y=\sqrt{x^2-25}$.

[**2.2**] 4. Given $3x+5y=15$. Write the equation in slope-intercept form and specify the slope. Write the equation in double-intercept form and specify both intercepts. Graph the equation.

5. Find the distance between the points $(1, 3)$ and $(8, -21)$, and the slope of the line determined by the points. Write a standard form equation of the line.

[**2.3**] *Graph each function.*
   6. $y=|x-4|$
   7. $y=x-[x]$

[**2.4**] 8. Graph $y=x^2+6x+4$. Specify the vertex and the axis.

9. Graph $y=4-x^2$ and $y=x+2$ on the same set of axes. Specify points of intersection of the graphs.

10. The height y (in feet) of a particular projectile above the ground is given by $y=400t-16t^2$, where t is the time in seconds. Determine: the domain over which t is nonnegative; the maximum height to which the projectile will rise.

[**2.5**] 11. If y varies jointly as x and the square of z, and $y=6$ when $x=3$ and $z=10$, find y when $x=10$ and $z=5$.

12. If P varies directly as T and inversely as V, and $P=360$ when $T=180$ and $V=0.5$, find P when $T=90$ and $V=3$.

SUPPLEMENT A

FAMILIES OF LINES

Consider the linear equation in x and y,

$$y = \tfrac{1}{3}x + b, \qquad\qquad \langle 1 \rangle$$

where b is an arbitrary constant. By comparison with the slope-intercept form of the equation of a line, we see that the graph of $\langle 1 \rangle$ is a line with slope $\tfrac{1}{3}$ and y-intercept b. Now, let us consider the effect of replacing b by different real numbers. For each such replacement for b we obtain a different linear equation and therefore a different line. This result suggests that if we view b as a variable, then equation $\langle 1 \rangle$ represents an *infinite set of lines* rather than a single line. From this point of view, the infinite set is called a **family of lines**; the number b is referred to as a *parameter*. Some of the lines that are members of the one-parameter family of lines that is the graph of equation $\langle 1 \rangle$ are shown in Figure A-1. One way to describe this set geometrically is "the family of all lines parallel to the line $y = \tfrac{1}{3}x$."

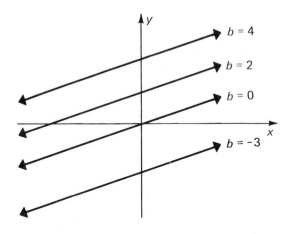

Figure A-1

EXAMPLE 1   The one-parameter family of nonvertical lines defined by

$$y - 3 = m(x - 1)$$

may be described as "the family of lines each of which includes the point

(1, 3) and has slope m," as indicated in Figure A–2. Note, however, that the vertical line $x=1$ is not a member of the family, even though it includes (1, 3), because a vertical line has no slope.

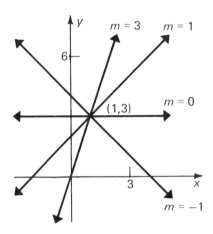

Figure A-2

EXAMPLE 2   Determine an equation of the family of lines perpendicular to the line $2x - 5y - 10 = 0$, and find line q, the member of the family that includes $P(8, -4)$. Sketch line q and two other members of the family.

SOLUTION   Solving for y, we have

$$2x - 5y - 10 = 0 \leftrightarrow y = \tfrac{2}{5}x - 2,$$

from which we see that the slope of the given line is $\tfrac{2}{5}$. Hence, the slope of each line in the required family is $-\tfrac{5}{2}$. With b as a parameter, an equation of the family is

$$y = -\tfrac{5}{2}x + b.$$

To find the line that includes $P(8, -4)$, we replace x by 8 and y by -4:

$$-4 = -\tfrac{5}{2}(8) + b \leftrightarrow b = 16.$$

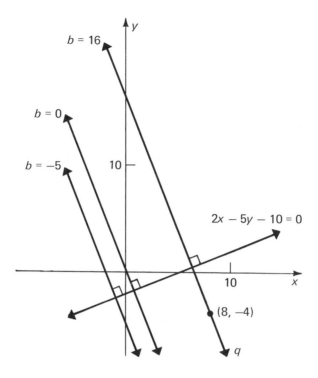

Figure A-3

The equation of line q is $y = -\frac{5}{2}x + 16$. Figure A-3 shows line q and two other members of the family.

Consider the following two lines that intersect in $P(2, 1)$, as shown in Figure A-4:

$$3x + y - 7 = 0 \qquad \text{and} \qquad 2x - 5y + 1 = 0.$$

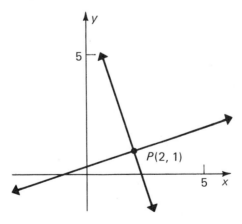

Figure A-4

With t as a parameter, the equation

$$3x + y - 7 + t(2x - 5y + 1) = 0 \qquad \langle 2 \rangle$$

defines a one-parameter family of lines each of which includes the point of intersection P. That is, for each real number t, the graph of equation $\langle 2 \rangle$ is a line through P. [You will find it instructive to substitute 2 for x and 1 for y in $\langle 2 \rangle$, and observe that $(2, 1)$ is a solution of equation $\langle 2 \rangle$ for *any* choice of t.] A major advantage of an equation such as $\langle 2 \rangle$ is that it can be used without having to find the coordinates of the point of intersection P.

EXAMPLE 3   Given the two lines intersecting in point P:

$$3x + y - 7 = 0 \quad \text{and} \quad 2x - 5y + 1 = 0.$$

Write an equation of the family of lines that include P, and then find an equation of the line that:

a. Includes point $Q(13, 2)$,   b. Has slope 2.

SOLUTION   a. The one-parameter family of lines through P is defined by

$$3x + y - 7 + t(2x - 5y + 1) = 0,$$

or, after removing the parentheses and regrouping, by

$$(3 + 2t)x + (1 - 5t)y + (t - 7) = 0. \qquad \langle 3 \rangle$$

To find the member of $\langle 3 \rangle$ that includes $(13, 2)$, we substitute 13 for x, 2 for y, and solve for t:

$$(3 + 2t)(13) + (1 - 5t)(2) + (t - 7) = 0 \quad \leftrightarrow \quad t = -2.$$

Replacing t by -2 in $\langle 3 \rangle$ and simplifying, we obtain the required equation:

$$x - 11y + 9 = 0.$$

b. Solving equation $\langle 3 \rangle$ for y, we have

$$y = -\frac{3 + 2t}{1 - 5t}x - \frac{t - 7}{1 - 5t}. \qquad \langle 4 \rangle$$

Because the slope of the line is equal to the coefficient of x, and is also to be equal to 2, we have

$$-\frac{3 + 2t}{1 - 5t} = 2 \quad \leftrightarrow \quad t = \tfrac{5}{8}.$$

Substituting $\tfrac{5}{8}$ for t in $\langle 3 \rangle$, and simplifying, we have the required equation:

$$2x - y - 3 = 0.$$

Exercise Set A

The symbol a names the x-intercept, b the y-intercept and m the slope of a line.

Write an equation for the family of lines that satisfies the given condition. (There may be more than one answer.) Sketch at least three lines of the family.

1. $b = -4$
2. $m = -4$
3. $a = -4$
4. $a + b = -4$
5. $m = 0$
6. m is not defined
7. Each line of the family includes $(5, 2)$.
8. Each line of the family includes the origin.
9. Each line of the family, together with the coordinate axes, determines a triangle of area 6.
10. The y-intercept of each line of the family is 3 times the x-intercept.

Rewrite each equation (if necessary) into one of the forms:

$$y = mx + b; \qquad y - y_1 = m(x - x_1); \qquad \frac{x}{a} + \frac{y}{b} = 1;$$

such that a single parameter appears in only one term. Sketch three members of the family.

11. $y = -2x + b$
12. $y = mx - 3$
13. $bx + 2y = 2b$
14. $-5x + ay + 5a = 0$
15. $y = mx + 2 - 3m$
16. $2mx - 12m = 2y - 1$

17. Write an equation of the family of lines parallel to $3x + 2y - 6 = 0$. Find the member of the family that includes $(3, 5)$.
18. Write an equation of the family of lines perpendicular to $3x + 2y = 6$. Find the member of the family that includes $(3, 5)$.
19. Given lines $x + 4y - 6 = 0$ and $3x + 2y - 9 = 0$ intersecting at P. Write an equation of the family of lines through P, in the form of equation $\langle 3 \rangle$, page 106.

In the family of lines of Exercise 19, find the particular line that satisfies the given condition.

20. The line that includes the origin.
21. The line through the point $(2, -4)$.
22. The line with y-intercept 4.
23. The line with x-intercept 4.
24. The line parallel to the x-axis.
25. The line parallel to the y-axis.
26. Use the results of Exercises 24 and 25 to find P.

SUPPLEMENT B

AVERAGE RATE OF CHANGE OF A FUNCTION

In Figure B-1, consider the graph of $y = \frac{1}{2}x^2$ and points $P(2, 2)$, $Q(4, 8)$ and $R(6, 18)$. If we start at P and imagine "moving along the curve" to Q, then the x-coordinate changes from 2 to 4, a change of 2 units. Using the symbol Δx to name such a change, we have that $\Delta x = 4 - 2 = 2$. Similarly, the y-coordinate changes from 2 to 8 and, using Δy to indicate the change in y, we have that $\Delta y = 8 - 2 = 6$. Now, our interest is in comparing the change in y to the change in x. Thus, we compute the fraction $\dfrac{\Delta y}{\Delta x}$ for those portions of the graph from P to Q, from P to R, and from Q to R.

$$P \text{ to } Q: \qquad \frac{\Delta y}{\Delta x} = \frac{8 - 2}{4 - 2} = \frac{3}{1}$$

$$P \text{ to } R: \qquad \frac{\Delta y}{\Delta x} = \frac{18 - 2}{6 - 2} = \frac{4}{1};$$

$$Q \text{ to } R: \qquad \frac{\Delta y}{\Delta x} = \frac{18 - 8}{6 - 4} = \frac{5}{1}.$$

The above equations indicate that: from P to Q, a 1-unit change in x produces a 3-unit change in y; from P to R, a 1-unit change in x produces a 4-unit change in y; from Q to R, a 1-unit change in x produces a 5-unit change in y. These results suggest what is intuitively evident from the graph—the amount of change in y depends upon *where we start* together with *how far we go*.

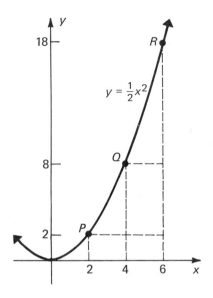

Figure B-1

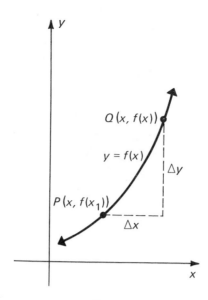

Figure B-2

To generalize the preceding discussion, in Figure B-2 consider the graph of a function defined by $y=f(x)$, together with points $P(x_1, f(x_1))$ and $Q(x, f(x))$. If we imagine starting at P and moving to Q, then x changes by an amount Δx and the coordinates of Q can be given in the form $(x_1+\Delta x, f(x_1+\Delta x))$. Thus,

$$\Delta y = f(x_1+\Delta x) - f(x_1)$$

and

$$\frac{\Delta y}{\Delta x} = \frac{f(x_1+\Delta x)-f(x_1)}{\Delta x}. \qquad \langle 1 \rangle$$

For a given function f, equation $\langle 1 \rangle$ measures *the average rate of change of y with respect to a unit change in x*, starting from the point $(x_1, f(x_1))$.

EXAMPLE 1   Given the function defined by $y=3x^2$, obtain an expression for $\frac{\Delta y}{\Delta x}$ in terms of x_1 and Δx. Find $\frac{\Delta y}{\Delta x}$ for $x_1=2$ and $\Delta x=5$.

SOLUTION   We use equation $\langle 1 \rangle$.

$$f(x_1+\Delta x) = 3(x_1+\Delta x)^2 = 3x_1^2 + 6x_1\Delta x + 3(\Delta x)^2; \qquad f(x_1) = 3(x_1)^2.$$

Hence,

$$\frac{\Delta y}{\Delta x} = \frac{3x_1^2 + 6x_1\Delta x + 3(\Delta x)^2 - 3x_1^2}{\Delta x}$$

$$\frac{\Delta y}{\Delta x} = \frac{6x_1\Delta x + 3(\Delta x)^2}{\Delta x} = 6x_1 + 3\Delta x.$$

For $x_1=2$ and $\Delta x=5$: $\dfrac{\Delta y}{\Delta x} = 6\cdot 2 + 3\cdot 5 = 27.$

EXAMPLE 2   Consider again the function of Example 1 and suppose that y represents the distance (in meters) traveled by a particle, and x represents the time (in seconds). Then $y = 3x^2$ relates the distance with the time. For example, the total distance traveled in 5 seconds is $3 \cdot 5^2 = 75$ meters. The fraction $\frac{\Delta y}{\Delta x}$ can now be interpreted as the average rate of change of distance with respect to time, or the *average speed*, of the particle. Interpreting the result of Example 1 from this point of view, we have the average speed between the time $x_1 = 2$ seconds and $x_1 + \Delta x = 7$ seconds is $\frac{\Delta y}{\Delta x} = 27$ meters per second.

Exercise Set B

For each function, use equation $\langle 1 \rangle$ to obtain an expression for $\frac{\Delta y}{\Delta x}$ in terms of x_1 and Δx.

1. $y = 4x^2 - 7x$ 2. $y = -x^2 + 2x + 5$ 3. $y = x^3 + 6x^2 - 8$

4. $y = 2x^3 + 9x - 8$ 5. $y = \dfrac{3}{x}$ 6. $y = \dfrac{x}{x+2}$

7. Given $y = 16x^2 + 5x + 12$, where y represents distance (in meters) and x represents time (in seconds). For $x_1 = 2$, compute the average speed for each Δx in the table.

Δx	5	1	0.1	0.01	0.001
$\dfrac{\Delta y}{\Delta x}$					

8. The results of Exercise 7 indicate that as Δx "approaches" 0 (as Δx is replaced by values closer and closer to 0), $\frac{\Delta y}{\Delta x}$ seems to "approach" 69. This suggests that 69 m/sec can be viewed as the *instantaneous speed* at $x = 2$. For the function of Exercise 7, find the instantaneous speed at $x = 1$; at $x = 3$.

9. The fraction $\frac{\Delta y}{\Delta x}$ may be interpreted as the *average slope* of a graph over the interval from x_1 to $(x_1 + \Delta x)$. Given $x_1 = 1$, find the average slope of the graph of $y = x^3$ for each Δx in the table.

Δx	5	1	0.1	0.01	0.001
$\dfrac{\Delta y}{\Delta x}$					

10. In Exercise 9, note that $\frac{\Delta y}{\Delta x}$ "approaches" 3 as Δx "approaches" 0. This suggests that 3 can be viewed as the *instantaneous slope* of the graph at $x = 1$. For the function of Exercise 9, find the instantaneous slope of the graph at $x = 2$; at $x = 5$.

11. For the function defined by $y = ax^2 + bx + c$, find an expression for the *average speed* in terms of x_1 and Δx. Find an expression for v, the *instantaneous speed*.

12. For the function defined by $y = ax^3 + bx^2 + cx + d$, find an expression for the *average slope* in terms of x_1 and Δx. Find an expression for m, the *instantaneous slope*.

13. For the function defined by $y = mx + b$, show that the average speed is equal to the instantaneous speed.

SUPPLEMENT C

ALGEBRA OF FUNCTIONS

If f and g are two functions defined over the same domain D, then $f(x)$ and $g(x)$ exist for any number x in D and we can use f and g to construct new functions, as described next.

Definition C-1   For any two functions f and g with a common domain D, if $x \in D$ then the functions $f+g$, $f-g$, $f \cdot g$, $\frac{f}{g}$ are defined by the following equations:

$$(f+g)(x)=f(x)+g(x);$$

$$(f-g)(x)=f(x)-g(x);$$

$$(f \cdot g)(x)=f(x) \cdot g(x);$$

$$\left(\frac{f}{g}\right)(x)=\frac{f(x)}{g(x)}, \qquad \text{provided } g(x) \neq 0.$$

EXAMPLE 1   Consider $f(x)=x^2$ and $g(x)=x+1$. The set R of real numbers is the common domain for f and g. Hence:

a. $(f+g)(x)=x^2+x+1;$

b. $(f-g)(x)=x^2-(x+1)=x^2-x-1;$

c. $(f \cdot g)(x)=x^2(x+1)=x^3+x^2;$

d. $\left(\dfrac{f}{g}\right)(x)=\dfrac{x^2}{x+1}, \qquad x \neq -1;$

e. $\left(\dfrac{g}{f}\right)(x)=\dfrac{x+1}{x^2}, \qquad x \neq 0.$

R is the domain of the first three functions. The domain for the function in (d) is $\{x: x \neq -1\}$; the domain for the function in (e) is $\{x: x \neq 0\}$.

If two functions do *not* have the same domain, then Definition C-1 is not immediately applicable. However, if we consider only those numbers x in the *intersection* of the two domains, then we may apply Definition C-1 accordingly.

EXAMPLE 2   Consider $f(x)=|x|$ and $g(x)=\sqrt{x-2}$. The domain of f is R, the domain of g is $\{x: x \geq 2\}$, the intersection of the two domains is $D=\{x: x \geq 2\}$. Hence, D is the domain of the functions $(f+g)$, $(f-g)$, $(f \cdot g)$ and $\left(\dfrac{g}{f}\right)$. For the function $\left(\dfrac{f}{g}\right)$, the domain is $\{x: x > 2\}$.

Another method of constructing functions from given functions involves the concept of "a function of a function." For instance, suppose that $f(x)=x+3$ and $g(x)=2x$, and let us compute $f(g(5))$. Because $g(5)=10$, we have

$$f(g(5))=f(10)=10+3=13.$$

It is customary to use the notation $(f\circ g)(5)$ to denote $f(g(5))$ and refer to the operation denoted by the symbol "$\circ$" as *composition*.

Definition C-2   Let f and g be two functions such that the range of g is a subset of the domain of f. The **composite of f with g**, denoted $f\circ g$, is given by

$$(f\circ g)(x)=f(g(x)),$$

for all x in the domain of g.

EXAMPLE 3   Let $f(x)=x+3$ and $g(x)=2x$. Because the set R of real numbers is the domain and the range of both functions, we may apply Definition C-2 as follows:

$$(f\circ g)(x)=f(g(x))=f(2x)=2x+3;$$
$$(g\circ f)(x)=g(f(x))=g(x+3)=2(x+3)=2x+6.$$

Note that $f\circ g$ and $g\circ f$ are not the same function.

If the range of g is *not* a subset of the domain of f, then Definition C-2 is not immediately applicable. However, if we consider only those numbers x in the domain of g for which $g(x)$ is in the domain of f, then Definition C-2 may be applied accordingly.

EXAMPLE 4   Let $f(x)=\sqrt{x}$ and $g(x)=\dfrac{1}{x}$. The domain of f is $\{x: x\geq 0\}$, the set of nonnegative numbers. The domain and the range of g is the set of all real numbers except zero, and so the range of g is not a subset of the domain of f. However, we note that if $x>0$, then $\dfrac{1}{x}>0$, which suggests that we restrict the domain of g to R^+, the set of positive numbers. With this restriction, the range of g is now R^+ which is a subset of the domain of f. Hence, for any x in R^+:

$$(f\circ g)(x)=f\left(\frac{1}{x}\right)$$

$$=\sqrt{\frac{1}{x}}, \quad \text{or} \quad \frac{1}{\sqrt{x}}.$$

Exercise Set C

For each of the following pairs of functions f and g, find: a. $f+g$; b. $f-g$; c. $f \cdot g$; d. $\dfrac{f}{g}$; e. $\dfrac{g}{f}$; f. $f \circ g$; g. $g \circ f$; *and specify the domain of each resulting function.*

1. $f(x)=x$, $\qquad g(x)=2x+1$
2. $f(x)=3-x$, $\qquad g(x)=x+3$
3. $f(x)=x^2-1$, $\qquad g(x)=3x$
4. $f(x)=\dfrac{1}{x}$, $\qquad g(x)=\dfrac{1}{x}$
5. $f(x)=3x-2$, $\qquad g(x)=\frac{1}{3}x+\frac{2}{3}$
6. $f(x)=\dfrac{1}{1-x}$, $\qquad g(x)=\dfrac{x-1}{x}$
7. $f(x)=x^2$, $\qquad g(x)=\sqrt{x}$
8. $f(x)=\sqrt{x-5}$, $\qquad g(x)=x^2+5$
9. $f(x)=2x^2+1$, $\qquad g(x)=3-x^2$
10. $f(x)=x^2+x+1$, $\qquad g(x)=x^2-x-1$
11. $f(x)=|x^2-4|$, $\qquad g(x)=x+2$
12. $f(x)=|x|$, $\qquad g(x)=-|x|$

13. Given $f(x)=\sqrt{x-3}$ and $g(x)=\sqrt{-x}$. Explain why $f+g$, $f-g$, $f \cdot g$ and $\dfrac{f}{g}$ do not exist.

14. Given the functions of Exercise 13: find $f \circ g$ and $g \circ f$ (if they exist) and specify the domain of each resulting function.

ELEMENTS OF ANALYTIC GEOMETRY

3.1   TECHNIQUES FOR GRAPHING

Sketching the graph of a linear function is a simple matter because we need only plot two points and then draw the line determined by these points. However, note the importance of knowing in advance that the graph is a *line*. Without such knowledge, we could plot a great number of points and still not be certain as to the actual form of the graph. Graphs of nonlinear relations, for example, quadratic functions, are generally more complicated than linear graphs; hence we try to obtain as much information concerning such graphs as we can before sketching them. In this section we consider certain concepts and techniques that provide information that is helpful when sketching graphs.

Continuous Curves   Roughly speaking, a graph that exhibits no "breaks" or "gaps" is described as a **continuous curve**. For example, lines and parabolas are continuous curves, whereas graphs of greatest integer functions and of the signum function (see Section 2.3) are not continuous. When necessary, we shall state without proof when a particular graph is, or is not, a continuous curve.

Discussing a Graph   We say "discuss the graph" of a given relation to mean "obtain information about:"

1. x-intercepts; y-intercepts.
2. Symmetry (to be considered next).
3. Domain and range of the relation.
4. Asymptotes (to be considered in Section 3.2).
5. Other points in the graph.

This listing is to be viewed only as a general guide to an efficient procedure for graphing; we do not expect that all the information indicated in the list can be obtained in every individual case.

Symmetry   In Section 2.4, from the standard form equation defining a quadratic function we found an axis of symmetry which was useful in graphing the function. Let us now consider the concept of **symmetry of a graph** in greater detail.

A graph is *symmetric with respect to the x-axis* if, whenever (x, y) is a point in the graph, then $(x, -y)$ is a point in the graph. (See Figure 3.1-1(a).)

A graph is *symmetric with respect to the y-axis* if, whenever (x, y) is a point in the graph, then $(-x, y)$ is a point in the graph. (See Figure 3.1-1(b).)

A graph is *symmetric with respect to the origin* if, whenever (x, y) is a point in the graph, then $(-x, -y)$ is a point in the graph. (See Figure 3.1-1(c).)

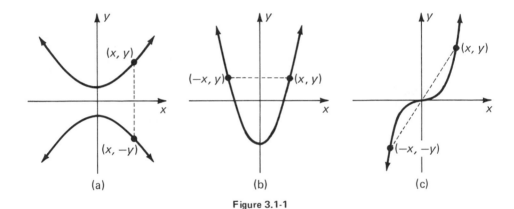

Figure 3.1-1

EXAMPLE 1   Given that $(1, 4)$, $(-2, -2)$ and $(4, 1)$ are points in a graph. If the graph is:

a. symmetric with respect to the *x*-axis, then $(1, -4)$, $(-2, 2)$ and $(4, -1)$ are also points in the graph. Note that each of these points is obtained by changing the sign of the *y*-coordinate. (See Figure 3.1-2(a).)

b. symmetric with respect to the *y*-axis, then $(-1, 4)$, $(2, -2)$ and $(-4, 1)$ are also points in the graph. Note that each of these points is obtained by changing the sign of the *x*-coordinate. (See Figure 3.1-2(b).)

c. symmetric with respect to the origin, then $(-1, -4)$, $(2, 2)$ and $(-4, -1)$ are also points in the graph. Note that each of these points is obtained by changing the signs of both coordinates. (See Figure 3.1-2(c).)

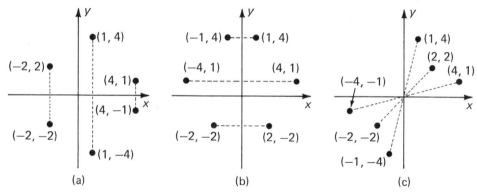

Figure 3.1-2

If a relation is defined by a given equation, certain straightforward **tests for symmetry** enable us to determine symmetries of its graph.

1. *Substitute $-y$ for y in the given equation*: if the resulting equation is equivalent to the original equation, then the graph is symmetric with respect to the x-axis.

2. *Substitute $-x$ for x in the given equation*: if the resulting equation is equivalent to the given equation, then the graph is symmetric with respect to the y-axis.

3. *Substitute $-x$ for x and $-y$ for y in the given equation*: if the resulting equation is equivalent to the given equation, then the graph is symmetric with respect to the origin.

EXAMPLE 2   Discuss and sketch the graph of $y^2 - x - 4 = 0$.

SOLUTION   (Some of the details are left for you to verify.)

Intercepts: Substituting 0 for y in the given equation, we find that -4 is an x-intercept. Substituting 0 for x we find that -2 and 2 are y-intercepts. Hence, $(-4, 0)$, $(0, -2)$ and $(0, 2)$ are points in the graph.

Symmetry: Substituting $-y$ for y we have

$$(-y)^2 - x - 4 = 0 \qquad \text{or} \qquad y^2 - x - 4 = 0.$$

Hence, the graph is symmetric with respect to the x-axis. Substituting $-x$ for x, we have

$$y^2 - (-x) - 4 = 0 \qquad \text{or} \qquad y^2 + x - 4 = 0.$$

The latter equation is not equivalent to the given equation, hence, the graph is not symmetric to the y-axis. Substituting $-y$ for y and $-x$ for x, we have

$$(-y)^2 - (-x) - 4 = 0 \qquad \text{or} \qquad y^2 + x - 4 = 0.$$

Hence, the graph is not symmetric with respect to the origin.

(Continued on next page)

Domain: Solving the given equation for y, we obtain $y^2 = x + 4$ or $y = \pm\sqrt{x+4}$, from which we see that $x + 4$ must be nonnegative. That is, $x + 4 \geq 0$ or $x \geq -4$. Hence, the domain is $\{x : x \geq -4\}$.

Range: Solving for x, we obtain $x = y^2 - 4$, from which we see that there are no restrictions on y. Hence, the range is $\{y : y \in R\}$.

Other points: As indicated by the domain $\{x : x \geq -4\}$, there are no points in the graph to the left of $x = -4$. Hence, we find some points to the right of $x = -4$. It is convenient to use the equation $y = \sqrt{x+4}$, to obtain $(-3, 1)$, $(2, \sqrt{6})$ and $(5, 3)$. By symmetry with respect to the x-axis, we also have $(-3, -1)$, $(2, -\sqrt{6})$ and $(5, -3)$. Using Table 4 on page 346, we approximate $\sqrt{6}$ by 2.4. The points determined in the preceding discussion are shown in Figure 3.1-3, together with the graph determined by these points. Note that the curve is continuous.

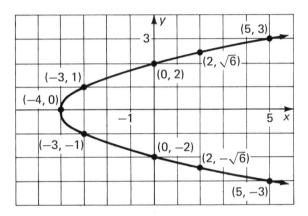

Figure 3.1-3

EXAMPLE 3   Discuss and sketch the graph of $y = 2\sqrt{1 - x^2}$.

SOLUTION   (Verify the details.)

Intercepts: 2 is a y-intercept; 1 and -1 are x-intercepts. Points $(0, 2)$, $(1, 0)$ and $(-1, 0)$ are points in the graph.

Symmetry: The graph is symmetric with respect to the y-axis only.

Domain: The replacement set for x must be such that $1 - x^2$ is nonnegative. That is,

$$1 - x^2 \geq 0 \quad \text{or} \quad (1 - x)(1 + x) \geq 0.$$

Solving the latter inequality, we find that the domain is $\{x : -1 \leq x \leq 1\}$.

Range: Solving the given equation for x, we have $x = \pm\frac{1}{2}\sqrt{4 - y^2}$, from which we see that $4 - y^2 \geq 0$ or $-2 \leq y \leq 2$.

However, because $2\sqrt{1-x^2}$ names a *nonnegative* number for each permissible replacement for x, it follows that y cannot be negative, and the range is $\{y: 0 \le y \le 2\}$.

Other points: Selecting arbitrary numbers between 0 and 1, we obtain such points as $\left(\frac{1}{2}, \sqrt{3}\right)$ and $\left(\frac{3}{4}, \frac{1}{2}\sqrt{7}\right)$. By symmetry with respect to the y-axis, we also have the points $\left(-\frac{1}{2}, \sqrt{3}\right)$ and $\left(-\frac{3}{4}, \frac{1}{2}\sqrt{7}\right)$. Using rational approximations as listed in Table 4, we obtain the points shown in Figure 3.1–4, and the continuous graph determined by these points.

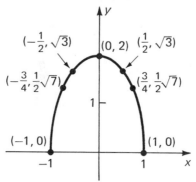

Figure 3.1-4

Exercise Set 3.1

(A) *For each point P, specify points P_1, P_2, and P_3 such that: P and P_1 are symmetric with respect to the x-axis; P and P_2 are symmetric with respect to the y-axis; P and P_3 are symmetric with respect to the origin. Graph each set of points. (See Example 1.)*

1. $P(4, -3)$
2. $P(-4, 3)$
3. $P(-6, 0)$
4. $P(6, 0)$
5. $P(3, \sqrt{7})$
6. $P(4, -\sqrt{11})$

*For each relation, use the **tests for symmetry** (page 000) to determine symmetries of its graph with respect to the: a. x-axis; b. y-axis; c. origin.*

7. $x^2 - y^2 = xy$
8. $x^2y^2 = x^2 + y^2$
9. $x^2 - 4y^2 = 6$
10. $y = 4x^3 - 2x$
11. $y = x^4 - 3x^2$
12. $x - y^2 = 4$

For each of the following relations, discuss the graph as in Examples 2 and 3, and sketch each graph.

13. $y^2 + x - 4 = 0$
14. $2y^2 - x + 8 = 0$
15. $y = \sqrt{x^2 - 9}$
16. $x = \sqrt{y^2 - 9}$
17. $x = \sqrt{81 - 4y^2}$
18. $y = \sqrt{81 - 4x^2}$
19. $xy = 8$
20. $xy = -8$

21. If a graph is symmetric with respect to the origin, is it necessarily symmetric with respect to the x-axis and the y-axis?
22. If a graph is symmetric with respect to the x-axis and to the y-axis, is it necessarily symmetric with respect to the origin?
23. Can the graph of a *function* be symmetric with respect to the x-axis?
24. Give an equation of a line that is symmetric with respect to the a. x-axis; b. y-axis; c. origin.

*A graph is **symmetric with respect to the line** $y=x$ if, whenever (x, y) is a point in the graph, then (y, x) is a point in the graph. Given points P, Q and R, specify points P_1, Q_1 and R_1 such that P and P_1, Q and Q_1, and R and R_1 are symmetric with respect to $y=x$. Graph all six points and show the line $y=x$.*

25. $P(3, -4)$, $Q(5, 5)$, $R(2, 3)$ 26. $P(6, \sqrt{2})$, $Q(0, 0)$, $R(-2, 3)$

*To test for symmetry with respect to $y=x$, **interchange x and y** in a given equation. If the resulting equation is equivalent to the given equation, then the graph is symmetric with respect to the line $y=x$. Test each of the following for symmetry to $y=x$.*

27. $xy=5$ 28. $x^2+y^2=9$ 29. $2x^2-y^2=4$

30. $x^2+y=8$ 31. $x^2-y^2=8$ 32. $4y^2+9x^2=36$

(B)   *For each equation, determine as much useful information as you can, and sketch the graph.*

33. $y=x^3-9x$ 34. $y=x^4-4x^2$ 35. $y=x^{3/2}$

36. $y^2=x^3$ 37. $y=|x^2-4|$ 38. $y=|4-x^2|$

3.2   CONIC SECTIONS

In Section 2.4 we introduced the equation

$$Ax^2+Bxy+Cx^2+Dx+Ey+F+0. \qquad \langle 1 \rangle$$

In general, equations such as $\langle 1 \rangle$ define relations. It can be shown that the graph (if it exists) of any such relation is the curve that is an intersection of a plane with a right circular cone, as shown and named in Figure 3.2–1. Consequently, such graphs are called **conic sections**. In this section we consider certain cases of equation $\langle 1 \rangle$, cases which we designate as *standard form equations*. Familiarity with these standard

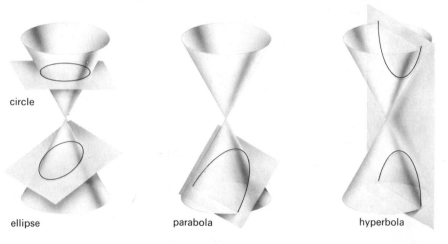

Figure 3.2-1

form equations enables us to readily identify which of the conic sections (parabola, circle, ellipse, or hyperbola) is the graph of a given equation.

Parabola with Horizontal Axis   From Section 2.4 we have that the graph of either of the equations

$$y = ax^2 + bx + c \quad \text{or} \quad y = a(x - h)^2 + k \qquad \langle 2 \rangle$$

is a parabola that opens upward if $a > 0$ and downward if $a < 0$. From the standard form equation (on the right in $\langle 2 \rangle$) we obtain the coordinates (h, k) of the vertex and the equation $x = h$ of the (vertical) axis of symmetry. The graph of the equation

$$x = ay^2 + by + c$$

is also a **parabola**. In this case the parabola opens *to the right* if $a > 0$ and *to the left* if $a < 0$. By completing the square in y we obtain the *standard form equation*

$$x = a(y - k)^2 + h,$$

from which we read the coordinates (h, k) of the vertex and the equation $y = k$ of the (horizontal) axis of symmetry.

EXAMPLE 1   Sketch the graph of $x = 2y^2 - 4y + 5$.

SOLUTION   (Not all of the details are shown.) There are no y-intercepts; 5 is the x-intercept and so $(5, 0)$ is a point of the graph. *Completing the square in y* we obtain

$$x = 2(y - 1)^2 + 3$$

from which we see that $a = 2$, $h = 3$ and $k = 1$. Hence, the graph is a parabola opening to the right, with vertex $(3, 1)$; the horizontal axis of symmetry is $y = 1$. It is convenient to select arbitrary *values for y* and use the given equation to compute associated values for x. For example, substituting -1 for y we find that $(11, -1)$ is a point of the graph. By counting units *above* and *below* the axis of symmetry, we find that $(5, 2)$ and $(11, 3)$ are also points of the graph, as shown in Figure 3.2–2. The vertical line test indicates that the relation defined by the given equation is not a function.

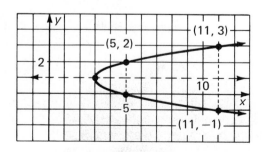

Figure 3.2-2

Central Conics   Conic sections that exhibit symmetry with respect to the origin are called *central* conics; the origin is called the *center*. In general, the graph of an equation such as

$$Ax^2 + Cy^2 + F = 0, \qquad\qquad <3>$$

where $A \neq 0$ and $C \neq 0$, is a central conic. From equation $<3>$, we note that central conics are symmetric with respect to the x-axis and to the y-axis, as well as to the origin. In order to decide which central conic is the graph of an equation such as $<3>$, we consider standard forms of the equation.

The graph of a quadratic relation defined by the *standard form* equation

$$x^2 + y^2 = r^2 \qquad\qquad <4>$$

is a **circle** with center at the origin and radius of length r, $r > 0$. It can be shown (See Exercise 3.2-43) that the domain and the range of the relation defined by equation $<4>$ is $\{x: -r \leq x \leq r\}$ and $\{y: -r \leq y \leq r\}$, respectively.

EXAMPLE 2   The graph of $x^2 + y^2 = 16$ is the circle with center at (0, 0) and radius of length $\sqrt{16}$, or 4, as shown in Figure 3.2-3.

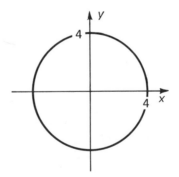

Figure 3.2-3

The graph of a quadratic relation defined by either of the *standard form* equations

$$\frac{x^2}{a^2} + \frac{y^2}{b^2} = 1 \qquad \text{or} \qquad \frac{y^2}{a^2} + \frac{x^2}{b^2} = 1, \qquad\qquad <5>$$

where $a > 0$, $b > 0$ and $a > b$, is called an **ellipse** with center at the origin. For the equation on the left in $<5>$, you can verify that a and $-a$ are x-intercepts; b and $-b$ are y-intercepts. Such an ellipse is shown in Figure 3.2-4(a). For the equation on the right, b and $-b$ are x-intercepts; a and $-a$ are y-intercepts. Such an ellipse is shown in Figure 3.2-4(b).

From Figure 3.2–4(a), we see that, for the relation on the left, the domain is $\{x: -a \leq x \leq a\}$; the range is $\{y: -b \leq y \leq b\}$. From Figure 3.2–4(b), for the relation on the right the domain is $\{x: -b \leq x \leq b\}$; the range is $\{y: -a \leq y \leq a\}$.

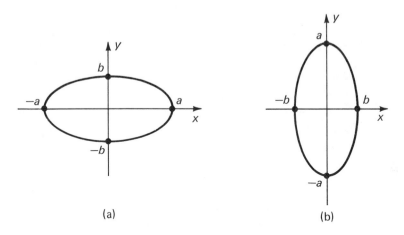

(a)   (b)

Figure 3.2-4

EXAMPLE 3   Sketch the graph of $x^2 + 4y^2 - 36 = 0$.

SOLUTION

$$x^2 + 4y^2 - 36 = 0 \quad \leftrightarrow \quad \frac{x^2}{36} + \frac{y^2}{9} = 1. \qquad \langle 6 \rangle$$

We recognize equation $\langle 6 \rangle$ as a standard form equation of an ellipse with $a = \sqrt{36} = 6$ and $b = \sqrt{9} = 3$. Hence, $(6, 0)$, $(-6, 0)$, $(0, 3)$ and $(0, -3)$ are four points of the graph. Choosing an arbitrary value for x between -6 and 6, say 4, and substituting into the given equation, we find that $y = \pm\sqrt{5}$, and so $(4, \sqrt{5})$ and $(4, -\sqrt{5})$ are two more points of the graph. By symmetry, we also have the points $(-4, \sqrt{5})$ and $(-4, -\sqrt{5})$. Figure 3.2–5 shows the above determined points, together with the ellipse that is the required graph.

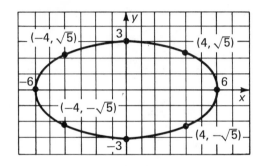

Figure 3.2-5

The graph of a quadratic relation defined by either of the *standard form* equations

$$\frac{x^2}{a^2} - \frac{y^2}{b^2} = 1 \quad \text{or} \quad \frac{y^2}{a^2} - \frac{x^2}{b^2} = 1, \qquad \langle 7 \rangle$$

where $a > 0$ and $b > 0$, is called a **hyperbola** with center at the origin. From the equation on the left you can verify that a and $-a$ are x-intercepts, but there are no y-intercepts. From the equation on the right you can verify that a and $-a$ are y-intercepts, but there are no x-intercepts. These results are illustrated in Figure 3.2–6, which shows two typical hyperbolas. Observe that each hyperbola consists of two *branches*—the curve is not continuous. Observe also that, for the relation on the left in $\langle 7 \rangle$, the domain is $\{x: x \le -a \text{ or } x \ge a\}$; the range is $\{y: y \in R\}$. For the relation on the right, the domain is $\{x: x \in R\}$; the range is $\{y: y \le -a \text{ or } y \ge a\}$.

Before considering a specific example of a hyperbola, we introduce a property that is helpful when sketching hyperbolas. In Figure 3.2–7(a), observe that the curve appears to be "approaching closer and closer" to

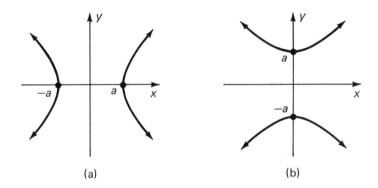

(a) (b)

Figure 3.2-6

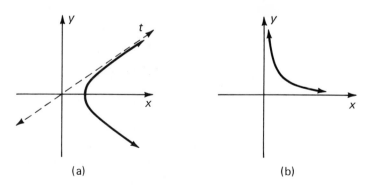

(a) (b)

Figure 3.2-7

line t; in Figure (b) the curve appears to be "approaching closer and closer" to the x-axis and to the y-axis. We say that line t is an **asymptote** to the graph in Figure (a); the x-axis and the y-axis are **asymptotes** to the graph in Figure (b). Of course, the preceding is a "rough" description of the somewhat subtle concept of an asymptote—a rigorous definition requires more advanced mathematics than we have available at this level. It can be shown that the hyperbola that is the graph of either of equations $\langle 7 \rangle$ has two asymptotes, as suggested by the following argument. First, the equation on the left in $\langle 7 \rangle$ can be solved for y to obtain

$$y = \pm \frac{b}{a} x \sqrt{1 - \frac{a^2}{x^2}}.$$

(You are asked to verify this in Exercise 49.) Next, as $|x|$ assumes successively greater values, $\frac{a^2}{x^2}$ assumes values successively closer to zero, and therefore y assumes values successively closer to $\pm \frac{b}{a} x \sqrt{1-0}$ or, simply, to $\pm \frac{b}{a} x$. Hence, as $|x|$ becomes very large, the graph of the given equation is approximated more and more closely by the lines

$$y = \frac{b}{a} x \quad \text{and} \quad y = -\frac{b}{a} x.$$

These two lines are the asymptotes.

Fortunately, a straightforward method (which we state without proof) is available for obtaining the equations of the asymptotes. We take the left member of either of the standard form equations $\langle 7 \rangle$, set that member equal to zero, and solve for y. For example, to obtain the asymptotes to the graph of

$$\frac{x^2}{a^2} - \frac{y^2}{b^2} = 1, \qquad\qquad \langle 8 \rangle$$

we set

$$\frac{x^2}{a^2} - \frac{y^2}{b^2} = 0 \quad \leftrightarrow \quad y = \pm \frac{b}{a} x.$$

The lines determined by the latter two equations (one with the $+$ sign, the other with the $-$ sign) are the asymptotes to the graph of $\langle 8 \rangle$.

EXAMPLE 4   Sketch the graph of $4x^2 - 9y^2 - 36 = 0$.

SOLUTION $$4x^2 - 9y^2 - 36 = 0 \quad \leftrightarrow \quad \frac{x^2}{9} - \frac{y^2}{4} = 1$$

We recognize the latter equation as the standard form equation of a hyperbola with $a = \sqrt{9} = 3$ and $b = \sqrt{4} = 2$. Substituting 0 for y, we find that 3 and -3 are x-intercepts. Substituting 0 for x, we find that there are no y-intercepts. For the asymptotes, we have

$$\frac{x^2}{9} - \frac{y^2}{4} = 0 \quad \leftrightarrow \quad y = \pm \frac{2}{3} x.$$

(Continued on next page)

To determine more points of the graph, we select arbitrary values for x greater than 3, say 4 and 5. Substituting appropriately, we obtain $\left(4, \frac{2}{3}\sqrt{7}\right), \left(4, -\frac{2}{3}\sqrt{7}\right), \left(5, \frac{8}{3}\right)$ and $\left(5, -\frac{8}{3}\right)$. By symmetry, we also have the points $\left(-4, \frac{2}{3}\sqrt{7}\right), \left(-4, -\frac{2}{3}\sqrt{7}\right), \left(-5, \frac{8}{3}\right)$ and $\left(-5, -\frac{8}{3}\right)$. Figure 3.2-8 shows the asymptotes (dashed lines), the points determined above, and the hyperbola that is the required graph.

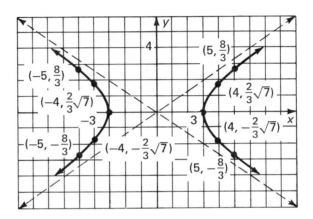

Figure 3.2-8

Exercise Set 3.2

(A) *Graph each equation. Specify the vertex and the axis of symmetry. (See Example 1.)*

1. $x = y^2 + 4y + 4$ 2. $x = y^2 + 8y + 16$
3. $x = -y^2 + 4y - 8$ 4. $x = -y^2 + 8y - 18$
5. $x = 3y^2 + 12y + 8$ 6. $x = -3y^2 + 12y - 8$

Graph each equation. Specify the length of a radius. (See Example 2.)

7. $2x^2 = 32 - 2y^2$ 8. $y^2 = 36 - x^2$

9. $\dfrac{x^2}{25} + \dfrac{y^2}{25} = 1$ 10. $\dfrac{x^2}{49} = 1 - \dfrac{y^2}{49}$

Graph each equation. Specify the intercepts. (See Example 3.)

11. $9x^2 + 16y^2 - 144 = 0$ 12. $16x^2 + 9y^2 - 144 = 0$
13. $36 - 9x^2 = y^2$ 14. $9y^2 + x^2 = 36$

Graph each equation. Specify the intercepts and the equations of the asymptotes. (See Example 4.)

15. $9x^2 - 16y^2 - 144 = 0$ 16. $16x^2 - 9y^2 - 144 = 0$
17. $16y^2 - 25x^2 = 400$ 18. $16y^2 - x^2 = 64$

*The graph of $xy = k$, $k \neq 0$, is called a **rectangular hyperbola**; the x-axis and y-axis are asymptotes. Graph each equation.*

19. $xy = 8$ 20. $xy = -8$ 21. $xy = -6$ 22. $xy = 12$

*Name and sketch the graph of each equation. If a **parabola**, specify the vertex and axis of symmetry; if an **ellipse**, specify the intercepts; if a **hyperbola**, specify intercepts and asymptotes.*

23. $49x^2 + 4y^2 - 196 = 0$ 24. $49y^2 = 196 + 4x^2$

25. $4x^2 - 49y^2 - 196 = 0$ 26. $4x^2 + 49y^2 = 196$

27. $y^2 = 4 - x^2$ 28. $x^2 = 4 - y$

29. $y^2 = 4 - x$ 30. $x^2 = 4 - y^2$

31. $xy = -12$ 32. $xy = 6$

33. $9y^2 - x^2 = 16$ 34. $9y^2 + x^2 = 16$

35. $4x^2 + 9y^2 = 25$ 36. $4x^2 - 9y^2 = 25$

37. Show that the graph of $\{(x, y): x^2 = h\}$, where $h > 0$, is the same as the graph of $\{(x, y): x = \sqrt{h} \text{ or } x = -\sqrt{h}\}$.

38. Show that the graph of $\{(x, y): y^2 = k\}$, where $k > 0$, is the same as the graph of $\{(x, y): y = \sqrt{k} \text{ or } y = -\sqrt{k}\}$.

Use Exercises 37 and 38 to sketch the graph of each relation.

39. $\{(x, y): x^2 = 9\}$ 40. $\{(x, y): y^2 = 9\}$

41. $\{(x, y): y^2 = 16\}$ 42. $\{(x, y): x^2 = 25\}$

(B) *Without graphing, find the domain and range of the quadratic relation defined by each of the following equations.* $(r > 0, a > 0, b > 0.)$

43. $x^2 + y^2 = r^2$ 44. $x^2 - y^2 = r^2$

45. $\dfrac{x^2}{a^2} + \dfrac{y^2}{b^2} = 1$ 46. $\dfrac{y^2}{a^2} + \dfrac{x^2}{b^2} = 1$

47. $\dfrac{x^2}{a^2} - \dfrac{y^2}{b^2} = 1$ 48. $\dfrac{y^2}{a^2} - \dfrac{x^2}{b^2} = 1$

49. Given $\dfrac{x^2}{a^2} - \dfrac{y^2}{b^2} = 1$, show that $y = \pm \dfrac{b}{a} x \sqrt{1 - \dfrac{a^2}{x^2}}$. (Hint: First show that

$y = \pm \dfrac{b}{a} \sqrt{x^2 - a^2}$, from which $y = \pm \dfrac{b}{a} \sqrt{x^2 \left(1 - \dfrac{a^2}{x^2}\right)}$, etc.)

3.3   TRANSLATED CONIC SECTIONS

In Section 3.2 we saw that the graph of a quadratic relation defined by an equation such as

$$Ax^2 + Cy^2 + F = 0 \qquad \langle 1 \rangle$$

is a conic section centered at the origin. Note that equation $\langle 1 \rangle$ does not include any linear (first degree) terms. In this section we consider graphs of quadratic relations defined by equations such as

$$Ax^2 + Cy^2 + Dx + Ey + F = 0, \qquad \langle 2 \rangle$$

where A and C are not both zero and linear terms *are* included. In general, such graphs are also conic sections, but they are not centered at the origin. In order to identify and sketch the particular conic that is the graph of an equation such as $<2>$, we use the technique of *completing the square*, together with another technique called *translation of axes*, to be considered next.

Translation of Axes   As shown in Figure 3.3–1, suppose that two Cartesian coordinate systems are introduced into a plane—an xy-system and an $\overline{xy}$-system—such that the x-axis and the $\overline{x}$-axis are parallel, and the y-axis and the $\overline{y}$-axis are parallel. Each point P in the plane will now have two pairs of coordinates, (x, y) and $(\overline{x}, \overline{y})$. From the figure, observe that these coordinates are related as follows:

$$\begin{bmatrix} x = \overline{x} + h \\ y = \overline{y} + k \end{bmatrix} \quad \text{or, equivalently,} \quad \begin{bmatrix} \overline{x} = x - h \\ \overline{y} = y - k. \end{bmatrix} \qquad <3>$$

Now, although we do not physically move lines in a plane, it is useful to visualize the $\overline{xy}$-system as the result of a "shift" or *translation of the axes* of the xy-system across the plane in such a manner that they remain parallel to their original positions, and the point with original coordinates (h, k) becomes the new origin $\overline{O}$. Equations $<3>$ are called the *equations of translation*.

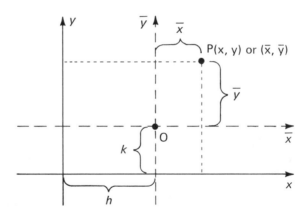

Figure 3.3-1

Graphing Noncentral Conics   To sketch the graph of a conic section not centered at the origin, we use the concept of translation of axes, together with the techniques developed in Section 3.2.

EXAMPLE 1   To identify the graph of

$$4x^2 + 9y^2 - 16x + 18y - 11 = 0, \qquad <4>$$

we first complete the square in x and in y:

$$4x^2 - 16x + 9y^2 + 18y = 11$$
$$4(x^2 - 4x\quad) + 9(y^2 + 2y\quad) = 11$$
$$4(x^2 - 4x + 4) + 9(y^2 + 2y + 1) = 11 + 16 + 9$$
$$4(x-2)^2 + 9(y+1)^2 = 36$$
$$\frac{(x-2)^2}{9} + \frac{(y+1)^2}{4} = 1 \qquad\qquad \langle 5 \rangle$$

Note that equation $\langle 5 \rangle$ resembles the standard form equation of an ellipse (see page 122). In fact, if we replace $x-2$ by $\bar{x}$ and $y+1$ by $\bar{y}$, the resulting equation appears as

$$\frac{\bar{x}^2}{9} + \frac{\bar{y}^2}{4} = 1, \qquad\qquad \langle 6 \rangle$$

which is a standard form equation of an ellipse centered at the origin of the $\overline{xy}$-system. Observe that we have used the translation equations

$$\begin{array}{ccc} \bar{x} = x - 2 & & \bar{x} = x - 2 \\ & \text{or,} & \\ \bar{y} = y + 1 & & \bar{y} = y - (-1), \end{array} \qquad \langle 7 \rangle$$

to translate the axes of the original xy-system into an $\overline{xy}$-system. Comparing equations $\langle 7 \rangle$ with equations $\langle 3 \rangle$, we see that $h = 2$ and $k = -1$. Hence, the origin of the $\overline{xy}$-system is at the point with *original* coordinates $(2, -1)$.

To sketch the ellipse that is the graph of equation $\langle 4 \rangle$ in the xy-system, we first draw in the $\bar{x}$-axis and the $\bar{y}$-axis as dashed lines intersecting at the point O with original coordinates $(2, -1)$, as shown in Figure 3.3-2. Next, using the procedures outlined in Section 3.2, we concentrate on sketching an ellipse with center at the origin of the $\overline{xy}$-system. Working from the standard form equation $\langle 6 \rangle$ and referring to the $\bar{x}$-axis and the $\bar{y}$-axis, we have that $a = 3$ and $b = 2$. Hence, 3 and -3 are $\bar{x}$-intercepts; 2 and -2 are $\bar{y}$-intercepts. Although we used an $\overline{xy}$-system as an aid in graphing, the resulting ellipse in Figure 3.3-2 is the graph of equation $\langle 4 \rangle$ *with respect to the original xy-system.*

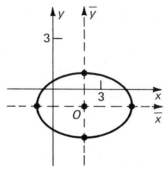

Figure 3.3-2

The procedures illustrated in the preceding example can be applied to show that the graph of each of the following is a standard form equation of a **conic section centered at the point** (h, k).

CIRCLE: $(x-h)^2 + (y-k)^2 = r^2$

ELLIPSE: $\dfrac{(x-h)^2}{a^2} + \dfrac{(y-k)^2}{b^2} = 1; \quad \dfrac{(y-k)^2}{a^2} + \dfrac{(x-h)^2}{b^2} = 1$

HYPERBOLA: $\dfrac{(x-h)^2}{a^2} - \dfrac{(y-k)^2}{b^2} = 1; \quad \dfrac{(y-k)^2}{a^2} - \dfrac{(x-h)^2}{b^2} = 1$

EXAMPLE 2 Identify and sketch the graph of

$$25x^2 - 4y^2 + 150x + 24y + 89 = 0.$$

SOLUTION Completing the square in x and in y, we have

$$25(x+3)^2 - 4(y-3)^2 = 100$$
$$\frac{(x+3)^2}{4} - \frac{(y-3)^2}{25} = 1. \qquad\qquad \langle 8 \rangle$$

By inspection, the equations of translation are

$$\bar{x} = x - (-3) \qquad \text{and} \qquad \bar{y} = y - 3,$$

from which $h = -3$ and $k = 3$. Hence, in the xy-system, the required graph is a hyperbola centered at $(-3, 3)$. In the $\bar{x}\bar{y}$-system, equation $\langle 8 \rangle$ is transformed into

$$\frac{\bar{x}^2}{4} - \frac{\bar{y}^2}{25} = 1.$$

To find equations for the asymptotes, we consider

$$\frac{\bar{x}^2}{4} - \frac{\bar{y}^2}{25} = 0 \quad \leftrightarrow \quad \bar{y} = \pm \frac{5}{2}\bar{x}.$$

These two lines are readily graphed in the $\bar{x}\bar{y}$-system. Finally, we sketch the hyperbola with respect to the $\bar{x}$-axis and the $\bar{y}$-axis. The resulting graph is shown in Figure 3.3-3.

The xy-system equations of the asymptotes of a noncentral hyperbola can be obtained from the $\bar{x}\bar{y}$-system equations.

EXAMPLE 3 In Example 2, the $\bar{x}\bar{y}$-system equations of the asymptotes are

$$\bar{y} = \pm \tfrac{5}{2}\bar{x}.$$

To obtain xy-system equations, we need only replace $\bar{x}$ by $x+3$, and $\bar{y}$ by $y-3$. Thus,

$$y - 3 = \pm \tfrac{5}{2}(x+3)$$

from which the required equations are

$$5x - 2y + 21 = 0 \qquad \text{and} \qquad 5x + 2y + 9 = 0.$$

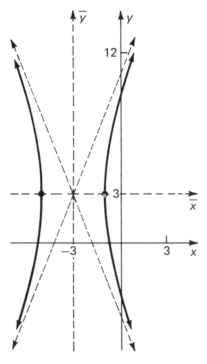

Figure 3.3-3

It is of interest to see how the standard form equations of parabolas fit into the scheme of translations of axes. For a parabola with vertical axis, consider

$$y = a(x - h)^2 + k \quad \leftrightarrow \quad y - k = a(x - h)^2,$$

which suggests the equations of translation

$$\bar{x} = x - h \quad \text{and} \quad \bar{y} = y - k.$$

Hence, in an $\overline{xy}$-system, we obtain the "simpler" equation

$$\bar{y} = a\bar{x}^2.$$

Similarly, reasoning from $x - h = a(y - k)^2$ leads to

$$\bar{x} = a\bar{y}^2$$

for the equation in an $\overline{xy}$-system of a parabola with horizontal axis. These results indicate that our procedure for obtaining the coordinates of the vertex of a parabola is equivalent to translating the x-axis and the y-axis so that the *new* origin is at the vertex of the parabola.

Translations for Other Graphs   The technique of translating axes is applicable to relations whose graphs are not necessarily conics.

EXAMPLE 4　　Sketch the graph of $y=|x-3|+2$.

SOLUTION　　We rewrite the given equation as

$$y-2=|x-3|,$$

which suggests $\bar{x}=x-3$ and $\bar{y}=y-2$ as equations of translation, with $h=3$ and $k=2$. Hence, we translate the xy-system so that the new origin is at the point with original coordinates $(3, 2)$, and then sketch the graph of

$$\bar{y}=|\bar{x}|,$$

as shown in Figure 3.3-4.

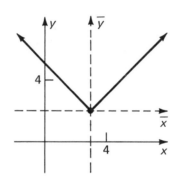

Figure 3.3-4

Exercise Set 3.3

(A)　*Name and sketch the graph of each equation. If the graph is: a **circle**, specify the center and the length of a radius; a **parabola**, specify the vertex and the axis; an **ellipse**, specify the center; a **hyperbola**, specify the center and, as in Example 3, the xy-system equations of asymptotes.*

1. $4x^2+9y^2-24x+18y+9=0$
2. $x^2+y^2+6x+4y+4=0$
3. $x^2+y^2-8x+2y-8=0$
4. $2x^2-12x-y+15=0$
5. $4x^2-9y^2+16x+18y-29=0$
6. $4x^2+16y^2+4x-48y+21=0$
7. $3y^2+x+6y+1=0$
8. $25x^2-4y^2-50x+32y-139=0$
9. $16x^2+16y^2-24x-64y-23=0$
10. $16x^2+9y^2+32x-36y-92=0$
11. $8x^2+8x-y+5=0$
12. $9x^2+9y^2+6x-12y-139=0$
13. $36x^2+4y^2-108x+45=0$
14. $4x^2-9y^2+16x+18y+43=0$

15. $16x^2 - y^2 + 64x - 6y + 71 = 0$

16. $y^2 - 2x - 4y + 2 = 0$

Write the standard form equation for each circle with center C(h, k) and radius of length r.

17. $C(2, -6)$, $r = 4$ 18. $C(-2, 6)$, $r = \sqrt{3}$

19. $C\left(5, \frac{3}{2}\right)$, $r = \sqrt{5}$ 20. $C\left(-\frac{2}{3}, -\frac{1}{4}\right)$, $r = 6$

Graph each equation. (Hint: Compare with Exercises 3.2–19 to 22.)

21. $(x-1)(y+3) = 1$ 22. $(x+2)(y+4) = -4$

23. $(x-1)(y+3) = -1$ 24. $(x+2)(y+4) = 4$

Transform each equation into an $\overline{xy}$-system equation and graph the resulting equation. (See Example 4.)

25. $y = |x-4| + 5$ 26. $y = |x| - 3$

27. $y = 3 - |x+4|$ 28. $y = -3 - |x+4|$

29. $y = \sqrt{x-1} - 4$ 30. $y = \sqrt{x+4} + 6$

31. $y = (x+2)^3 + 1$ 32. $y = (x-5)^3 - 2$

3.4   LINEAR AND QUADRATIC INEQUALITIES

An inequality that can be written equivalently as a first degree inequality in two variables x and y such as

$$Ax + By + C > 0, \qquad\qquad \langle 1 \rangle$$

where A and B are not both zero, is called a **linear inequality**. Inequality $\langle 1 \rangle$ is also said to be a linear inequality if the $>$ symbol is replaced by $\geq$, $<$, or $\leq$. Relations defined by linear inequalities are called *linear relations*.

Graphs of Linear Inequalities   From geometry, we have the fact that any line in a plane separates that plane into three disjoint sets of points—the line itself and two *regions* called *half-planes*. The separating line is called the *edge* of either half-plane. The union of a half-plane and its edge is called a *closed half-plane*; a half-plane without its edge is called an *open half-plane*. As an example, the x-axis separates a plane into two opposite open half-planes:

$$\{(x, y): y > 0\} \qquad \text{and} \qquad \{(x, y): y < 0\}.$$

In general, graphs of linear inequalities are either closed or open half-planes. To sketch such graphs, we follow a straightforward procedure.

1.   First, graph the line $Ax + By + C = 0$ that is the edge of the half-plane. Sketch a *solid line* if the half-plane is *closed*; a *dashed line* if the half-plane is *open*.

2. Chose any convenient point in either half-plane as a "test" point. If the coordinates of that point satisfy the given inequality, then the half-plane that includes the test point is the required region. If the coordinates do *not* satisfy the given inequality, then the *opposite* half-plane is the required region.

EXAMPLE 1 The graph of $x - 2y - 4 > 0$ is obtained as follows. First, sketch the graph of the line

$$x - 2y - 4 = 0,$$

as in Figure 3.4-1(a). The dashed line indicates that this line is not a part of the graph but is the edge of the required region. Second, (for simpler arithmetic) we choose the origin (0, 0) as a test point. Substituting 0 for x and 0 for y in the given inequality, we have

$$0 - 2 \cdot 0 - 4 > 0, \quad \text{or} \quad -4 > 0,$$

a *false* statement. Hence, the required half-plane is the one that does not include the origin. The graph of the given inequality is the shaded region in Figure 3.4-1(b).

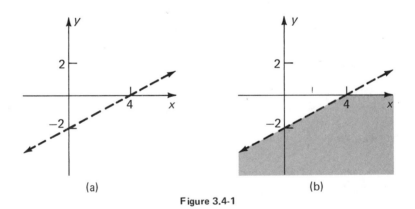

(a)   (b)

Figure 3.4-1

Graphs of Quadratic Inequalities An inequality that can be written equivalently as a second degree inequality in two variables x and y such as

$$Ax^2 + Bxy + Cy^2 + Dx + Ey + F > 0, \qquad \langle 2 \rangle$$

where A and C are not both zero, is called a **quadratic inequality**. Relations defined by quadratic inequalities are called **quadratic relations**.

In general, the graph of a relation defined by a quadratic inequality in two variables is a region in the plane, with a conic section as its edge. To sketch such graphs, we follow an approach very much like that used for linear inequalities, as illustrated next.

EXAMPLE 2   The graph of $x^2+y^2+2x-6y+6\geq0$ is obtained as follows. First, after completing the square in x and in y we consider the equation of the edge:

$$(x+1)^2+(y-3)^2=4,$$

which we recognize as a circle with center at $(-1, 3)$ and radius of length 2. The "solid" circle indicates that the edge is part of the graph [see Figure 3.4-2(a)]. Second, we select an arbitrary test point, say $(0, 0)$. Substituting into the given inequality, we obtain

$$0^2+0^2+2\cdot0-6\cdot0+6\geq0, \quad \text{or} \quad 6\geq0,$$

a *true* statement. Hence, the required graph is the shaded region of the plane that includes the origin, together with the circular edge, as shown in Figure 3.4-2(b).

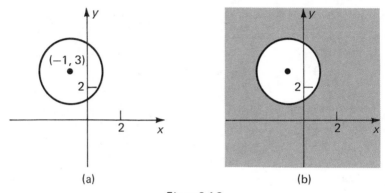

(a)   (b)

Figure 3.4-2

It may sometimes be necessary to use more than one test point when graphing a quadratic inequality.

EXAMPLE 3   The graph of $x^2-y^2\leq9$ is obtained as follows. First, we sketch the hyperbola $x^2-y^2=9$, as in Figure 3.4-3, page 136. Next, we choose a test point in each of three regions: to the left of the left branch; between left and right branches; to the right of the right branch. For convenience, let us take $(-4, 0)$, $(0, 0)$ and $(4, 0)$.

$$\text{At } (-4, 0): \quad 16-0\leq9, \text{ false.}$$

$$\text{At } (0, 0): \quad 0-0\leq9, \text{ true.}$$

$$\text{At } (4, 0): \quad 16-0\leq9, \text{ false.}$$

Hence, the shaded region shown in the figure, including the two branches, is the required graph.

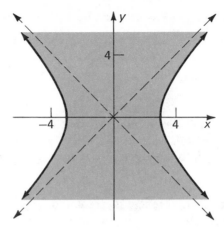

Figure 3.4-3

Exercise Set 3.4

(A) *Graph each linear inequality. (See Example 1.)*

1. $y > 3 - x$
2. $y < 4 + x$
3. $2y < 3 - 2x$
4. $4y > x + 5$
5. $y \geq x$
6. $y \leq -x$
7. $5y \leq 3x + 2$
8. $6x - y + 4 \geq 0$

Graph each quadratic inequality. (See Examples 2 and 3.)

9. $x^2 + y^2 - 2x + 4y \leq 0$
10. $x^2 + y^2 + 6x + 8y > 0$
11. $y^2 - 2x > 4y - 2$
12. $2x^2 - 12x < y - 15$
13. $y^2 - x^2 < 16$
14. $x^2 - y^2 \geq 16$
15. $x^2 + 16y^2 + 240 \geq 128y$
16. $4x^2 + 49y^2 - 16x \geq 294y - 261$
17. $16x^2 + y^2 < 128x - 240$
18. $49x^2 + 4y^2 \leq 294x + 16y - 261$

Graph each relation.

19. $\{(x, y): x > 3\}$
20. $\{(x, y): y < -2\}$
21. $\{(x, y): |x| \geq 3\}$
22. $\{(x, y): |y| \leq 4\}$
23. $\{(x, y): |y| \leq 2\}$
24. $\{(x, y): |x| \geq 1\}$

(Hint: For Exercises 25–29, see Exercise Set 3.2–39 to 42.)

25. $\{(x, y): x^2 \geq 9\}$
26. $\{(x, y): x^2 < 9\}$
27. $\{(x, y): y^2 < 16\}$
28. $\{(x, y): y^2 \geq 16\}$

(B) *Graph each inequality. (Hint: See Exercise Set 3.3–25 to 32.)*

29. $y < |x - 4| + 5$
30. $y > |x| - 3$
31. $y \geq (x + 2)^3 + 1$
32. $y \leq (x - 5)^3 - 2$
33. $y \leq \sqrt{x - 1} - 4$ (Reminder: all graphs are in $R \times R$.)
34. $y \geq \sqrt{x + 4} + 6$

Sketch the graph of the intersection of each pair of relations.

35. $\{(x, y): x^2 + y^2 \geq 36\} \cap \{(x, y): |x| \leq 3\}$
36. $\{(x, y): x^2 < 25 + y^2\} \cap \{(x, y): |x + 1| \leq 4\}$

37. $\{(x, y): y \leq x^2 + 2\} \cap \{(x, y): |y - 4| \leq 2\}$

38. $\{(x, y): xy \geq 6\} \cap \{(x, y): |y| \leq 9\}$

3.5   INVERSE RELATIONS AND FUNCTIONS

Much of mathematics is concerned with the construction and study of various types of functions. In this section we consider a concept that is sometimes used to construct a new function from a given function. First, we introduce another kind of symmetry.

In Figure 3.5-1, observe that the x-coordinate of P is equal to the y-coordinate of Q; the y-coordinate of P is equal to the x-coordinate of Q. We say that P and Q are *symmetric with respect to* $y = x$. More generally, a graph is *symmetric with respect to the line* $y = x$ if, whenever (a, b) is a point in the graph, then (b, a) is a point in the graph. (See also Exercises 3.1–25, 26.)

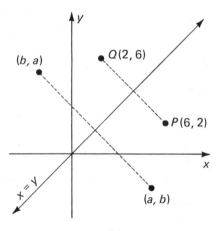

Figure 3.5-1

Inverse Relations   Consider the relations

$$F = \{(1, 2), (3, 4), (5, 4)\} \quad \text{and} \quad G = \{(2, 1), (4, 3), (4, 5)\}$$

and note that either relation can be constructed from the other by interchanging the components of each ordered pair. Relation G is called the *inverse* of F; each relation is said to be the inverse of the other. Note that the domain of F, $\{1, 3, 5\}$, is the range of G; the range of F, $\{2, 4\}$, is the domain of G. Symbols such as F^{-1}, G^{-1}, . . . are often used to name the inverses of relations F, G, . . . , respectively.

Definition 3.1   The **inverse of a relation** F is the relation F^{-1} obtained by interchanging the components in the ordered pairs of F. The domain of F^{-1} is the range of F, the range of F^{-1} is the domain of F.

If a relation F is defined by an equation in which the variables x and y represent members in the domain and range, respectively, then the inverse relation F^{-1} is defined by the equation that results when x and y are interchanged. Furthermore, if F includes point (a, b), then F^{-1} includes point (b, a). It follows that if the graphs of F and F^{-1} are drawn on the same set of axes, the resulting graph is symmetric with respect to $y=x$. We say that the graphs of a relation and its inverse are "mirror images" or "reflections" of each other with respect to the line $y=x$.

EXAMPLE 1   A relation F is defined by each equation. Find a defining equation(s) in the form $y=F^{-1}(x)$ for the inverse F^{-1}.

a. $y=3x-6$ b. $y=x^2+2$

SOLUTION   In each case, we first *interchange variables* x and y in the given equation, and then solve the resulting equation for y in terms of x.

a. $x=3y-6 \quad \leftrightarrow \quad y=\frac{1}{3}x+2.$

The latter equation defines the inverse relation F^{-1}.

b. $x=y^2+2 \quad \leftrightarrow \quad y=\pm\sqrt{x-2}.$

The latter equations define the inverse relation F^{-1}.

EXAMPLE 2   Graph each pair of inverse relations from Example 1 on the same set of axes, together with the line $y=x$. Specify the domain and the range of F and of F^{-1}.

a. $F(x)=3x-6; \quad F^{-1}(x)=\frac{1}{3}x+2.$

b. $F(x)=x^2+2; \quad F^{-1}(x)=\pm\sqrt{x-2}.$

SOLUTION

a. Figure 3.5-2(a) shows the lines that are the graphs of F and F^{-1}. The set R of real numbers is the domain and the range of both F and F^{-1}.

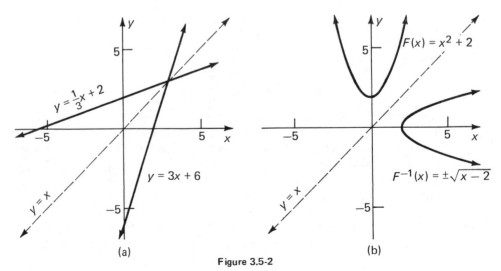

(a)   (b)

Figure 3.5-2

b. Figure 3.5-2(b) shows the parabolas that are the graphs of F and F^{-1}. The domain of F is $\{x: x \in R\}$, the range is $\{y: y \geq 2\}$. By Definition 3.1 (with the appropriate interchange of x and y), the domain of F^{-1} is $\{x: x \geq 2\}$, the range is $\{y: y \in R\}$.

Inverse Functions   Because each function is a relation, it follows that every function has an inverse relation. However, the inverse of a function is not necessarily itself a function.

EXAMPLE 3   Consider Figure 3.5-2(b) again. By the vertical line test, we see that F *is* a function but F^{-1} *is not* a function.

Given a function f, if each element in the domain is paired with exactly one element in the range, and if each element in the range is paired with exactly one element in the domain, then f is called a **one-to-one function**. It can be shown that the inverse of each one-to-one function is itself a function. That is, *if f is a one-to-one function, then f^{-1} is a function.* The graph of a function can be visually checked to determine whether or not the function is one-to-one. *If no horizontal line intersects the graph more than once, then the function is one-to-one.* This check is referred to as the **horizontal line test**.

EXAMPLE 4   Consider each of the following graphs.

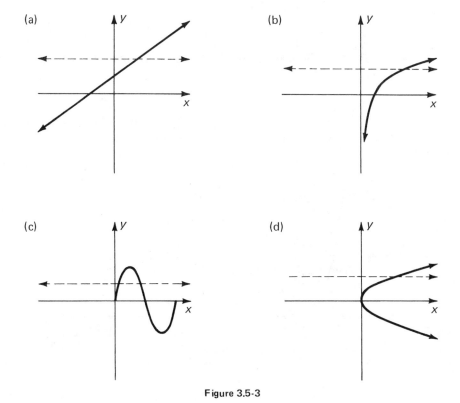

(a)

(b)

(c)

(d)

Figure 3.5-3

By the horizontal line test, the graphs in Figures 3.5–3(a) and (b) are graphs of one-to-one functions, and so their respective inverses are also functions. The graph in Figure (c) is not the graph of a one-to-one function. The graph in Figure (d) is not that of a one-to-one function, although it appears to "pass" the horizontal line test, because it is not the graph of a function to begin with.

If a function f is not one-to-one, then its inverse is not a function. However, given a function f, we can sometimes *restrict the domain* of f in such a manner that the resulting function is one-to-one and it will therefore have an inverse that is also a function.

EXAMPLE 5   The quadratic function defined by $y=x^2$ with $x \in R$ is not one-to-one, as you can see by checking the graph in Figure 3.5–4(a). Consider, instead, the function f defined by

$$y=x^2, \qquad x \le 0, \qquad\qquad \langle 1 \rangle$$

where we have restricted the domain to the set of nonpositive numbers. Function f is one-to-one with range $\{y: y \ge 0\}$; its graph is the solid portion of the parabola. It follows that f^{-1}, the inverse of f, is also a function. To determine f^{-1}, we interchange x and y in $\langle 1 \rangle$ to obtain $x=y^2$. Solving for y, we have:

$$y=\pm\sqrt{x}. \qquad\qquad \langle 2 \rangle$$

Now, by Definition 3.1, the range of f^{-1} is the domain of f. Hence, the range of f^{-1} is the set of nonpositive numbers, and so we must choose the *minus sign* to precede the radical expression in $\langle 2 \rangle$. That is, f^{-1} is defined by

$$y=-\sqrt{x},$$

with domain $\{x: x \ge 0\}$ and range $\{y: y \le 0\}$. Figure 3.5–4(b) shows the graph of f^{-1}.

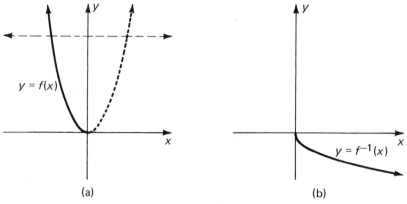

(a)   (b)

Figure 3.5-4

Exercise Set 3.5

(A) *A relation F is defined by each of the following equations.*

 a. *Find a defining equation(s) in the form $y = F^{-1}(x)$ for the inverse relation F^{-1}. (See Example 1.)*

 b. *Graph each pair of inverse relations on the same set of axes together with the line $y = x$. Specify the domain and the range of relations F and F^{-1}. (See Example 2.)*

 c. *Specify whether or not F^{-1} is a function. (See Example 3.)*

1. $y = 3x - 4$	2. $y = 2x + 7$	3. $x - 3y = 5$
4. $y - 3x = 5$	5. $y = 2$	6. $y = -3$
7. $x = -3$	8. $x = 2$	9. $y = x^2 - 4$
10. $y = -4x^2 + 3$	11. $y = x^3$	12. $y = (x - 2)^3$
13. $y = \sqrt{x}$	14. $y = \sqrt{x - 4}$	

Specify whether each of the following is, or is not, the graph of a function whose inverse is a function. (See Example 4.)

15.

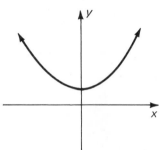

16.

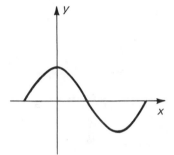

17.

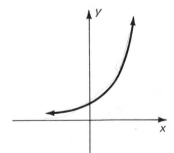

18.

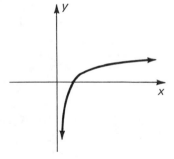

19.

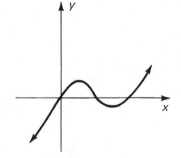

20.

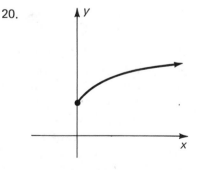

Each of the following equations defines a function f.

a. *Sketch the graph of f; specify its domain and range.*

b. *If f has an inverse function f^{-1}; find a defining equation for the inverse in the form $y=f^{-1}(x)$; sketch the graph of f^{-1}; specify its domain and range. If f has no inverse function, so state. (See Example 5.)*

21. $y=x^2,\quad x\geq 0$ 22. $y=x^2,\quad x\leq 2$

23. $y=x^2,\quad x\geq -2$ 24. $y=x^2,\quad x\leq -2$

25. $y=3x^2+2,\quad x\leq 0$ 26. $y=3x^2+2,\quad x\geq 0$

27. $y=|2x|-4,\quad x\geq 0$ 28. $y=|2x|-4,\quad x\leq 0$

(B)   *Inverse functions may be defined as follows: "Functions f and g are inverse functions if and only if $f[g(x)]=x$ for each x in the domain of g and $g[f(x)]=x$ for each x in the domain of f." Use this definition to determine whether the functions in each of the following pairs are, or are not, inverses.*

EXAMPLE $f(x)=\frac{1}{3}x-2;\qquad g(x)=3x+6.$

SOLUTION $f[g(x)]=\frac{1}{3}(3x+6)-2=x;\qquad g[f(x)]=3(\frac{1}{3}x-2)+6=x.$

Hence, f and g are inverse functions.

29. $f(x)=x+3;\qquad g(x)=x-3$

30. $f(x)=-2x+5;\qquad g(x)=\dfrac{5}{2}-\dfrac{x}{2}$

31. $f(x)=3x+6;\qquad g(x)=\dfrac{x}{3}+2$

32. $f(x)=-5x-15;\qquad g(x)=x+3$

33. $f(x)=x^2, x\geq 0;\qquad g(x)=\sqrt{x},\ x\geq 0$

34. $f(x)=ax+b;\qquad g(x)=\dfrac{1}{a}x-\dfrac{b}{a}$

Chapter 3   Self-Test

[**3.1**]   *Test each relation for symmetry with respect to the: x-axis; y-axis; origin.*

     1. $x^2-xy+y^2=0$ 2. $x^2y^2+x^2=0$

     3. Discuss (intercepts, symmetry, domain and range) and sketch the graph of $y=\sqrt{x^2-1}$.

[**3.2,**   *Name and graph each equation. If the graph is: a **parabola**, specify the*
3.3]   *vertex and axis of symmetry; an **ellipse**, specify the center; a **circle**, specify the center and length of a radius; a **hyperbola**, specify the center and the equations of the asymptotes.*

     4. $9x^2=36(1-y^2)$ 5. $3x^2+12y=6x-3y^2-3$

     6. $x^2=36(1+y^2)$ 7. $2x^2-y+4=2x$

     8. $9x^2+4y^2-18x+16y=11$ 9. $9x^2-4y^2-18x-16y=43$

[**3.4**] *Graph each inequality or relation.*

    10. $3x - 5y - 15 \leq 0$ 11. $2x - 4 > y^2 + 2y$

    12. $\{(x, y): |x| \geq 3\}$ 13. $\{(x, y): |y| \leq 3\}$

[**3.5**] *For each function f, specify the equation of the inverse f^{-1} in the form $y = f^{-1}(x)$. Specify the domain and range of f and f^{-1}. State whether f^{-1} is, or is not, a function.*

    14. $f\!: 3x - 2y = 6$ 15. $f\!: y = x^2 - 1, \ x \leq 0$

SUPPLEMENT D

RELATED GRAPHS

In Section 3.3 we saw that the graph of $y=f(x)$ is *related* to the graph of $y-k=f(x-h)$ in the sense that one graph can be used as an aid in constructing the other. In this supplement we consider other ways of relating graphs. *In the figures, a "dashed" curve represents the graph of $y=f(x)$, a "solid" curve represents the related graph.*

$f(x)$ **AND** $cf(x)$ For each x in the domain of f, the number $cf(x)$ is equal to c times the number $f(x)$. Thus, if (a, b) is a point in the graph of $y=f(x)$, then there is a corresponding point (a, cb) in the graph of $y=cf(x)$. Hence, points in the graph of $y=cf(x)$ can be obtained by multiplying the y-coordinates of (arbitrarily) selected points in the graph of $y=f(x)$ by c, as indicated in Figure D–1(a). If $|c|>1$, the resulting graph will appear to be "stretched" in the y-direction; if $-1<c<1$, $c\neq 0$, the resulting graph will appear to be "squeezed" in the y-direction.

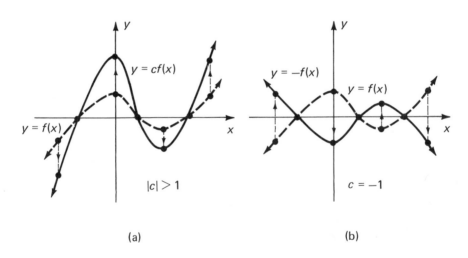

(a)   (b)

Figure D-1

If $c=-1$, then we want the graph of $y=-f(x)$. Now, to each point (a, b) in the graph of $y=f(x)$ there corresponds the point $(a, -b)$ in the graph of $y=-f(x)$. Thus, the two graphs are symmetric with respect to the x-axis and the graph of $y=-f(x)$ can be constructed by *reflecting* the graph of $y=f(x)$ across the x-axis, as indicated in Figure D–1(b). Note that points of intersection with the x-axis are the same for both graphs.

EXAMPLE 1   Graph: a. $y=2(x-3)^2$;   b. $y=\frac{1}{2}(x-3)^2$.

SOLUTION   In each case we begin with the graph of $y=(x-3)^2$.

a. For the graph of $y=2(x-3)^2$, we first select points in the graph of $y=(x-3)^2$ and then multiply y-coordinates by 2, sending: (1, 4) into (1, 8); (2, 1) into (2, 2); (3, 0) into (3, 0); etc.; to obtain the required graph shown in Figure D-2(a).

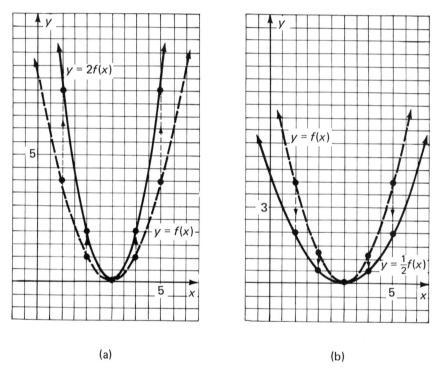

(a)   (b)

Figure D-2

b. For the graph of $y=\frac{1}{2}(x-3)^2$, we multiply y-coordinates of selected points in the graph of $y=(x-3)^2$ by $\frac{1}{2}$, sending: (1, 4) into (1, 2); (2, 1) into $\left(2, \frac{1}{2}\right)$; (3, 0) into (3, 0); etc.; to obtain the required graph shown in Figure D-2(b).

EXAMPLE 2   Graph $y = -\big((x-2)^2 - 3\big)$.

SOLUTION   We first sketch the graph of $y + 3 = (x-2)^2$ (think of $\bar{y} = \bar{x}^2$). For the graph of $y = -\big((x-2)^2 - 3\big)$, we reflect selected points in the graph of $y = (x-2)^2 - 3$ across the x-axis: $(0, 1)$ into $(0, -1)$; $(1, -2)$ into $(1, 2)$; $(2, -3)$ into $(2, 3)$; etc.; to obtain the required graph shown in Figure D-3.

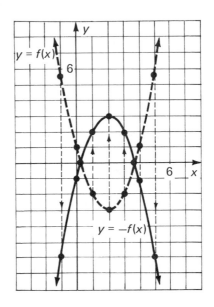

Figure D-3

$f(x)$ and $|f(x)|$ For each x in the domain of f:

$$|f(x)| = f(x) \text{ if } f(x) \ge 0; \qquad |f(x)| = -f(x) \text{ if } f(x) < 0.$$

Hence, to construct the graph of $y = |f(x)|$ from the graph of $y = f(x)$, we need only reflect all those points with *negative y-coordinates* (all those portions of the graph below the x-axis) across the x-axis, as indicated in Figure D-4. All other points remain "fixed."

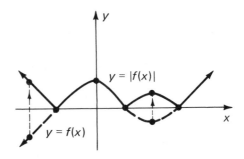

Figure D-4

EXAMPLE 3   Graph   a. $y=|x-2|$ b. $y=|(x-2)^2-3|$.

SOLUTION

a.   Figure D-5(a) shows the graph of $y=x-2$ with the dashed portion of the graph *below* the x-axis. By reflecting only the dashed portion of the graph across the x-axis, we obtain the required graph, as shown.

b.   Figure D-5(b) shows the graph of $y=(x-2)^2-3$ with the dashed portion of the graph *below* the x-axis. By reflecting the dashed portion of the graph across the x-axis, we obtain the required graph, as shown.

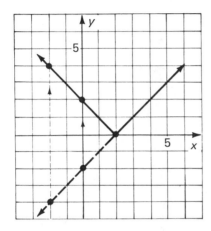

(a)   Figure D-5   (b)

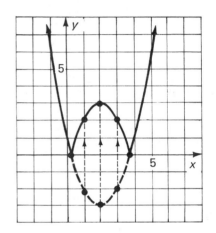

$f(x)$ and $\dfrac{1}{f(x)}$ Points in the graph of $y=\dfrac{1}{f(x)}$ can be obtained from the graph of $y=f(x)$ by taking reciprocals of y-coordinates of selected points (see Figure D-6). Special attention is needed for those values of x for which the denominator $f(x)$ is equal to zero. At such points we expect to find *vertical* asymptotes, as indicated by the dashed lines in the figure.

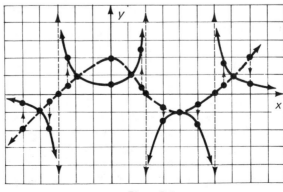

Figure D-6

EXAMPLE 4   Graph $y = \dfrac{1}{x^2 - 4}$.

SOLUTION   First, we sketch the graph of $y = x^2 - 4$ and observe that there are two x-intercepts: 2 and -2. Hence, there are two vertical asymptotes: $x = 2$ and $x = -2$, as shown in Figure D-7. Points in the required graph are now obtained by taking reciprocals of y-coordinates: $(-3, 5)$ into $\left(-3, \frac{1}{5}\right)$; $(-1, -3)$ into $\left(-1, -\frac{1}{3}\right)$; $(0, -4)$ into $\left(0, -\frac{1}{4}\right)$; etc.

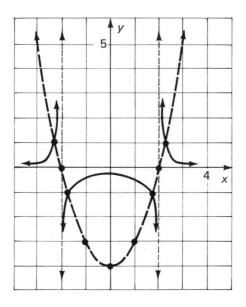

Figure D-7

Addition of Ordinates   The graph of an equation in a form such as

$$y = f(x) \pm g(x)$$

can be constructed by a technique called the *Addition of Ordinates* method, as illustrated next.

EXAMPLE 5   Graph $y = x^2 + 2x$.

SOLUTION   First, we sketch the graphs of $y = x^2$ and $y = 2x$ on the same set of axes, as in Figure D-8. Now, to each arbitrarily selected value of x there corresponds a *pair* of points, one in each of the two sketched graphs. The associated point in the required graph is obtained by (graphically) adding the y-coordinates of those two points. Some of the pairings and their associated points are:

$(-2, 4)$ and $(-2, -6)$ into $(-2, -2)$;
$(-1, 1)$ and $(-1, -2)$ into $(-1, -1)$;
$(1, 1)$ and $(1, 2)$ into $(1, 3)$; etc.

The solid curve in Figure D-8 is the required graph.

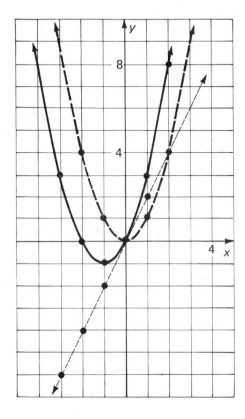

Figure D-8

The graph of an equation in a form such as $y = f(x) - g(x)$ can be constructed by applying the Addition of Ordinates method to the graphs of $y = f(x)$ and $y = -g(x)$.

Exercise Set D

In Exercises 1 through 4, copy each graph as closely as you can. Given the graph of $y = f(x)$, construct the graph of:

a. $y = 3f(x)$

b. $y = \frac{1}{3}f(x)$

c. $y = -f(x)$

d. $y = |f(x)|$

e. $y = \dfrac{1}{f(x)}$

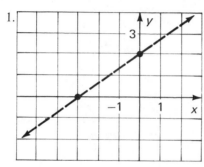

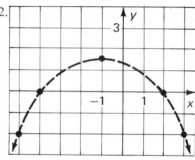

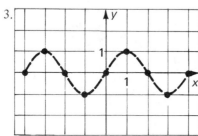

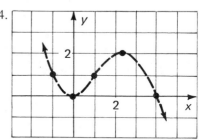

Use the graph of $y=x^2$ to construct the graph of each of the following equations. (You may wish to review translations, Section 3.3.)

5. $y=3x^2$

6. $y=-\frac{1}{3}x^2$

7. $y=-\frac{1}{2}(x+1)^2$

8. $y=2(x-1)^2$

9. $y=|x^2-3|$

10. $y=|(x-4)^2-6|$

11. $y=\dfrac{1}{(x-1)^2-4}$

12. $y=\dfrac{-1}{(x-3)^2}$

Use the Addition of Ordinates method to graph each equation.

13. $y=x+\frac{1}{2}x^2$

14. $y=x-\frac{1}{2}x^2$

15. $y=x-\dfrac{4}{x}$

16. $y=x+\dfrac{4}{x}$

POLYNOMIAL FUNCTIONS

4.0   INTRODUCTION

In section 0.9 we described a polynomial as a monomial, or a finite sum of monomials. A more formal definition follows.

Definition 4.1   A **polynomial** of degree n in one variable x is an expression of the form

$$a_n x^n + a_{n-1} x^{n-1} + \cdots + a_1 x + a_0, \qquad a_n \neq 0,$$

where $x \in C$, n is a natural number and the coefficients named by $a_0, a_1, \ldots, a_{n-1}, a_n$ are complex numbers.

The term of greatest degree, $a_n x^n$, is called the **leading term**; its coefficient a_n is the **leading coefficient**. If all of the coefficients are real numbers, the polynomial is called a **real polynomial**; if imaginary coefficients are included, then the polynomial is called a **complex polynomial**. We name polynomials by symbols such as $A(x)$, $Q(x)$, etc., where, unless stated otherwise, it is understood that such named polynomials are arranged in *descending* powers of the variable, as indicated by Definition 4.1.

If $A(x)$ is a polynomial, then the equation $y = A(x)$, that is

$$y = a_n x^n + a_{n-1} x^{n-1} + \cdots + a_1 x + a_0$$

defines a **polynomial function** with the set C of complex numbers as its domain, and a subset of C as its range. If c is a complex number such that $A(c) = 0$, then c is called a **zero of the function** defined by $y = A(x)$ or, more simply, a **zero of $A(x)$**. Zeros of a function are sometimes further classified as *real zeros*, *imaginary zeros*, *rational zeros*, etc. Thus, if

$$A(x) = 2x^4 - x^3 - 4x^2 - x - 6,$$

then $-\frac{3}{2}$ is a real zero and i is an imaginary zero of $A(x)$ because (as you may verify),

$$A\left(-\tfrac{3}{2}\right) = 2\left(-\tfrac{3}{2}\right)^4 - \left(-\tfrac{3}{2}\right)^3 - 4\left(-\tfrac{3}{2}\right)^2 - \left(-\tfrac{3}{2}\right) - 6 = 0;$$

$$A(i) = 2i^4 - i^3 - 4i^2 - i - 6 = 0.$$

4.1   BASIC POLYNOMIAL THEOREMS

In this chapter, our basic goal is that of determining zeros of polynomials, which, in turn, will enable us to solve polynomial equations. As you will see, this goal can be reached if we can factor polynomials. Towards this end, we first introduce a special technique for finding quotients, and then consider some theorems that relate zeros of polynomials with factors of polynomials.

Synthetic Division   Consider the expression $\dfrac{A(x)}{x-c}$, where $A(x)$ is a polynomial. By the Division Algorithm for Polynomials (Theorem 0.24), there exists a unique *quotient* polynomial $Q(x)$ and a unique *remainder* polynomial $R(x)$ such that

$$A(x) = (x-c)Q(x) + R(x), \qquad\qquad <1>$$

where either $R(x) = 0$, or $R(x)$ is a polynomial of degree less than the degree of $x-c$. But $x-c$ is a polynomial of degree one. Hence, either $R(x) = 0$, or $R(x) = r$, where r is a constant, and equation $<1>$ takes the form

$$A(x) = (x-c)Q(x) + r. \qquad\qquad <2>$$

EXAMPLE 1   Consider the expression $\dfrac{x^3 - 7x^2 + 18x - 20}{x-4}$. By the "long" division process we obtain:

$$
\begin{array}{r}
x^2 - 3x + 6 \\
x-4\overline{)x^3 - 7x^2 + 18x - 20} \\
\underline{x^3 - 4x^2} \qquad\qquad\qquad \\
-3x^2 + 18x \qquad \\
\underline{-3x^2 + 12x} \qquad \\
6x - 20 \\
\underline{6x - 24} \\
4,
\end{array}
$$

from which we see that $x^2 - 3x + 6$ is the quotient $Q(x)$; 4 is the remainder r. In the form of equation $<2>$, we have

$$x^3 - 7x^2 + 18x - 20 = (x-4)(x^2 - 3x + 6) + 4. \qquad <3>$$

The long division process shown in Example 1 can be considerably simplified by an abbreviated method called **synthetic division**, which we can use whenever the divisor is in the form $x-c$. The procedure is demonstrated below for

$$A(x) = x^3 - 7x^2 + 18x - 20 \qquad \text{and} \qquad x-4.$$

1. List the coefficients of $A(x)$ in the order of descending powers of x:

$$1 \qquad -7 \qquad 18 \qquad -20$$

Note: Be sure to enter 0 for any "missing" powers of x. (See Example 2.)

2. Prefix this listing by the replacement value of x for which

$$x - c = 0.$$

In this case, $x - 4 = 0$ if x is replaced by 4.

$$4 \underline{|} \quad 1 \quad -7 \quad 18 \quad -20$$

3. Draw a line under the list of coefficients and "bring down" the leading coefficient:

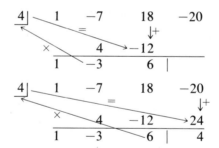

4. Multiply the prefixed 4 by the "brought down" coefficient 1 and add the product to the next coefficient -7:

5. Repeat the routine of Step 4 until each of the coefficients has been used.

The final result appears as

$$
\begin{array}{r|rrrr}
4 & 1 & -7 & 18 & -20 \\
  &   & 4 & -12 & 24 \\
\hline
  & 1 & -3 & 6 & \,|\, 4
\end{array}
$$

6. All but the last of the entries in the third row are the coefficients of the quotient $Q(x)$ arranged in descending powers. That is,

$$Q(x) = 1x^2 - 3x + 6.$$

The last entry of the third row is the remainder: $r = 4$. Thus,

$$x^3 - 7x^2 + 18x - 20 = (x - 4)(x^2 - 3x + 6) + 4,$$

which equation is identical to equation $\langle 3 \rangle$.

EXAMPLE 2   Given $A(x) = 3x^4 - 24x^2 + 11x - 4$. For each of the following divisors, use the synthetic division process to find $Q(x)$ and r such that $A(x) = (x - c)Q(x) + r$.

    a. $x + 3$ b. $x - i$.

SOLUTION   We view $A(x)$ as $3x^4 + 0x^3 - 24x^2 + 11x - 4$.

    a. $x + 3 = 0$ if x is replaced by -3. Hence:

$$
\begin{array}{r|rrrrr}
-3 & 3 & 0 & -24 & 11 & -4 \\
   &   & -9 & 27 & -9 & -6 \\
\hline
   & 3 & -9 & 3 & 2 & \,|\; -10.
\end{array}
$$

From the entries in the last line we see that

$$A(x) = (x + 3)(3x^3 - 9x^2 + 3x + 2) - 10.$$

    b. $x - i = 0$ if x is replaced by i. Hence:

$$
\begin{array}{r|rrrrr}
i & 3 & 0 & -24 & 11 & -4 \\
  &   & 3i & -3 & -27i & 27 + 11i \\
\hline
  & 3 & 3i & -27 & 11 - 27i & \,|\; 23 + 11i.
\end{array}
$$

From the entries in the last line we see that

$$A(x) = (x - i)[3x^3 + 3ix^2 - 27x + (11 - 27i)] + (23 + 11i).$$

For the remainder of this chapter, unless specified otherwise, *all divisions are to be done by synthetic division.*

The Basic Theorems   Let us compute $A(-3)$ for the polynomial

$$A(x) = 3x^4 - 24x^2 + 11x - 4: \qquad\qquad \langle 4 \rangle$$
$$A(-3) = 3(-3)^4 - 24(-3)^2 + 11(-3) - 4 = -10.$$

In Example 2, we divided polynomial $\langle 4 \rangle$ by $x + 3$ and found the remainder $r = -10$, which we now note is the same number as $A(-3)$. The fact that $r = A(-3)$ is not a coincidence!

Theorem 4.1 (The Remainder Theorem)   If a polynomial $A(x)$ is divided by $x - c$, where c is a complex number, then the remainder r is equal to $A(c)$.   **(P)**

EXAMPLE 3   If $A(x) = 2x^4 - 5x^3 - 7x + 4$, use Theorem 4.1 to compute $A(3)$.

SOLUTION   By Theorem 4.1, if $A(x)$ is divided by $x - 3$, then $r = A(3)$. Thus,

$$
\begin{array}{r|rrrrr}
3 & 2 & -5 & 0 & -7 & 4 \\
  &   & 6 & 3 & 9 & 6 \\
\hline
  & 2 & 1 & 3 & 2 & \,|\; 10,
\end{array}
$$

from which $r = 10$. Hence, $A(3) = 10$.

Two important consequences of the Remainder Theorem follow in the case that the remainder $A(c)$ is equal to zero. Consider the following division of $A(x) = x^3 - 31x + 30$ by $x - 1$:

$$
\begin{array}{r|rrrr}
1 & 1 & 0 & -31 & 30 \\
  &   & 1 & 1 & -30 \\
\hline
  & 1 & 1 & -30 \;| & 0.
\end{array}
$$

From the last line we see that

$$A(x) = (x - 1)(x^2 + x - 30) + 0$$

or, after factoring $x^2 + x - 30$ and dropping the 0 term,

$$A(x) = (x - 1)(x - 5)(x + 6). \tag*{$\langle 5 \rangle$}$$

From the right member of $\langle 5 \rangle$ it is clear that

    1. $x - 1$, $x - 5$ and $x + 6$ are *factors* of $A(x)$.

Furthermore, by successively substituting 1, 5 and -6 for x in $\langle 5 \rangle$, it is also clear that

    2. 1, 5 and -6 are *zeros* of $A(x)$.

These two results suggest the next theorem.

Theorem 4.2   If $A(x)$ is a polynomial of degree $n \geq 1$ and c is a complex number, then $x - c$ is a factor of $A(x)$ if and only if $A(c) = 0$.   **(P)**

Essentially, Theorem 4.2 implies that: If we can factor a polynomial $A(x)$, then we can find the zeros of $A(x)$; conversely, if we know the zeros of $A(x)$, then we can find the factors of $A(x)$.

EXAMPLE 4   Given that 2 is a zero of $A(x) = 2x^3 + x^2 - 13x + 6$. Completely factor $A(x)$ over R and specify zeros of $A(x)$.

SOLUTION   By Theorem 4.2, if 2 is a zero of $A(x)$, then $x - 2$ is a factor of $A(x)$. That is,

$$A(x) = (x - 2)Q(x),$$

where $Q(x)$ can be determined by division:

$$
\begin{array}{r|rrrr}
2 & 2 & 1 & -13 & 6 \\
  &   & 4 & 10 & -6 \\
\hline
  & 2 & 5 & -3 \;| & 0.
\end{array}
$$

Hence, $2x^3 + x^2 - 13x + 6 = (x - 2)(2x^2 + 5x - 3)$. After factoring the quotient $2x^2 + 5x - 3$, we obtain the completely factored form

$$A(x) = (x - 2)(2x - 1)(x + 3) = 2(x - 2)(x - \tfrac{1}{2})(x + 3),$$

where $(2x - 1)$ has been rewritten as $2(x - \tfrac{1}{2})$. By Theorem 4.2, the zeros of $A(x)$ are 2, $\tfrac{1}{2}$, and -3.

Synthetic Division Tables   In our later work it will be convenient to list, in a table, the results of more than one synthetic division with respect to a given polynomial. To accomplish this, we agree not to write the products obtained at each step. For example, the synthetic division of Example 4 would be written as

$$\begin{array}{r|rrrr} & 2 & 1 & -13 & 6 \\ \hline 2 & 2 & 5 & -3 & |\ 0. \end{array}$$

The next example illustrates one of the uses of such a table.

EXAMPLE 5   Consider $A(x) = x^3 - 2x^2 - x + 2$. The following table lists the results of successively dividing $A(x)$ by $x-3$, $x-2$, $x-1$, x and $x+1$.

$$\begin{array}{r|rrrr} & 1 & -2 & -1 & 2 \\ \hline 3 & 1 & 1 & 2 & 8 \\ 2 & 1 & 0 & -1 & 0 \\ 1 & 1 & -1 & -2 & 0 \\ 0 & 1 & -2 & -1 & 2 \\ -1 & 1 & -3 & 2 & 0 \end{array}$$

Examining the entries in the last column, which we call the **remainder column**, we see that $r=0$ when $A(x)$ is divided by $x-2$, $x-1$ and $x+1$. By Theorem 4.2, it follows that 2, 1 and -1 are zeros of $A(x)$ and that $x-2$, $x-1$ and $x+1$ are factors of $A(x)$.

Exercise Set 4.1

(A)   *Given a polynomial and two divisors. For each divisor, use synthetic division to write the polynomial in the form $(x-c)Q(x)+r$. (See Example 2.)*

1. $x^3 - 6x^2 + 2x + 4$:   a. $x-2$;   b. $x+2$.
2. $x^3 - 9x^2 + 14x + 24$:   a. $x-5$;   b. $x+5$.
3. $2x^4 - 3x^3 - 8x^2 - 5x - 3$:   a. $x-1$;   b. $x+1$.
4. $2x^4 + 2x^3 - 9x^2 - x - 5$:   a. $x-3$;   b. $x+3$.
5. $3x^5 - 46x^3 - 33x - 2$:   a. $x-4$;   b. $x+4$.
6. $2x^5 + 7x^4 - 64x + 1$:   a. $x-2$;   b. $x+2$.
7. $3x^3 - 2x^2 + 12x - 5$:   a. $x-2i$;   b. $x+2i$.
8. $x^3 + 4x^2 - x + 36$:   a. $x-3i$;   b. $x+3i$.
9. $4x^3 - 6x^2 + 3x + 9$:   a. $x-\frac{1}{2}$;   b. $x-\frac{3}{2}$.
10. $9x^3 - 12x^2 + 8x - 7$:   a. $x-\frac{2}{3}$;   b. $x-\frac{4}{3}$.
11. $16x^4 + 63x^2 - 4$:   a. $x-\frac{1}{4}$;   b. $x+2i$.
12. $16x^4 + 63x^2 - 4$:   a. $x+\frac{1}{4}$;   b. $x-2i$.

Use synthetic division and the Remainder Theorem (Theorem 4.1) to find the indicated values. (See Example 3.)

13. $A(x) = 2x^3 - 6x^2 + 6x - 18$: $A(-2)$; $A(\frac{1}{2})$.

14. $A(x) = 5x^3 - 11x^2 - 14x - 10$: $A(3)$; $A(\frac{1}{5})$.

15. $P(x) = 2x^4 + 5x^3 + 2x^2 + 7x + 5$: $P(-3)$; $P(i)$.

16. $P(x) = 3x^4 - 2x^3 + x^2 - x + 7$: $P(-1)$; $P(-i)$.

Given a polynomial and one of its zeros. Use Theorem 4.2 to factor the polynomial completely over R, and specify its zeros. (See Example 4.)

17. $x^3 + 2x^2 - 9x - 18$; -2.   18. $x^3 + 4x^2 - 9x - 36$; -4.

19. $3x^3 + x^2 - 8x + 4$; 1.   20. $2x^3 - 5x^2 - x + 6$; -1.

21. $4x^3 + 28x^2 - x - 7$; -7.   22. $4x^3 + 24x^2 - 3x - 18$; -6.

Construct a synthetic division table to determine which of the members of $\{-3, -2, -1, 0, 1, 2, 3\}$ are zeros of each polynomial. Specify the factor that corresponds to each such zero. (See Example 5.)

23. $x^3 + 3x^2 - 6x - 8$ 24. $x^3 - 3x^2 - 4x + 12$

25. $x^4 - 4x^3 - 7x^2 + 10x$ 26. $x^4 + 4x^3 + x^2 - 6x$

(B)   *Determine a value for k so that the second polynomial is a factor of P(x).*

27. $P(x) = x^3 - kx^2 + 3x + 7k$; $x - 3$

28. $P(x) = 3x^4 - 40x^3 + 118x^2 + kx + 27$; $x - 9$

29. Prove that $x - 1$ is a factor of $x^{1000} - 1$.

30. Is $x + 2$ a factor of $x^{10} - 1024$? Justify your answer.

(P)   *The following modification of Theorem 0.24 is helpful in constructing proofs of the theorems in this section.*

Theorem A   If $A(x)$ is a polynomial of degree $n \geq 1$ and $c \in C$, then there exists exactly one polynomial $Q(x)$ of degree $n - 1$ and exactly one complex number r such that $A(x) = (x - c)Q(x) + r$.

   I   Prove Theorem 4.1. [Hint: Begin with Theorem A and then consider $A(c)$.]

Theorem 4.2 is an "if and only if" statement. Hence, a proof must be constructed for two parts. See Exercises II and III.

   II   Prove this part of Theorem 4.2: If $A(x)$ is a polynomial of degree $n \geq 1$, $c \in C$ and $A(c) = 0$, then $x - c$ is a factor of $A(x)$. [Hint: Begin with Theorem A and apply Theorem 4.1.]

   III   Prove this part of Theorem 4.2: If $A(x)$ is a polynomial of degree $n \geq 1$, $c \in C$ and $x - c$ is a factor of $A(x)$, then $A(c) = 0$. [Hint: If $x - c$ is a factor of $A(x)$, then $A(x) = (x - c)Q(x)$.]

4.2 ZEROS OF POLYNOMIALS

The theorems of this section answer the key question "How many zeros does a given polynomial function have?" Additionally, they furnish a substantial amount of information as to the nature of the zeros—real, imaginary, positive, negative, etc.

Number of Zeros   The first of our theorems asserts that each polynomial has *at least one* zero.

Theorem 4.3 (The Fundamental Theorem of Algebra)   If $A(x)$ is a polynomial of degree $n \geq 1$, then there exists at least one complex number c such that $A(c) = 0$.

An important consequence of Theorem 4.3 is the fact that each polynomial $A(x)$ can (theoretically) be factored into a product of linear factors each in the form $x - c$, and a constant factor a_n, as indicated by the following argument. By Theorem 4.3, the polynomial $A(x)$ has at least one zero; call it c_1. By Theorem 4.2, it follows that $x - c_1$ is a factor of $A(x)$. That is,

$$A(x) = (x - c_1)Q_1(x).$$

But the quotient polynomial $Q_1(x)$ must also have at least one zero; call it c_2. Then $Q_1(x) = (x - c_2)Q_2(x)$ and so

$$A(x) = (x - c_1)(x - c_2)Q_2(x).$$

By repeated application of the same reasoning, we can continue this process until we reach a stage where the quotient polynomial $Q(x)$ is already in the form $a_n(x - c_n)$, at which point we stop. This line of reasoning motivates the next theorem.

Theorem 4.4   If $A(x)$ is a polynomial of degree $n \geq 1$, then there exist n complex numbers, $c_1, c_2, \ldots, c_n$, not necessarily distinct, such that

$$A(x) = a_n(x - c_1)(x - c_2) \cdots (x - c_n),$$

where a_n is the leading coefficient of $A(x)$.

The phrase "not necessarily distinct" in the statement of Theorem 4.4 means that two or more of the numbers from c_1 to c_n may be equal to each other. In such cases we say that the number c is a **multiple zero** of the polynomial. For example, if

$$A(x) = 5(x - 2)(x - 2)(x - 2)(x + 6)(x + 6),$$

then 2 is a zero of *multiplicity three* (a *triple zero*) and -6 is a zero of *multiplicity two* (a *double zero*) of $A(x)$.

By Theorem 4.2, each of the n numbers $c_1, c_2, \ldots, c_n$ is a zero of $A(x)$, and so $A(x)$ has at least n zeros. It can be shown that these n numbers are the *only* zeros of $A(x)$, from which we conclude that each polynomial has exactly n zeros.

Theorem 4.5 Every polynomial of degree $n \geq 1$ has exactly n complex zeros, if zeros of multiplicity k are counted k times.

EXAMPLE 1 Given $A(x) = x^4 + 4x^3 - 12x^2 - 32x + 64$. If 2 is a double zero, factor $A(x)$ completely over C and specify all of its zeros.

SOLUTION If 2 is a zero, then $x - 2$ is a factor of $A(x)$. Dividing $A(x)$ by $x - 2$, we have

$$\begin{array}{r|rrrrr} & 1 & 4 & -12 & -32 & 64 \\ 2 & 1 & 6 & 0 & -32 & 0 \end{array}$$

from which

$$A(x) = (x - 2)(x^3 + 6x^2 - 32).$$

Because 2 is a double zero, $x - 2$ is a repeated factor, and so it must be a factor of $x^3 + 6x^2 - 32$, which we next divide by $x - 2$:

$$\begin{array}{r|rrrr} & 1 & 6 & 0 & -32 \\ 2 & 1 & 8 & 16 & 0. \end{array}$$

Hence,

$$A(x) = (x - 2)(x - 2)(x^2 + 8x + 16),$$
$$A(x) = (x - 2)(x - 2)(x + 4)(x + 4),$$

The zeros of $A(x)$ are: 2 (double) and -4 (double).

Imaginary Zeros Recall from Section 0.8 that, if $a, b \in R$, then $a + bi$ and $a - bi$ are *conjugate* complex numbers. The next theorem asserts that if a real polynomial has imaginary zeros, then these zeros always occur in pairs of conjugate complex numbers.

Theorem 4.6 If $A(x)$ is a real polynomial of degree $n \geq 1$, and if the complex number c is a zero of $A(x)$, then the conjugate of c is also a zero of $A(x)$. **(P)**

EXAMPLE 2 We can verify that both i and its conjugate, $-i$, are imaginary zeros of $A(x) = x^3 - 2x^2 + x - 2$, by dividing $A(x)$ by $x - i$ and by $x + i$ and examining the remainders:

$$\begin{array}{r|rrrr} & 1 & -2 & 1 & -2 \\ i & 1 & -2 + i & -2i & 0 \\ -i & 1 & -2 - i & 2i & 0. \end{array}$$

From the table we see that $A(i) = 0$ and $A(-i) = 0$.

EXAMPLE 3 Given $A(x) = 2x^3 - 3x^2 + 18x - 27$. If $3i$ is an imaginary zero of $A(x)$, factor $A(x)$ completely over C.

(Continued on next page)

SOLUTION If $3i$ is a zero of $A(x)$ then, by Theorem 4.6, $-3i$ is also a zero of $A(x)$. By Theorem 4.2 it follows that $x-3i$ and $x+3i$ are factors of $A(x)$. That is,

$$2x^3 - 3x^2 + 18x - 27 = (x-3i)(x+3i)Q(x), \qquad \langle 1 \rangle$$

where $Q(x)$ remains to be determined. From $\langle 1 \rangle$, we have

$$Q(x) = \frac{2x^3 - 3x^2 + 18x - 27}{x^2 + 9}.$$

By long division (check the details), $Q(x) = 2x - 3$. Hence,

$$A(x) = (x-3i)(x+3i)(2x-3)$$

or, in the form of Theorem 4.4,

$$A(x) = 2(x-3i)(x+3i)(x-\tfrac{3}{2}).$$

Note that $3i$, $-3i$ and $\tfrac{3}{2}$ are *all* of the zeros of $A(x)$.

Real Zeros If the set of zeros of a real polynomial includes real numbers, then there is a simple test that enables us to determine the *possible* number of positive and negative zeros in the set. Before stating the test, we need a new concept. Let $A(x)$ be a polynomial arranged in descending powers of x. A **variation in sign** is said to occur if, as we read from left to right, successive coefficients are opposite in sign. For example, in

$$A(x) = 4x^5 - 9x^4 - 3x^3 + 7x^2 + 2x - 3.$$

we count *three variations* in sign. We shall also need to count variations in sign in $A(-x)$. Thus

$$A(-x) = 4(-x)^5 - 9(-x)^4 - 3(-x)^3 + 7(-x)^2 + 2(-x) - 3,$$

$$A(-x) = -4x^5 - 9x^4 + 3x^3 + 7x^2 - 2x - 3,$$

from which we count *two variations* in sign.

Theorem 4.7 (Descartes' Rule of Signs) If $A(x)$ is a real polynomial in the form

$$a_n x^n + a_{n-1} x^{n-1} + \cdots + a_1 x + a_0,$$

then the number of *positive* zeros of $A(x)$ is either equal to the number of variations in signs of the coefficients of $A(x)$, or is less than this number by an even natural number. The number of *negative* zeros of $A(x)$ is either equal to the number of variations in the signs of the coefficients of $A(-x)$, or is less than this number by an even natural number.

Although Theorem 4.7 may be formidable to read, it is not difficult to apply.

EXAMPLE 4   There are *three* variations in sign in

$$A(x) = 4x^5 - 9x^4 - 3x^3 + 7x^2 + 2x - 3.$$

Hence, by Theorem 4.7, there are either 3, or 1, positive zeros of $A(x)$. There are *two* variations in sign in

$$A(-x) = -4x^5 - 9x^4 + 3x^3 + 7x^2 - 2x - 3.$$

Hence, there are either 2, or 0, negative zeros of $A(x)$.

A useful technique when finding zeros of polynomials, as well as when solving polynomial equations, is that of making a list of all possible combinations of real and imaginary zeros of a given polynomial.

EXAMPLE 5   Determine the possible number of positive, negative and imaginary zeros of $A(x) = 4x^5 - 9x^4 - 3x^3 + 7x^2 + 2x - 3$.

SOLUTION   First, by Theorem 4.5, we know that there are five complex zeros. Next, by Theorem 4.6, imaginary zeros occur in conjugate pairs. Hence, there must always be an even number of imaginary zeros, or none at all. Finally, using the results of Example 4, we have

POSSIBLE NUMBER OF ZEROS

Positive	Negative	Imaginary
3	2	0
3	0	2
1	2	2
1	0	4

Upper and Lower Bounds   Consider a set S of real numbers. If m and n are real numbers such that S includes no number greater than n and no number less than m, then n is an *upper bound* on S, m is a *lower bound* on S. Thus, if $S = \{-1, 0, 2, 3\}$, then 4 is an upper bound and -2 is a lower bound on S. The next theorem provides a test for lower and upper bounds on the set of real zeros of a polynomial. Again, we have a theorem that is formidable to read but not difficult to apply.

Theorem 4.8   If $A(x)$ and $Q(x)$ are real polynomials, c is a real number and

$$A(x) = (x - c)Q(x) + A(c),$$

then

I   If $c \geq 0$ and if $A(c)$ and the coefficients of the terms of $Q(x)$ are all of the same sign, then $A(x)$ can have no real zeros greater than c.

II   If $c \leq 0$ and if $A(c)$ and the coefficients of the terms of $Q(x)$ alternate in sign when the terms are arranged in descending powers of x, then $A(x)$ can have no real zeros less than c.

Theorem 4.8-I describes conditions under which the number c *is an upper bound* on the set of real zeros of $A(x)$; Theorem 4.8-II describes conditions under which c *is a lower bound* on the set of real zeros of $A(x)$.

EXAMPLE 6   Given $A(x) = 2x^3 + 4x^2 - 2x - 5$, show that 2 is an upper bound and -3 is a lower bound on the set of real zeros of $A(x)$.

SOLUTION   We divide $A(x)$ by $x - 2$ and by $x + 3$ to obtain

$$A(x) = (x - 2)(2x^2 + 8x + 14) + 23 \qquad \langle 2 \rangle$$

and

$$A(x) = (x + 3)(2x^2 - 2x + 4) + (-17). \qquad \langle 3 \rangle$$

In $\langle 2 \rangle$, the coefficients 2, 8, and 14 and the remainder 23 have like signs $(+, +, +, +)$. Hence, by Theorem 4.8-I, $A(x)$ has no zeros greater than 2. In $\langle 3 \rangle$, the coefficients 2, -2, and 4, and the remainder -17, alternate in sign $(+, -, +, -)$. Hence, by Theorem 4.8-II, $A(x)$ has no zeros less than -3.

In actual practice, to find upper and lower bounds we construct a table of synthetic divisions, dividing successively by $x - c$, with $c = 1, 2, 3, \ldots$, until we obtain a pattern of *all signs alike*. Then we divide successively by $x - c$ with $c = -1, -2, -3, \ldots$ until we obtain a pattern of *alternating signs*. When applying Theorem 4.8, a zero entry may be viewed as preceded by either a $+$ sign or a $-$ sign.

EXAMPLE 7   Find an upper and a lower bound on the set of real zeros of $A(x) = 3x^4 - 8x^3 + 9x + 5$.

SOLUTION   For an upper bound, we divide by $x - c$ with $c = 1, 2, 3, \ldots$:

	3	-8	0	9	5
1	3	-5	-5	4	9
2	3	-2	-4	1	7
3	3	1	3	18	59.

Because the entries in the last line have like signs (all positive), we conclude that 3 is an upper bound on the set of real zeros. For a lower bound, we let $c = -1, -2, -3, \ldots$:

	3	-8	0	9	5
-1	3	-11	11	-2	7.

After only one division the entries alternate in sign $(+, -, +, -, +)$ and we conclude that -1 is a lower bound on the set of real zeros.

The last theorem in this section furnishes a method for determining intervals in which real zeros of a polynomial are located.

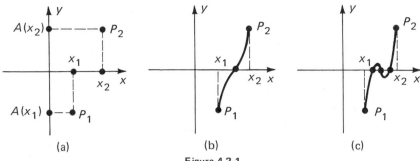

Figure 4.2-1

In Figure 4.2-1(a), P_1 and P_2 are points in the graph of $A(x)$, with x-coordinates x_1 and x_2, respectively. Because $A(x)$ is a continuous function, it follows that the graph of $A(x)$ must cross the x-axis at least once between x_1 and x_2, as indicated in Figures 4.4-1(b) and (c). Hence, there is at least one real zero of $A(x)$ located between x_1 and x_2, as stated in the next theorem.

Theorem 4.9 (Location Theorem)   If $A(x)$ is a real polynomial, x_1 and x_2 are real numbers with $x_1 < x_2$, and $A(x_1)$ and $A(x_2)$ are opposite in sign, then there is at least one real number c such that $x_1 < c < x_2$ and $A(c) = 0$.

When applying the Location Theorem, we usually take x_1 and x_2 as a pair of consecutive integers.

EXAMPLE 8   Verify that $A(x) = 2x^4 + 3x^3 - 14x^2 - 15x + 9$ has a real zero between 2 and 3; a second between 0 and 1; a third between -2 and -1; the fourth between -3 and -2.

SOLUTION   We construct a synthetic division table and look for remainders that are *opposite in sign*:

	2	3	-14	-15	9
→ 3	2	9	13	24	81 ←
2	2	7	0	-15	-21
→ 1	2	5	-9	-24	-15 ←
0	2	3	-14	-15	9 ←
→ -1	2	1	-15	0	9
→ -2	2	-1	-12	9	-9 ←
→ -3	2	-3	-5	0	9. ←

Remainders that are opposite in sign are indicated on the right side of the table; associated pairs of consecutive integers are indicated on the left side. By Theorem 4.9, we have verified the location of each of the four zeros.

Exercise Set 4.2

(A)   *Given a polynomial and a multiple zero. Factor the polynomial completely over C and specify all of its zeros. (See Example 1.)*

1. $x^4 + 3x^3 - 7x^2 - 15x + 18$; -3 is a double zero.
2. $x^4 - 11x^3 + 42x^2 - 64x + 32$: 4 is a double zero.
3. $x^4 + 4x^3 + 5x^2 + 4x + 4$: -2 is a double zero.
4. $x^5 - 3x^4 + 4x^3 - 4x^2 + 3x - 1$: 1 is a triple zero.

Given a polynomial and one of its imaginary zeros. Factor the polynomial completely over C and specify all of its zeros. (See Examples 2 and 3.)

5. $x^3 - 5x^2 + x - 5$; i 6. $x^3 + 7x^2 + x + 7$; $-i$

7. $x^3 - 5x^2 + 8x - 6$; $1 + i$ 8. $x^3 - 2x^2 - 3x + 10$; $2 - i$

Specify the possible number of positive, negative and imaginary zeros of each polynomial. (See Examples 4 and 5.)

9. $3x^3 - 2x^2 - 11x + 4$ 10. $x^3 - x^2 - 4x + 9$

11. $x^4 + 4x^3 + 4x^2 + 2x + 1$ 12. $x^4 + 4x^2 + 2x + 1$

13. $2x^5 + 4x^3 + 3x + 2$ 14. $x^6 + 1$

Find an upper bound and a lower bound for the set of real zeros for each polynomial. (See Example 7.)

15. $2x^3 - x^2 - 4x + 2$ 16. $3x^3 + 2x^2 - 26x + 4$

17. $x^4 - 2x^3 - 7x^2 + 10x + 10$ 18. $2x^4 + 3x^3 - 14x^2 - 15x + 9$

19. $x^5 - 3x^3 + 21$ 20. $x^7 + 9$

Verify that each polynomial has a zero between the given pair of consecutive integers. (See Example 8.)

21. $2x^3 + 7x^2 + 2x - 6$; $-2, -1$

22. $x^3 + x^2 - 5x - 5$; $2, 3$

23. $x^4 - 2x^2 + 12x + 7$; $-3, -2$

24. $2x^4 + 3x^3 + 15x + 9$; $-2, -3$

25. $2x^5 - 5x^4 - x^3 + x^2 + x - 8$; $2, 3$

26. $x^5 - 8x^3 - 4x^2 + 3x - 2$; $3, 4$

Each polynomial has three real zeros. Locate each of the three real zeros between a pair of consecutive integers.

27. $x^3 - x^2 - 2x + 1$ 28. $x^3 + 3x^2 - 6x - 3$

(B)   *Given $A(x) = a_n x^n + a_{n-1} x^{n-1} + \cdots + a_1 x + a_0$.*

29. Prove: If a_0 and each coefficient of $A(x)$ is positive, then $A(x)$ has no positive zeros.

30. Prove: If $A(x)$ includes only even powers of x, and a_0 and each coefficient of $A(x)$ is positive, then $A(x)$ has no real zeros.

(C)   I   Here are two BASIC computer programs for obtaining approximations to the real zeros of $y = 3x^3 - x^2 + 2x + 10$, using the Location Theorem (Theorem 4.9). Run the programs and discuss the differences in the two programs. How could the second program be revised in order to approximate the zero between -2 and -1 to the nearest one-thousandth?

Program 1	Program 2

```
  5  FOR X=-3 TO 3
 10  LET Y=3*X↑3-X↑2+2*X+10
 15  PRINT "("X","Y")"
 20  NEXT X
 25  END
```

```
  5  FOR X=-2 TO -1 STEP 0.1
 10  LET Y=3*X↑3-X↑2+2*X+10
 15  IF Y>2 THEN 25
 20  PRINT "("X","Y")"
 25  NEXT X
 30  END
```

(P) *A proof of Theorem 4.6 can be constructed with the aid of Theorem A, which you may want to prove first. (The overbar means "the conjugate of.")*

Theorem A If z_1, z_2 and z are complex numbers, then

i $\overline{z_1 + z_2} = \overline{z_1} + \overline{z_2}$

ii $\overline{z_1 z_2} = \overline{z_1} \cdot \overline{z_2}$

iii $\overline{z^n} = \overline{z}^n$

iv $z = \overline{z}$ if and only if z is a real number.

I Prove Theorem 4.6. (Hint: If $A(c) = 0$, then

$$a_n c^n + a_{n-1} c^{n-1} + \cdots + a_1 c + a_0 = 0.$$

Now, apply Theorem A to

$$\overline{a_n c^n + a_{n-1} c^{n-1} + a_1 c + a_0} = \overline{0}$$

to show that $A(\overline{c}) = 0$, where $\overline{c}$ is the conjugate of c.)

4.3   SOLVING POLYNOMIAL EQUATIONS

We are now able to obtain a significant amount of information with respect to the zeros of a polynomial function. With the introduction of one more theorem, we will be ready to consider techniques for solving polynomial equations.

Rational Zeros of Polynomials   If all of the coefficients of a polynomial $A(x)$ are integers, then the set S of all possible *rational* zeros of $A(x)$ can be found by use of the next theorem.

Theorem 4.10   If a_n, a_{n-1}, . . . , a_0 are integers and if the rational number $\dfrac{p}{q}$ in lowest terms is a zero of the polynomial

$$a_n x^n + a_{n-1} x^{n-1} + \cdots + a_1 x + a_0, \qquad a_n \neq 0,$$

then p is a factor of a_0 and q is a factor of a_n.

Theorem 4.10 implies that we need consider only integral factors of a_0 and a_n in order to find possible rational zeros of $A(x)$.

EXAMPLE 1   Specify the set S of possible rational zeros of

$$A(x) = 6x^3 + 11x^2 - 4x - 4.$$

SOLUTION   By Theorem 4.10, each rational zero of $A(x)$ is of the form $\dfrac{p}{q}$, where p is a factor of the constant term -4 and q is a factor of the leading coefficient 6. That is,

$$p \in \{\pm 4, \pm 2, \pm 1\} \quad \text{and} \quad q \in \{\pm 6, \pm 3, \pm 2, \pm 1\},$$

from which it follows that

$$\frac{p}{q} \in S \quad \text{where} \quad S = \{\pm 4, \pm 2, \pm 1, \pm \tfrac{4}{3}, \pm \tfrac{2}{3}, \pm \tfrac{1}{2}, \pm \tfrac{1}{3}, \pm \tfrac{1}{6}\}.$$

A word of caution—Theorem 4.10 does not assert that $A(x)$ must have rational zeros. It does assert that *if* $A(x)$ has rational zeros, then those zeros must be members of the set S obtained by use of the theorem. Hence, to determine any rational zeros of a given polynomial, we need only test members of S.

EXAMPLE 2   Given $A(x) = 6x^3 + 11x^2 - 4x - 4$. Find any positive rational zeros of $A(x)$.

SOLUTION   From Example 1, the set of all possible rational zeros of $A(x)$ is:

$$S = \{\pm 4, \pm 2, \pm 1, \pm \tfrac{4}{3}, \pm \tfrac{2}{3}, \pm \tfrac{1}{2}, \pm \tfrac{1}{3}, \pm \tfrac{1}{6}\}.$$

Testing positive numbers in S:

	6	11	-4	-4
$\frac{1}{6}$	6	12	-2	$-\frac{13}{3}$
$\frac{1}{3}$	6	13	$\frac{1}{3}$	$-\frac{35}{9}$
$\frac{1}{2}$	6	14	3	$-\frac{5}{2}$
$\frac{2}{3}$	6	15	6	0

From the last line we see that $A(\tfrac{2}{3}) = 0$. Furthermore, by Theorem 4.8, $\tfrac{2}{3}$ is an upper bound on the positive zeros of $A(x)$. Hence, the only positive rational zero of $A(x)$ is $\tfrac{2}{3}$.

Theorem 4.10 can also be used in the case that the coefficients of a polynomial include any rational fractions, not just all integers. For example, consider

$$P(x) = x^3 + \tfrac{11}{6}x^2 - \tfrac{2}{3}x - \tfrac{2}{3}.$$

The lowest common denominator is 6, and so we can "factor out" $\tfrac{1}{6}$ and rewrite $P(x)$ as

$$P(x) = \tfrac{1}{6}(6x^3 + 11x^2 - 4x - 4) = \tfrac{1}{6}A(x).$$

Consequently, the zeros of $P(x)$ are the same as the zeros of $A(x)$, where $A(x) = 6x^3 + 11x^2 - 4x - 4$ (see Example 2).

Polynomial Equations   Associated with each polynomial function $y = A(x)$ is the polynomial equation $A(x) = 0$, or

$$a_n x^n + a_{n-1} x^{n-1} + \cdots + a_1 x + a_0 = 0. \qquad \langle 1 \rangle$$

If c is a zero of $A(x)$, then $A(c) = 0$, which implies that c is also a solution of equation $\langle 1 \rangle$. Hence, the *zeros* of $A(x)$ are the *solutions* of $A(x) = 0$, and the problem of solving polynomial equations is essentially the same as that of finding zeros of polynomial functions. Thus, we may use "zero of a polynomial function" and "solution of a polynomial equation" synonymously, and we continue to use the theorems and techniques introduced heretofore. For example, the rational zeros of a polynomial $A(x)$ are also the rational solutions of the polynomial equation $A(x) = 0$. Our method for solving polynomial equations, where $A(x)$ is of degree greater than 2, depends mainly upon a search for rational solutions, together with the concept of *depressed equations*, introduced next.

Suppose that c is a solution of the polynomial equation $A(x) = 0$. Then c is a zero of $A(x)$ and it follows that $x - c$ is a factor of $A(x)$. That is, $A(x) = (x - c)Q(x)$, and the given equation can be written as

$$(x - c)Q(x) = 0.$$

The remaining solutions of the given equation are found by solving the **depressed equation**

$$Q(x) = 0.$$

EXAMPLE 3   Solve: $2x^4 - 5x^3 - x^2 - 5x - 3 = 0.$ $\qquad\qquad \langle 2 \rangle$

SOLUTION

1. Applying Descartes' Rule (Theorem 4.7) we consider:

    $$A(x) = 2x^4 - 5x^3 - x^2 - 5x - 3; \quad 1 \text{ sign change};$$

    $$A(-x) = 2x^4 + 5x^3 - x^2 + 5x - 3; \quad 3 \text{ sign changes};$$

    and list possible solutions as

POSITIVE	NEGATIVE	IMAGINARY
1	3	0
1	1	2

2. By Theorem 4.10, the set of all possible rational solutions is

    $$S = \{\pm 3, \pm 1, \pm \tfrac{3}{2}, \pm \tfrac{1}{2}\}.$$

3. Testing the positive numbers in S, we have

	2	-5	-1	-5	-3
$\frac{1}{2}$	2	-4	-3	$-\frac{13}{2}$	$-\frac{25}{4}$
1	2	-3	-4	-9	-12
$\frac{3}{2}$	2	-2	-4	-11	$-\frac{39}{2}$
3	2	1	2	1	0.

From the last line we see that 3 is a solution. Hence, $x-3$ is a factor of $A(x)$, $2x^3+x^2+2x+1$ is the other factor, and equation $\langle 2\rangle$ takes the form

$$(x-3)(2x^3+x^2+2x+1)=0.$$

4. We next consider the depressed equation

$$2x^3+x^2+2x+1=0. \qquad\qquad \langle 3\rangle$$

Because there can only be one positive solution (See Step 1) and we have found it, we now test for negative solutions. Thus,

$$
\begin{array}{r|cccc}
 & 2 & 1 & 2 & 1 \\
\hline
-\frac{1}{2} & 2 & 0 & 2 & \mid\ 0
\end{array}
$$

from which we see that $-\frac{1}{2}$ is a solution and equation $\langle 3\rangle$ can be written as

$$(x+\tfrac{1}{2})(2x^2+2)=0.$$

5. Again we consider a depressed equation:

$$2x^2+2=0 \quad\leftrightarrow\quad 2(x^2+1)=0.$$

Factoring over C, we have

$$2(x-i)(x+i)=0,$$

from which the remaining two solutions are i and $-i$. Equation $\langle 2\rangle$ can now be written in the factored form

$$2(x-3)(x+\tfrac{1}{2})(x-i)(x+i)=0,$$

from which we see that 3, $-\frac{1}{2}$, i and $-i$ are the zeros of $A(x)$. It follows that the solution set of the given equation is $\{3,\ -\frac{1}{2},\ i,\ -i\}$.

Multiple Solutions   If $A(x)$ is a polynomial of degree n then, counting multiplicities, $A(x)$ has exactly n zeros, and the equation $A(x)=0$ has exactly n solutions (including multiple solutions). However, the solution set of the equation will show *at most n* solutions, because we do not list the same number more than once. For example, although the (factored form) polynomial equation

$$(x-1)(x-4)^3(x+2)^2=0$$

has exactly six solutions, the solution set appears as $\{1, 4, -2\}$ because 4 is a triple solution and -2 is a double solution.

Suppose that, while solving a polynomial equation, you determine that the number c is a solution of the given equation. If you have any reason to conjecture that c is a multiple solution, then you can confirm, or reject, your conjecture simply by testing c in any or all of the subsequent depressed equations.

EXAMPLE 4 Show that 3 is a double solution of the equation

$$x^4 - 6x^3 + 7x^2 + 12x - 18 = 0. \qquad \langle 4 \rangle$$

SOLUTION First we show that 3 is a solution of equation $\langle 4 \rangle$:

$$\begin{array}{r|rrrrr} & 1 & -6 & 7 & 12 & -18 \\ \hline 3 & 1 & -3 & -2 & 6 & 0. \end{array}$$

Hence, 3 is a solution and $\langle 4 \rangle$ can be written as

$$(x-3)(x^3 - 3x^2 - 2x + 6) = 0.$$

Next, we must show that 3 is a solution of the depressed equation

$$x^3 - 3x^2 - 2x + 6 = 0: \qquad \langle 5 \rangle$$

$$\begin{array}{r|rrrr} & 1 & -3 & -2 & 6 \\ \hline 3 & 1 & 0 & -2 & 0. \end{array}$$

We see that 3 is a solution of the depressed equation $\langle 5 \rangle$. Hence, 3 is a double solution of the given equation $\langle 4 \rangle$.

As you have now seen, the process of solving a polynomial equation, as illustrated above, hinges upon the equation having at least one rational solution. If a polynomial equation has no rational solutions, other methods can be employed to approximate any real solutions. (Supplement E considers one such method.)

Theorem 4.10 can also be used to determine when a polynomial equation has no rational solutions.

EXAMPLE 5 Show that $2x^3 - 5x^2 + 3x - 7 = 0$ has no rational solutions.

SOLUTION The set of all possible rational solutions is

$$S = \{\pm 7, \pm 1, \pm \tfrac{7}{2}, \pm \tfrac{1}{2}\}.$$

Testing positive numbers in S, we have

$$\begin{array}{r|rrrr} & 2 & -5 & 3 & -7 \\ \hline \tfrac{1}{2} & 2 & -4 & 1 & -\tfrac{13}{2} \\ 1 & 2 & -3 & 0 & -7 \\ \tfrac{7}{2} & 2 & 2 & 10 & 28 \end{array}$$

From the last line we see that $\tfrac{7}{2}$ is an upper bound on the set of positive solutions and so we need not test 7. Next, testing negative numbers in S, we have

$$\begin{array}{r|rrrr} & 2 & -5 & 3 & -7 \\ \hline -\tfrac{1}{2} & 2 & -6 & 6 & -10 \end{array}$$

and $-\tfrac{1}{2}$ is a lower bound on the set of negative solutions. Hence, we need test no further. We conclude that the given equation has no rational solutions.

(A) *For each polynomial equation: List the number of possible positive, negative and imaginary solutions; specify the set S of possible rational solutions (see Example 1); solve the equation (see Example 3).*

1. $x^3 - x^2 - 4x + 4 = 0$ 2. $x^3 - 7x + 6 = 0$

3. $x^3 + 6x^2 + 11x + 6 = 0$ 4. $x^3 - 9x^2 + 23x - 15 = 0$

5. $3x^3 + 8x^2 - 7x - 12 = 0$ 6. $4x^3 + 16x^2 - x - 4 = 0$

7. $3x^3 + 2x^2 + 12x + 8 = 0$ 8. $3x^3 + 8x^2 - x - 20 = 0$

9. $x^3 - \frac{1}{3}x^2 - \frac{1}{4}x + \frac{1}{12} = 0$ 10. $x^3 - \frac{1}{6}x^2 - \frac{11}{9}x - \frac{4}{9} = 0$

11. $x^4 + 2x^3 - 7x^2 - 8x + 12 = 0$ 12. $x^4 + 5x^3 - 10x^2 - 20x + 24 = 0$

13. $3x^4 - 11x^3 + 9x^2 + 13x - 10 = 0$ 14. $x^4 + x^3 - 5x^2 - 15x - 18 = 0$

Each of the following equations has at least one multiple solution. Solve each equation. (See Example 4.)

15. $x^4 + x^3 - 10x^2 - 4x + 24 = 0$

16. $x^4 - 10x^3 + 32x^2 - 38x + 15 = 0$

17. $x^5 + 5x^4 + 3x^3 - 13x^2 - 8x + 12 = 0$

18. $x^5 + x^4 - 5x^3 - x^2 + 8x - 4 = 0$

Prove that each of the following equations has no rational solutions. (See Example 5.)

19. $2x^4 - 3x^3 - 8x^2 - 5x - 3 = 0$ 20. $2x^4 + 2x^3 + 9x^2 - x - 5 = 0$

(B) 21. A rectangular box with a volume of 48 cubic inches is to be made from a 10-inch square of tin by cutting a small square from each of the four corners, and bending up the projecting sides through 90°. Find the length of the side of the small square to be cut out. How many solutions are possible?

22. How long is the edge of a cube if its volume could be doubled by increasing one dimension by 6 inches, decreasing one dimension by 4 inches, and increasing the third dimension by 12 inches? How many solutions are possible?

23. Prove that $\sqrt{3}$ is not a rational number. (Hint: Let $x = \sqrt{3}$ and consider the equation $x^2 = 3$.)

24. Prove that $\sqrt[3]{3}$ is an irrational number.

25. Prove: If r is a prime number, $r > 1$, then $\sqrt[3]{r}$ is an irrational number.

26. Factor over C: $A(x) = x^5 + 3x^4 + 6x^3 + 10x^2 + 9x + 3$.

27. Find the three cube roots of -1 by solving the equation $x^3 = -1$.

28. Find the three cube roots of -8.

4.4   GRAPHING POLYNOMIAL FUNCTIONS

A reasonably efficient method for sketching the graph of a polynomial function involves the construction of a synthetic division table. Such a table furnishes information about x-intercepts (the same as the *zeros* of the function); y-intercepts; upper and lower bounds on the real zeros; and a sufficient number of points to enable us to sketch the graph. Whenever we refer to a specific line of entries in a synthetic division table, we do not count the top row of listed coefficients as such a line. Thus, "the first line" is to be understood to mean "the first line below the listing of coefficients."

EXAMPLE 1 Sketch the graph of $y = x^3 - 9x^2 + 14x + 24$.

SOLUTION By synthetic division, we have

	1	−9	14	24
0	1	−9	14	24
1	1	−8	6	30
2	1	−7	0	24
3	1	−6	−4	12
4	1	−5	−6	0
5	1	−4	−6	−6
6	1	−3	−4	0
7	1	−2	0	24
8	1	−1	6	72
9	1	0	14	150

From the first line of entries we see that 24 is a y-intercept. The zero entries in the remainder column indicate that 4 and 6 are x-intercepts. By Theorem 4.8, the last line in the table indicates that 9 is an upper bound on the real zeros of the function. Hence, there are no x-intercepts to the right of 9, and we now consider negative numbers.

	1	−9	14	24
−1	1	−10	24	0
−2	1	−11	36	−48.

From the first line, we see that -1 is an x-intercept, and we now have all three zeros of the function: -1, 4, and 6. Also, -2 is a lower bound on the real zeros of the function, so there are no x-intercepts to the left of -2. Because $(8, 72)$, $(9, 150)$ and $(-2, -48)$ are relatively far from the x-axis, we will not graph these three points. The remaining points are used to obtain the graph shown in Figure 4.4−1, page 172. Note that, for convenience, *different scales* are used on the x-axis and the y-axis.

In Example 1, all of the x-intercepts are rational numbers. If some or all of the zeros of a polynomial function are irrational, we apply the Location Theorem (Theorem 4.9) to locate the x-intercepts between a pair of consecutive integers.

EXAMPLE 2 Sketch the graph of $y = x^3 - 6x^2 + 2x + 4$. Specify a pair of successive integers between which each irrational x-intercept is located.

(Continued on next page)

By synthetic division, we have

	1	−6	2	4
0	1	−6	2	4
1	1	−5	−3	1 ←
→2	1	−4	−6	−8
3	1	−3	−7	−17
4	1	−2	−6	−20
5	1	−1	−3	−11
→6	1	0	2	16 ←

From the first line we see that 4 is a y-intercept. Now, observe that $A(1)=1>0$ and $A(2)=-8<0$. Hence, there is at least one x-intercept between 1 and 2. Similarly, there is at least one x-intercept between 5 and 6. Observing that 6 is an upper bound on the real zeros, we next consider negative numbers.

	1	−6	2	4
−1	1	−7	9	−5

From the table we see that −1 is a lower bound on the real zeros. Also, because $A(0)=4>0$ and $A(-1)=-5<0$, there is at least one x-intercept between 0 and −1. We have now located all three x-intercepts, as follows: between 5 and 6; between 1 and 2; between 0 and −1. The required graph is shown in Figure 4.4−2.

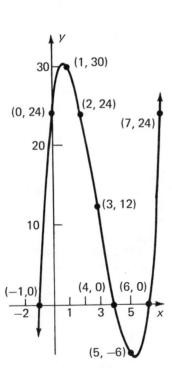

Figure 4.4-1

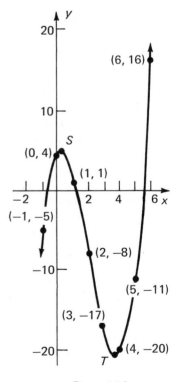

Figure 4.4-2

In Figure 4.4–2, observe that at point S the curve stops "rising to the right" and begins "falling to the right." At point T, the curve stops "falling to the right" and begins "rising to the right." Points such as S and T are called *turning points*. Methods of higher mathematics are needed to determine the coordinates of such points. Consequently, when sketching graphs, we only estimate the location of turning points as best we can. It is of interest to know that if a polynomial is of degree n, then its graph can have *at most* $n-1$ turning points. Thus, the function in Example 2 is of degree 3, and its graph has no more than two turning points.

Exercise Set 4.4

(A) *Graph each polynomial function. Specify the x-intercepts of the graph. (See Example 1.)*

1. $y = x^3 - 4x^2 - x + 4$ 2. $y = x^3 - x^2 - 16x + 16$
3. $y = x^3 - 6x^2 - x + 30$ 4. $y = x^3 - x^2 - 9x + 9$

Graph each polynomial function. Specify a pair of consecutive integers between which each x-intercept falls. (See Example 2.)

5. $y = x^3 - 2x^2 - 2x + 2$ 6. $y = 2x^3 + x^2 - 6x - 3$
7. $y = x^3 + x - 1$ 8. $y = x^3 + 3x^2 - 3$
9. $y = 2x^3 + x^2 - 6$ 10. $y = 3x^3 - 4x^2 + 3$
11. $y = x^4 - 8x^2 + 2$ 12. $y = x^4 - x^3 - 5x^2 + 3x + 1$
13. $y = x^4 + 5x^3 + 4x^2 + 6x - 4$ 14. $y = x^4 - 2x^3 - 7x^2 - 10x + 10$

The graphs of $y = A(x)$ and $y = -A(x)$ are symmetric with respect to the x-axis; the x-intercepts of both graphs are the same. Use this concept to help sketch the graph of each polynomial.

15. $y = -x^3 + 4x^2 + x - 4$ (See Exercise 1.)
16. $y = -x^3 + x^2 + 16x - 16$ (See Exercise 2.)
17. $y = -x^4 + 8x^2 - 2$ (See Exercise 11.)
18. $y = -x^4 + x^3 + 5x^2 - 3x - 2$ (See Exercise 12.)

(B) *A polynomial in factored form can often be quickly (but "roughly") sketched by plotting single points to the left of, between, and to the right of the points of intersection of the graph with the x-axis. Graph each of the following polynomials.*

EXAMPLE   Sketch a "rough" graph of $y = (x - 4)(x + 3)(x + 6)$.

SOLUTION   The x-intercepts are 4, -3 and -6. We find other selected points as follows:

$$
\begin{aligned}
\text{at } x = -7: \quad & y = (-11)(-4)(-1) = -44; \\
\text{at } x = -4: \quad & y = (-8)(-1)(2) = 16; \\
\text{at } x = 0: \quad & y = (-4)(3)(6) = -72; \\
\text{at } x = 5: \quad & y = (1)(8)(11) = 88.
\end{aligned}
$$

(Continued on next page)

Points $(-7, -44)$, $(-6, 0)$, $(-4, 16)$, $(-3, 0)$, $(0, -72)$, $(4, 0)$ and $(5, 88)$ suffice to obtain the required graph, as shown.

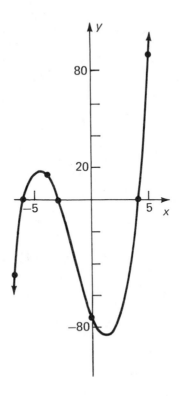

19. $y = x(x-3)(x+3)$
20. $y = 2x(x+2)(x-4)$
21. $y = (x-5)(x-3)(x+1)(x+2)$
22. $y = (x-1)(x-2)(x+5)(x+3)$

Factor each polynomial and "roughly" sketch its graph.
23. $y = x^3 - 9x^2 + 20x$
24. $y = x^3 + 8x^2 + 15x$
25. $y = x^4 - 13x^2 + 36$
26. $y = x^4 - 17x^2 + 16$
27. $y = x^4 + x^3 - 7x^2 - x + 6$
28. $y = x^4 - x^3 - 8x^2 + 2x + 12$

(C) *The following BASIC program can be used to determine coordinates of points in the graph of $y = a_3x^3 + a_2x^2 + a_1x + a_0$ over the interval with left endpoint XO and right endpoint XF. Run the program with the test data provided and select some different values of XO, XF and Z, where Z is the increment in x. How can the program be revised for use with a fourth degree polynomial?*

```
10 PRINT "ENTER THE LEAST AND GREATEST VALUES OF X AND THE INCREMENT"
20 INPUT X0,XF,Z
30 FOR J=X0 TO XF STEP Z
40 LET Y=0
50 FOR K=1 TO 4
60 READ A
70 LET Y=(Y+A)*X
80 NEXT K
90 PRINT X,Y
100 RESTORE
110 NEXT J
120 PRINT "DO YOU WISH TO CONTINUE? 1=YES,0=NO"
130 PRINT "YOUR DECISION";
140 INPUT W
150 IF W=1 THEN 10
160 DATA 1,0,-1,0
170 END
```

4.5   RATIONAL FUNCTIONS

A **rational function** is a function defined by an equation such as

$$y = \frac{A(x)}{B(x)}, \qquad [B(x) \neq 0],$$

where $A(x)$ and $B(x)$ are real polynomials. Rational functions are continuous functions, except for those values of x for which the denominator is equal to zero. For each such value of x the graph may be expected to exhibit a "break." We say that the function is *discontinuous*, or there is a *discontinuity*, at each such value of x. The behavior of graphs of rational functions over intervals that include discontinuities is of particular interest.

Asymptotes   In our study of hyperbolas (Section 3.2) we introduced the concept of a curve "approaching" a line called an *asymptote*. In general, graphs of rational functions also approach asymptotes. If the equation of an asymptote is of the form $x = h$, or $y = k$, then the line is called a **vertical asymptote**, or a **horizontal asymptote**, respectively. Although the theory underlying the concept of asymptotes requires mathematics not usually available at this level, methods for obtaining asymptotes of rational functions are quite straightforward. The following two theorems furnish a basis for these methods.

Theorem 4.11   If $\dfrac{A(x)}{B(x)}$ is a rational expression in lowest terms, then the graph of $y = \dfrac{A(x)}{B(x)}$ has a vertical asymptote for each value of x for which $B(x)$ is equal to zero.

We can reason that no point of a vertical asymptote can be a point of the graph, as follows. By Theorem 4.11, vertical asymptotes occur for values of x for which the denominator is equal to zero. Inasmuch as the function is not defined (does not exist) at such values of x, it follows that the graph of a rational function cannot intersect a vertical asymptote.

EXAMPLE 1   Consider the rational function defined by

$$y = \frac{4x^2 - 9}{x^2 - 16}.$$

Setting the denominator $B(x)$ equal to zero, we have

$$x^2 - 16 = 0 \quad \leftrightarrow \quad x = \pm 4.$$

Hence, $B(x) = 0$ if $x = 4$ or if $x = -4$, and there are two vertical asymptotes: $x = 4$ and $x = -4$.

The next theorem establishes conditions for the existence of horizontal asymptotes.

Theorem 4.12   If $A(x)$ and $B(x)$ are polynomials of degree m and n respectively, and if a and b are the coefficients of the terms of greatest degree in $A(x)$ and $B(x)$ respectively, then the graph of

$$y = \frac{A(x)}{B(x)}$$

   I   has the horizontal asymptote $y = 0$ if $m < n$;

   II   has the horizontal asymptote $y = \frac{a}{b}$ if $m = n$;

   III   has no horizontal asymptote if $m > n$.

To use Theorem 4.12, we need only compare the respective degrees of the numerator and denominator polynomials.

EXAMPLE 2   For each rational function, determine any horizontal asymptotes.

   a. $y = \dfrac{3}{x^2 + 4}$ b. $y = \dfrac{4x^2 - 9}{x^2 - 16}$ c. $y = \dfrac{x^3 + 1}{4x^2 - x}$

SOLUTION

   a.   The numerator is of degree 0; the denominator is of degree 2; $0 < 2$. By Theorem 4.12–I, $y = 0$ (the x-axis) is a horizontal asymptote.

   b.   Numerator and denominator are of degree 2. By Theorem 4.12–II, $y = \frac{4}{1}$ or $y = 4$ is a horizontal asymptote.

   c.   The numerator is of degree 3; the denominator is of degree 2; $3 > 2$. By Theorem 4.12–III, there is no horizontal asymptote.

Recall that the asymptotes to the hyperbolas of Section 3.2 were neither vertical nor horizontal. Such asymptotes are called **oblique asymptotes**. In the case of rational functions, oblique asymptotes occur when the degree of the numerator is exactly one more than the degree of the denominator. An equation for such an asymptote can be found by division.

EXAMPLE 3   Consider the rational function defined by

$$y = \frac{x^2 + 2x - 3}{x + 4}.$$ $\langle 1 \rangle$

By division, $\langle 1 \rangle$ can also be written as

$$y = x - 2 + \frac{5}{x + 4}.$$ $\langle 2 \rangle$

Now, consider the effect of successively substituting greater and greater values for x in the remainder term:

x	1	6	46	496	4996	49996
$\dfrac{5}{x+4}$	1	0.5	0.1	0.01	0.001	0.0001

From the table, it is clear that the remainder "approaches" zero in value. It follows that, for increasing values of x, the number y "approaches" the quotient $x - 2$ more and more closely. (Similar results follow if decreasing values of x are substituted in $\langle 2 \rangle$.) We say that $y = x - 2$ is an oblique asymptote to the graph.

Graph of a Rational Function   To sketch the graph of a rational function, we first discuss the graph (see Section 3.1), determine any asymptotes, and then sketch the curve.

EXAMPLE 4   Discuss and sketch the graph of $y = \frac{4x^2 - 9}{x^2 - 16}$.

SOLUTION   (Verification of intercepts and symmetry is left to you.)

x-intercepts: $\frac{3}{2}, -\frac{3}{2}$ **y-intercept:** $\frac{9}{16}$

Symmetry: With respect to the y-axis only.

Domain: Values for which the denominator is equal to zero are to be excluded. Hence, the domain is $\{x: x \neq 4, \ x \neq -4\}$.

Range: Solving the given equation for x in terms of y, we obtain

$$x^2 = \frac{16y - 9}{y - 4} \quad \text{or} \quad x = \pm\sqrt{\frac{16y - 9}{y - 4}},$$

from which we see at once that $y \neq 4$. Because we require that x be a real number, it follows that

$$\frac{16y - 9}{y - 4} \geq 0.$$ $\langle 3 \rangle$

Inequality $\langle 3 \rangle$ can be solved to obtain the range $\{y: y \leq \frac{9}{16} \text{ or } y > 4\}$.

Asymptotes: From Example 1, $x = 4$ and $x = -4$ are vertical asymptotes. From Example 2(b), $y = 4$ is a horizontal asymptote.

Point plotting: The information obtained in the preceding steps is shown in Figure 4.5–1(a). It remains to determine a sufficient number of points to help sketch the rest of the graph. We note that all the points of intersection of the graph with the axes are in the region between the vertical asymptotes. Hence, we compute coordinates of some other points in this region; for example, $\left(2, -\frac{7}{12}\right)$ and $\left(3, -\frac{27}{7}\right)$. By symmetry with respect to the y-axis we also have points $\left(-2, -\frac{7}{12}\right)$ and $\left(-3, -\frac{27}{7}\right)$. Plotting these four points (of course, we approximate fractional coordinates) and drawing a smooth curve through them, as shown in Figure 4.5–1(b), we obtain one branch of the curve. Observe how the graph approaches each vertical asymptote.

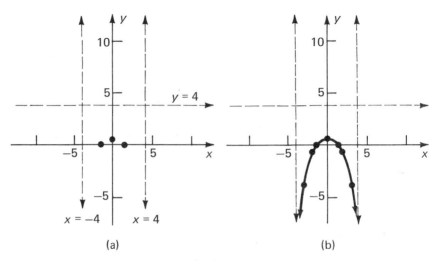

(a)   (b)

Figure 4.5-1

Next, we consider points in the region to the right of $x=4$: say $\left(5, \frac{91}{9}\right), \left(7, \frac{187}{33}\right)$ and $\left(9, \frac{315}{65}\right)$. Plotting these points and drawing a smooth curve through them, we obtain a second branch of the curve. Note that this branch approaches the horizontal asymptote $y=4$ and the vertical asymptote $x=4$, as shown in Figure 4.5–2(a). Finally, by symmetry with respect to the y-axis, a third branch of the curve is determined by the points $\left(-5, \frac{91}{9}\right), \left(-7, \frac{187}{33}\right), \left(-9, \frac{315}{65}\right)$. Sketching in this branch, we obtain the completed graph, as shown in Figure 4.5–2(b). It is of interest to relate this graph to the domain and range, as specified in the discussion. Note that, as indicated by the domain, there are no points in the graph with x-coordinate 4 or -4. Further, as indicated by the range, there are no points in the graph in the region between the lines $y=4$ and $y=\frac{9}{16}$ (the latter line is not shown on the graph.)

178   POLYNOMIAL FUNCTIONS

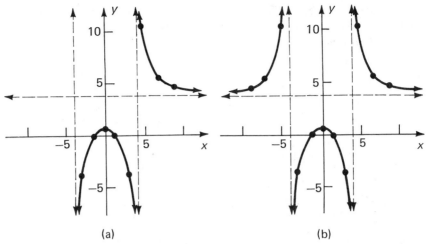

(a)   (b)

Figure 4.5-2

EXAMPLE 5   Discuss and sketch the graph of $y = \dfrac{x^2+1}{x-1}$.

SOLUTION   (Verify the details.)

   x-intercepts: none;   **y-intercept**: -1

   Symmetry: none

   Domain: $\{x: x \neq 1\}$;   **Range**: $\{y: y \leq 2-2\sqrt{2} \text{ or } y \geq 2+2\sqrt{2}\}$

   Asymptotes: (Vertical) $x=1$;   (Oblique) $y=x+1$

   Point-plotting: $\left(-2, -\frac{5}{3}\right)$, $(-1, -1)$, $(0, -1)$, $\left(\frac{1}{2}, -\frac{5}{2}\right), \left(\frac{3}{2}, \frac{13}{2}\right)$, $(2, 5)$, $(3, 5), \left(4, \frac{17}{3}\right)$.

The graph is shown in Figure 4.5-3.

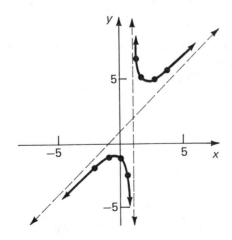

Figure 4.5-3

Exercise Set 4.5

(A) *Find equations for any vertical, horizontal, or oblique asymptotes. (See Examples 1, 2, 3.)*

1. $y = \dfrac{3x+2}{x^2-9}$

2. $y = \dfrac{x+5}{4x^2-1}$

3. $y = \dfrac{4x^2-1}{x+5}$

4. $y = \dfrac{x^2-9}{3x+2}$

5. $y = \dfrac{5x^2+3}{6x^2+1}$

6. $y = \dfrac{6x^2+1}{5x^2+3}$

7. $y = \dfrac{x^3+6x^2}{x^2-1}$

8. $y = \dfrac{x^3+2x+4}{x^2+5x+6}$

Discuss and sketch the graph of each function. Show asymptotes, if any. (See Examples 4, 5.)

9. $y = \dfrac{1}{x}$ rect hyperbole

10. $y = \dfrac{3}{x^2-1}$

11. $y = \dfrac{x+4}{x+3}$

12. $y = \dfrac{2x-1}{3x+6}$

13. $y = \dfrac{2x-1}{x^2-1}$

14. $y = \dfrac{5}{2x^2+5x-3}$

15. $y = \dfrac{3x-6}{x^2+2}$

16. $y = \dfrac{6x^2+5}{x^2+5}$

17. $y = \dfrac{x^2-4x+9}{x-4}$

18. $y = \dfrac{x^2+x+1}{x+2}$

19. $y = \dfrac{x^3-4x}{x^2-1}$

20. $y = \dfrac{x^2+x-12}{x^2-6x}$

21. $x^2y - 9y = 8 - 2x^2$

22. $yx^2 + 4y = 8$

(B) *Discuss and sketch the graph of each of the following equations. Note: Although each equation is of the form $y = \dfrac{A(x)}{B(x)}$, none of the equations defines a rational function. Nevertheless, solving the equation $B(x)=0$ will help to determine any vertical asymptotes. If you can solve for x in terms of y, then a similar consideration of the resulting denominator will help to determine any horizontal asymptotes.*

23. $y = \dfrac{x}{\sqrt{x^2-1}}$

24. $y = \dfrac{x}{\sqrt{9-x^2}}$

25. $y = \dfrac{x}{\sqrt{x^2+1}}$

26. $y = \dfrac{x+1}{\sqrt{x^2-6x+9}}$

Chapter 4 Self-Test

[4.1] 1. If $x-3$ is a divisor of $A(x) = x^4 - 2x^3 - 13x^2 + 14x + 24$, use synthetic division to write the polynomial in the form $(x-c)Q(x)+r$.

2. For the polynomial of Exercise 1, use synthetic division to find $A(-3)$ and $A(2)$.

3. If -4 and 1 are zeros of $A(x)=x^4+2x^3-13x^2-14x+24$, factor $A(x)$ completely over C and specify all of its zeros.

4. Find a least integer upper bound and a greatest integer lower bound for the set of real zeros of $x^4+x^3-3x^2-3x-3$.

5. Specify the possible number of positive, negative and complex zeros of $x^4+x^3-3x^2-3x-3$.

6. Locate each of the three real zeros of x^3-4x+1 between a pair of consecutive integers.

7. Solve $4x^4-12x^3-7x^2+18x+9=0$.

8. Graph the polynomial function $y=2x^3+7x^2-6x-6$; specify a pair of consecutive integers between which each x-intercept falls.

9. Find equations for any vertical, horizontal or oblique asymtotes to the graph of:

a. $y=\dfrac{2x^3+x^2+3}{x^2-1}$ b. $y=\dfrac{x^2}{x^2+3x-10}$.

10. Discuss and sketch the graph of $y=\dfrac{9}{x^2-4}$.

SUPPLEMENT E

IRRATIONAL ZEROS OF POLYNOMIALS

There are several numerical methods available for obtaining *rational approximations* to irrational zeros of polynomials (or irrational solutions of polynomial equations). One relatively straightforward method depends upon repeated synthetic division together with applications of Theorem 4.9, the Location Theorem.

Consider the polynomial

$$A(x) = x^3 - 6x^2 + 2x + 4.$$

As you can verify, $A(1) = 1$ and $A(2) = -8$. Hence, by Theorem 4.9, there is at least one real zero of $A(x)$ between 1 and 2—let us call it x_1. By Theorem 4.10, the only possible *rational* zeros of $A(x)$ are the members of $\{\pm 4, \pm 2, \pm 1\}$, from which we conclude that x_1 is irrational. Now consider Figure E-1 which shows two points in the graph of $A(x)$, $P_1(1, 1)$ and $P_2(2, -8)$, together with the line segment $\overline{P_1 P_2}$ intersecting the x-axis at Q_1. By inspection, the x-coordinate of Q_1 is approximately 1.1. Hence, we use 1.1 as a *first approximation* to x_1. By synthetic division, we have

	1	−6	2	4
1.1	1	−4.9	−3.39	0.271
1.2	1	−4.8	−3.76	−0.512

Because $A(1.1) = 0.271 > 0$ and $A(1.2) = -0.512 < 0$, by Theorem 4.9 it follows that

$$1.1 < x_1 < 1.2.$$

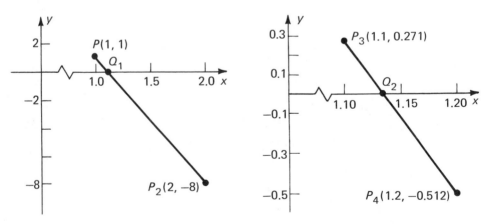

Figure E-1   Figure E-2

Figure E-2 shows $P_3(1.1, 0.271)$, $P_4(1.2, -0.512)$ and Q_2, the intersection of $\overline{P_3 P_4}$ with the x-axis. By inspection, the coordinate of Q_2

appears to be a little more than 1.1. We try 1.13 and 1.14 as *second approximations* to x_1, and test by synthetic division:

	1	−6	2	4
1.13	1	−4.87	−3.5031	0.041497
1.14	1	−4.86	−3.5404	−0.036056

Because $A(1.13) > 0$ and $A(1.14) < 0$, by Theorem 4.9 it follows that

$$1.13 < x_1 < 1.14.$$

Now, the actual y-coordinate of the intersection of the graph of $A(x)$ and the x-axis is, of course, zero. Hence, because the difference between zero and 0.041497 is more than the difference between zero and −0.036056, we conclude that x_1 is closer to 1.14. That is, $x_1 = 1.14$ to two decimal places.

By repeated applications of the above process, an irrational zero of a polynomial function can be approximated to any required number of decimal places.

Exercise Set E

For each polynomial function, compute the indicated zero to the specified number of places.

1. $y = x^2 + 3x - 5$; one positive zero, two decimal places.
2. $y = x^2 + 3x - 5$; one negative zero, two decimal places.
3. $y = x^3 - 3x + 1$; between 0 and 1, one decimal place.
4. $y = x^3 - 3x + 1$; between −1 and −2, one decimal place.
5. $y = x^3 + x^2 + x + 2$; the greatest negative zero, one decimal place.
6. $y = x^3 + 3x^2 - 6x - 3$; the greatest positive zero, one decimal place.
7. $y = x^4 - x^3 - 4x^2 - 5x + 5$; the greatest positive zero, one decimal place.
8. $y = x^4 - x^3 - 4x^2 - 5x + 5$; between 0 and 1, one decimal place.
9. Approximate $\sqrt[3]{2}$ to two decimal places. (Hint: Locate the positive zero of $A(x)$, where $A(x) = x^3 - 2$.)
10. Approximate $\sqrt[4]{6}$ to two decimal places. (Hint: See Exercise 9.)

SYSTEMS OF EQUATIONS

INTRODUCTORY CONCEPTS

A **linear equation in n variables** is an equation of the form

$$a_1x_1 + a_2x_2 + \cdots + a_{n-1}x_{n-1} + a_nx_n = k,$$

where not all of the coefficients $a_1, a_2, \ldots, a_n$ are equal to zero. The **replacement set** for the variables is a *set of ordered n-tuples* of real numbers (ordered pairs, ordered triples, etc.). A **solution** of a linear equation in n variables is an ordered n-tuple $(x_1, x_2, \ldots, x_n)$ that satisfies the equation. For equations in two or three variables it is customary to use x, y and z as variables and a, b, c and d as coefficients. Thus,

$$ax + by = c \qquad \text{and} \qquad ax + by + cz = d$$

are linear equations in two and three variables, respectively; ordered pairs (x, y) and ordered triples (x, y, z) denote members of their respective replacement sets. The equations

$$3x + 4y = 9 \qquad \text{and} \qquad x + 2y - 7z = 4$$

are examples of linear equations in two and three variables, respectively. You can verify that $(3, 0)$ and $(7, 3)$ are solutions of the equation in two variables; $(4, 0, 0)$ and $(3, 4, 1)$ are solutions of the equation in three variables.

A set of two linear equations in two variables defines a **2 × 2 system of linear equations** or, simply, a **2 × 2 linear system**. A set of three linear equations in three variables defines a **3 × 3 linear system**; a set of m linear equations in n variables defines an **$m \times n$ linear system**. The **solution set** of a 2×2 linear system is the set of all ordered pairs that are solutions of both equations; the **solution set** of a 3×3 linear system is the set of all ordered triples that are solutions of all three equations; the **solution set** of an $m \times n$ linear system is the set of all ordered n-tuples that are solutions of all m equations. Thus, the *solution set of a system is the intersection of the solution sets of the equations of the system.* To *solve a system* means to *find the solution set of the system.* A **consistent**

system is a system that has at least one solution; a system that has no solutions is **inconsistent**. Hence, the solution set of a consistent system includes at least one member; the solution set of an inconsistent system is the empty set, $\emptyset$.

5.1   2×2 AND 3×3 SYSTEMS

Let us consider a general 2×2 linear system in *standard form*

$$a_1x + b_1y = c_1$$
$$a_2x + b_2y = c_2,$$

where a_1 and b_1 are not both zero; a_2 and b_2 are not both zero. Because the graph of each equation is a line, the graph of the system is a pair of lines and the existence of solutions of the system can be determined by whether or not the lines intersect. The three possible cases are illustrated in Figure 5.1-1, and described geometrically and algebraically as follows:

GEOMETRY	*ALGEBRA*
a. The lines intersect in exactly one point.	a. The system is *consistent*; there is *exactly one solution*.
b. The lines coincide.	b. The system is *consistent*; there is an *infinite number of solutions*.
c. The lines are parallel— they do not intersect.	c. The system is *inconsistent*; there are *no solutions*.

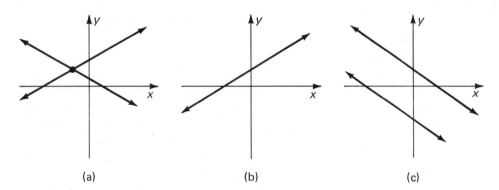

(a)   (b)   (c)

Figure 5.1-1

EXAMPLE 1   The graph of the 2×2 linear system

$$x - 4y = -8$$
$$3x - 2y = 6$$

(Continued on next page)

is shown in Figure 5.1-2. By inspection, we see that the lines intersect at the point (4, 3). Hence, the solution set of the system is $\{(4, 3)\}$.

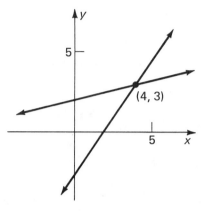

Figure 5.1-2

Equivalent Systems   Consider the two 2×2 linear systems

$$x = 4 \qquad\qquad x - 4y = -8$$
$$y = 3; \qquad\qquad 3x - 2y = \ \ 6.$$

It is clear that (4, 3) is the only solution of the system on the left, and so $\{(4, 3)\}$ is its solution set. From Example 1 we have that $\{(4, 3)\}$ is also the solution set of the second system. Systems with the *same* solution set are of particular interest.

Definition 5.1   Systems of equations are **equivalent systems** if and only if they have equal solution sets.

The concept of equivalent systems is the basis of the methods used for solving systems. If the solution set of a system is not obvious, we replace it by an equivalent system, repeatedly if need be, until the solution set becomes obvious.

Generating Equivalent Systems   The following two examples illustrate two methods for generating equivalent systems. For each of these methods, our strategy is to generate an equivalent system in which at least one of the equations includes only one variable, in which case we say that we have *eliminated* a variable. The choice of which variable to eliminate is entirely arbitrary. The first of the two methods is referred to as the **substitution method** because it is based upon the Substitution Axiom (Axiom E–4).

EXAMPLE 2   Use the substitution method to solve the system

$$①\quad 4x + 3y = 9$$
$$②\quad x - \ y = -3.$$

SOLUTION   Solving equation ② for x in terms of y, we have

$$x = y - 3.$$

Hence, each solution of equation ② is an ordered pair of the form $(y - 3, y)$. Because each solution of the system must be a solution of *both* equations, it follows that we must determine a value for y such that $(y - 3, y)$ is also a solution of ①. Hence, we substitute $y - 3$ for x in ① to obtain

$$4(y - 3) + 3y = 9 \quad \leftrightarrow \quad y = 3.$$

We now have the equivalent system

$$③ \qquad y = 3$$
$$④ \quad x - y = -3$$

obtained by replacing equation ① with $y = 3$. Substituting 3 for y in ④, we have

$$x - 3 = -3 \quad \leftrightarrow \quad x = 0.$$

Thus, we have still another equivalent system

$$y = 3$$
$$x = 0,$$

for which there is exactly one solution, $(0, 3)$. Hence, the solution set of the given system is $\{(0, 3)\}$.

To introduce the second method, let us reconsider the system of Example 1,

$$① \quad x - 4y = -8$$
$$② \quad 3x - 2y = 6,$$

<1>

with solution set $\{(4, 3)\}$. Suppose that each member of ① is multiplied by a nonzero number s, each member of ② is multiplied by a nonzero number t, and the corresponding members of the resulting equations are added to obtain

$$s(x - 4y) + t(3x - 2y) = -8s + 6t. \qquad <2>$$

Equation <2> is called a **linear combination** of equations ① and ②. We now assert that the solution $(4, 3)$ of system <1> is also a solution of linear combination equation <2>. For verification, we need only substitute 4 for x and 2 for y in <2> to obtain

$$s(4 - 4 \cdot 3) + t(3 \cdot 4 - 2 \cdot 3) = -8s + 6t$$

$$-8s + 6t = -8s + 6t,$$

which last equation is certainly true for *any* numbers s and t. More generally, if (r_1, r_2) is a common solution of a 2×2 system, then (r_1, r_2) is also a solution of any linear combination of the two equations of the system (see Exercise 5.1–33). The next theorem makes use of the linear combination concept.

Theorem 5.1   For any two nonzero real numbers s and t, if either equation of the given system

$$a_1 x + b_1 y = c_1 \quad (a_1,\ b_1 \text{ not both zero})$$
$$a_2 x + b_2 y = c_2 \quad (a_2,\ b_2 \text{ not both zero.})$$

is replaced by the linear combination equation

$$s(a_1 x + b_1 y) + t(a_2 x + b_2 y) = s c_1 + t c_2,$$

then the resulting system is equivalent to the given system.

Application of Theorem 5.1 to solving systems is referred to as the **method of linear combinations**. The numbers s and t are so chosen that the resulting linear combination equation will involve only one variable.

EXAMPLE 3   Solve the system by the method of linear combinations.

$$\textcircled{1} \quad 6x - 5y = 17$$
$$\textcircled{2} \quad 9x + 2y = -3.$$

SOLUTION   We note that if each member of $\textcircled{1}$ is multiplied by 2 and each member of $\textcircled{2}$ is multiplied by 5, the terms involving y will sum to zero. Hence, we choose $s = 2$, $t = 5$, and multiply accordingly, to obtain

$$\textcircled{3} \quad 12x - 10y = 34$$
$$\textcircled{4} \quad 45x + 10y = -15.$$

Adding corresponding members of $\textcircled{3}$ and $\textcircled{4}$, we have the linear combination equation

$$57x - 19 = 0, \quad \text{or} \quad x = \tfrac{1}{3}.$$

Replacing equation $\textcircled{2}$ by $x = \tfrac{1}{3}$, we turn to the equivalent system

$$\textcircled{5} \quad 6x - 5y = 17$$
$$\textcircled{6} \quad x = \tfrac{1}{3}.$$

Substituting $\tfrac{1}{3}$ for x in $\textcircled{5}$, we find that $y = -3$. Hence, the only solution is $(\tfrac{1}{3}, -3)$ and the solution set of the given system is $\{(\tfrac{1}{3}, -3)\}$.

Each of the two preceding examples displays a consistent system with exactly one solution. The next two examples illustrate how to recognize a consistent system with an infinite number of solutions or an inconsistent system.

EXAMPLE 4   Solve:

$$\textcircled{1} \quad x - 2y = 3$$
$$\textcircled{2} \quad 2x - 4y = 6.$$

<3>

SOLUTION   We multiply ① by -2, $(s=-2)$ and ② by 1, $(t=1)$ to obtain

$$③ \quad -2x+4y=-6$$
$$④ \quad 2x-4y=6.$$

Adding corresponding members of ③ and ④, we have

$$0x+0y=0.$$

Applying Theorem 5.1, we next consider the equivalent system

$$⑤ \quad x-2y=3$$
$$⑥ \quad 0x+0y=0. \qquad \langle 4 \rangle$$

Theorem 5.1 is not applicable to system $\langle 4 \rangle$ because a_2 and b_2 are both zero. However, we can complete the job of solving the given system by reasoning as follows. *Any* pair of real numbers satisfies equation ⑥. Hence, any pair of real numbers that is a solution of ⑤ is also a solution of ⑥, and so the solution set of $\langle 4 \rangle$ is the infinite set of solutions of equation ⑤. But system $\langle 4 \rangle$ is equivalent to the given system $\langle 3 \rangle$. It follows that the solution set of system $\langle 3 \rangle$ is the same as the solution set of equation ①. That is, the solution set of the given system is $\{(x, y): x-2y=3\}$. This is an example of case (b) on page 185.

EXAMPLE 5   Solve:

$$① \quad x-2y=3.$$
$$② \quad 2x-4y=4.$$

SOLUTION   We multiply ① by -2 and ② by 1 to obtain

$$③ \quad -2x+4y=-6$$
$$④ \quad 2x-4y=4.$$

Adding corresponding members of ③ and ④, we have

$$0x+0y=-2,$$

and we now consider the equivalent system

$$⑤ \quad x-2y=3$$
$$⑥ \quad 0x+0y=-2. \qquad \langle 5 \rangle$$

But equation ⑥ has no solutions. Hence, system $\langle 5 \rangle$, and therefore the given system, has no solutions. That is, the solution set of the given system is $\emptyset$. This is an example of case (c) on page 185.

3×3 Systems   The methods illustrated above can be extended to solving systems involving more than two equations in two variables.

EXAMPLE 6   Solve the 3×3 linear system

$$\begin{array}{r} \text{(1)} \quad 3x+2y+\ z=7 \\ \text{(2)} \quad x+\ y-3z=3 \\ \text{(3)} \quad -2x+\ y-\ z=0. \end{array}$$

SOLUTION   Our plan is to first eliminate one of the variables, say x, and then to solve the resulting 2×2 system. To accomplish this, we form linear combinations of two equations at a time, choosing s and t for each such combination so that the sum of the terms involving x will be zero. To start, we multiply each member of ① by 1 and each member of ② by -3 to obtain

$$\begin{array}{r} 3x+2y+\ z=7 \\ -3x-3y+9z=-9. \end{array}$$

Adding corresponding members, we have

$$\text{(4)} \quad -y+10z=-2.$$

Next, we multiply each member of ② by 2 and each member of ③ by 1 to obtain

$$\begin{array}{r} 2x+2y-6z=6 \\ -2x+\ y-\ z=0. \end{array}$$

Adding corresponding members, we have

$$\text{(5)} \quad 3y-7z=6.$$

Equations ④ and ⑤ form the 2×2 system

$$\begin{array}{r} -y+10z=-2 \\ 3y-\ 7z=\ \ 6. \end{array}$$

which can be solved by any method to find that $y=2$ and $z=0$. Substituting 2 for y and 0 for z in equation ① (or ② or ③), we find that $x=1$. Hence, the solution set of the given system is $\{(1, 2, 0)\}$.

Exercise Set 5.1

(A)   *Solve each system by the substitution method. (See Example 2.)*

1. $x+3y=4$
   $2x+\ y=3$

2. $2x+\ y=3$
   $x-3y=5$

3. $2x+3y=-14$
   $x+3y=13$

4. $5x-\ y=11$
   $2x+9y=42$

5. $x-2y=11$
   $2x-4y=22$

6. $6x-3y=9$
   $2x-\ y=3$

Solve each system by the linear combination method. (See Examples 3, 4 and 5.)

7. $2x-\ y=12$
   $3x-6y=-9$

8. $4x+\ y=3$
   $8x-3y=-6$

9. $11x+14y=1$
   $3x-\ 5y=-2$

10. $-3x-7y=4$
    $2x+5y=12$

11. $3x+6y=14$
    $2x+4y=-3$

12. $2x-\ 7y=9$
    $-4x+14y=20$

Solve each system. (See Example 6.)

13. $x+y+z=1$
$x-y+2z=7$
$x+y-z=-3$

14. $x+2y+3z=9$
$2x-y+2z=11$
$3x+4y-2z=-4$

15. $x+2y-z=0$
$2x-y+z=5$
$4x+2y+5z=6$

16. $3x+2y-z=4$
$x+y+z=1$
$-4x-3y+2z=-5$

17. $3x-2y-3z=-1$
$6x+y+2z=7$
$9x+3y+4z=9$

18. $4x-y+3z=1$
$7x+7y+z=2$
$2x+y-5z=-11$

View each system as a 2×2 linear system in $\dfrac{1}{x}$ and $\dfrac{1}{y}$, and solve.

19. $\dfrac{2}{x}-\dfrac{3}{y}=2$

$\dfrac{3}{x}+\dfrac{2}{y}=16$

20. $\dfrac{1}{x}+\dfrac{4}{y}=3$

$-\dfrac{6}{x}+\dfrac{1}{y}=7$

In Exercises 21 through 24, use the concept: "If a point is on a graph, then the coordinates of the point must satisfy the equation of the graph."

The graph of $x^2+y^2+Dx+Ey+F=0$ is a circle. Find an equation of the circle that includes the three given points.

21. $(1, 1), (4, 0), (2, -1)$

22. $(-3, 1), (2, 2), (3, 0)$

The graph of $y=ax^2+bx+c$ is a parabola. Find an equation of the parabola that includes the three given points.

23. $(0, 6), (3, -6), (8, 14)$

24. $(3, 5), (-3, 2), (5, 10)$

25. Find two numbers such that the sum of their reciprocals is 8 and the difference of their reciprocals is 5.

26. A chemist has two bottles of the same acid, one containing a 10% solution and the other a 4% solution. How much of each solution is to be mixed in order to obtain 120 cubic centimeters of a 5% solution?

27. An investment of \$15,000 earns an annual interest of \$840 when part of it is invested at 6% and the balance at 5%. Find the amount invested at each rate.

28. One auto can travel the same distance in 2 hours that a second auto can travel in 3 hours. If they both travel for 4 hours, the first car will travel 80 miles further than the second. Find the speed of each car.

(B) *Solve each system.*

29. $|x-3y|=5$
$2x-y=-5$

30. $|2x+5y|=13$
$3x+y=-13$

31. $|x|+|y|=2$
$x+y=0$

32. $|x|-|y|=1$
$x-2y=0$

33. Given any two nonzero real numbers s and t. If (r_1, r_2) is a solution of

$$a_1x+b_1y=c_1$$
$$a_2x+b_2y=c_2,$$

show that (r_1, r_2) is a solution of the linear combination

$$s(a_1x+b_1y)+t(a_2x+b_2y)=sc_1+tc_2.$$

(C)   *The following BASIC program can be used to solve 2×2 linear systems. Run the program with the test data provided. Use the program to check some of your answers to Exercises 1–12. Why does this program work? How could the program be changed to solve 3×3 linear systems?*

```
10 DIM A(2,3)
20 MAT READ A
30 LET D=A(1,1)*A(2,2)-A(2,1)*A(1,2)
40 IF D<>0 THEN 100
50 IF A(1,1)*A(2,3)-A(1,3)*A(2,1)<>0 THEN 80
60 PRINT "EQUATIONS INCONSISTENT, LINES PARALLEL"
70 GO TO 10
80 PRINT "EQUATIONS CONSISTENT, LINES COINCIDENT"
90 GO TO 10
100 LET X=(A(1,3)*A(2,2)-A(1,2)*A(2,3))/D
110 LET Y=(A(1,1)*A(2,3)-A(1,3)*A(2,1))/D
120 PRINT "EQUATIONS CONSISTENT, ONE SOLUTION"
130 PRINT "("X","Y")"
140 GO TO 10
150 DATA 1,2,3,2,4,6,1,2,3,1,-1,10,1,2,3,2,4,10
160 END
```

5.2   ELEMENTARY TRANSFORMATIONS; MATRICES

In this section we introduce procedures that enable us to solve linear systems—particularly those with more than two equations—in a methodical and efficient manner.

Elementary Transformations   Each of the following changes in the form of a linear system is called an **elementary transformation**:

1. Interchanging two equations.

2. Multiplying each member of an equation by the same nonzero real number.

3. Adding the same multiple of each member of an equation to the corresponding member of another equation.

Each application of an elementary transformation to a given system results in a system with the same solution set as the given system, as asserted by the following theorem.

Theorem 5.2   If a system of linear equations is obtained from a given system of linear equations as a result of a finite sequence of elementary transformations, the resulting system is equivalent to the given system.

When applying Theorem 5.2, our goal is to transform a given system into a form such as

$$1x + 0y = k_1 \qquad 1x + 0y + 0z = k_1 \qquad \text{etc.,}$$
$$0x + 1y = k_2, \qquad 0x + 1y + 0z = k_2$$
$$0x + 0y + 1z = k_3,$$

from which the respective solution sets are readily seen to be $\{(k_1, k_2)\}$, $\{(k_1, k_2, k_3)\}$, etc. In the following examples, we use notation such as:

$$(\tfrac{1}{3} \times A\,①); \qquad (2 \times B\,②) + B\,①;$$

to mean "multiply each member of equation ① of system A by $\tfrac{1}{3}$"; "add two times each member of equation ② of system B to equation ① of system B," respectively.

EXAMPLE 1   Solve system A by using elementary transformations.

$$A\begin{cases} ① & 2x + 3y = -2 \\ ② & x - 5y = 12 \end{cases}$$

SOLUTION

Objective: to obtain a first equation with $1x$ as the first term.

$$B\begin{cases} ① & x - 5y = 12 \\ ② & 2x + 3y = -2 \end{cases} \qquad (\text{Interchange } A\,① \text{ and } A\,②)$$

Objective: to eliminate the x-term in $B\,②$.

$$C\begin{cases} ① & x - 5y = 12 \\ ② & 0 + 13y = -26 \end{cases} \qquad (-2 \times B\,①) + B\,②$$

Objective: to obtain the coefficient 1 for the y-term in $C\,②$.

$$D\begin{cases} ① & x - 5y = 12 \\ ② & 0 + y = -2 \end{cases} \qquad (\tfrac{1}{13} \times C\,②)$$

Objective: to eliminate the y-term in $D\,①$.

$$E\begin{cases} ① & x + 0 = 2 \\ ② & 0 + y = -2 \end{cases} \qquad (5 \times D\,②) + D\,①$$

From E it is clear that $x = 2$ and $y = -2$. Hence, $(2, -2)$ is the only solution of system E and, by Theorem 5.2, the solution set of equivalent system A is $\{(2, -2)\}$.

Matrices   Solving systems by elementary transformations often necessitates a considerable amount of repetitive writing. The procedure can be much simplified by use of a symbol to be introduced next—a symbol in which only the coefficients and constants of the system are listed—thereby relieving us of the need to write the variables over and over again. Although we shall use 2×2 and 3×3 systems as examples, the techniques are applicable to $m \times n$ systems as well.

Consider the standard form 3×3 system A and the symbols labeled M and G:

$$A\begin{cases} ① & a_1 x + b_1 y + c_1 z = d_1 \\ ② & a_2 x + b_2 y + c_2 z = d_2 \\ ③ & a_3 x + b_3 y + c_3 z = d_3 \end{cases}$$

$$M:\begin{bmatrix} a_1 & b_1 & c_1 \\ a_2 & b_2 & c_2 \\ a_3 & b_3 & c_3 \end{bmatrix} \qquad G:\left[\begin{array}{ccc:c} a_1 & b_1 & c_1 & d_1 \\ a_2 & b_2 & c_2 & d_2 \\ a_3 & b_3 & c_3 & d_3 \end{array}\right]$$

Rectangular arrays of entries, such as M and G, are called **matrices** (singular: **matrix**); the horizontal listings of entries are called **rows**; the vertical listings are called **columns**. A matrix with m rows and n columns is called an **$m \times n$ matrix**. The 3×3 matrix M, in which the entries are the coefficients of the variables of system A, is called the **coefficient matrix** of the system. The 3×4 matrix G is obtained from matrix M by inserting into M an additional (last) column in which the entries are the constant terms d_1, d_2, d_3 of the equations in system A. We call matrix G the **augmented matrix** of the system.

EXAMPLE 2 Each system is followed by its coefficient matrix and its augmented matrix:

a. $\begin{aligned} 2x+3y&=5 \\ x-4y&=-2; \end{aligned}$ $\begin{bmatrix} 2 & 3 \\ 1 & -4 \end{bmatrix}$; $\left[\begin{array}{cc:c} 2 & 3 & 5 \\ 1 & -4 & -2 \end{array}\right]$

b. $\begin{aligned} x-2y+\ z&=0 \\ y+4z&=5 \\ 3x-2z&=-2; \end{aligned}$ $\begin{bmatrix} 1 & -2 & 1 \\ 0 & 1 & 4 \\ 3 & 0 & -2 \end{bmatrix}$; $\left[\begin{array}{ccc:c} 1 & -2 & 1 & 0 \\ 0 & 1 & 4 & 5 \\ 3 & 0 & -2 & -2 \end{array}\right]$

Row Operations The preceding discussion implies that a given $m \times n$ matrix may be viewed as the augmented matrix of a linear system in standard form.

EXAMPLE 3 Using x, y and z as variables, we may associate a matrix and a linear system as follows:

$\left[\begin{array}{cc:c} 1 & 2 & 3 \\ 4 & 5 & 6 \end{array}\right]$ with $\begin{aligned} x+2y&=3 \\ 4x+5y&=6; \end{aligned}$

$\left[\begin{array}{ccc:c} 1 & 2 & -1 & -3 \\ 2 & -1 & 1 & 5 \\ 3 & 2 & -2 & -3 \end{array}\right]$ with $\begin{aligned} x+2y-\ z&=-3 \\ 2x-\ y+\ z&=5 \\ 3x+2y-2z&=-3; \end{aligned}$

$\left[\begin{array}{ccc:c} 1 & 0 & 0 & 7 \\ 0 & 1 & 0 & -4 \\ 0 & 0 & 1 & -2 \end{array}\right]$ with $\begin{aligned} x&=\ 7 \\ y&=-4 \\ z&=-2. \end{aligned}$

This association between matrices and linear systems is developed further by the introduction of certain processes involving matrices alone. Corresponding to each of the three elementary transformations on systems (see page 192) there is associated an operation on matrices called an **elementary row operation**. These three elementary row operations are:

1. Interchanging two rows.

2. Multiplying each entry in a row by the same nonzero real number.

3. Adding the same multiple of each entry in a row to the corresponding entry in another row.

If a matrix M is obtained from a matrix A by a finite sequence of elementary row operations, we say matrices M and A are **row-equivalent**.

Matrix Solution of Linear Systems   The preceding discussion and Theorem 5.2 imply that linear systems of equations are equivalent systems if their augmented matrices are row-equivalent. It follows that linear systems can be solved by applying row operations to their augmented matrices. Our overall goal in the process is to transform the augmented matrix of the system (in standard form) into a row-equivalent matrix in a form such as

$$\begin{bmatrix} 1 & 0 & \vdots & k_1 \\ 0 & 1 & \vdots & k_2 \end{bmatrix}, \qquad \begin{bmatrix} 1 & 0 & 0 & \vdots & k_1 \\ 0 & 1 & 0 & \vdots & k_2 \\ 0 & 0 & 1 & \vdots & k_3 \end{bmatrix} \qquad \text{etc.,}$$

if possible. Note that we want 1 as the first entry in the first row. This can be accomplished either by interchanging rows:

$$\begin{bmatrix} 3 & 2 & \vdots & 5 \\ 1 & -3 & \vdots & 4 \end{bmatrix} \quad \text{into} \quad \begin{bmatrix} 1 & -3 & \vdots & 4 \\ 3 & 2 & \vdots & 5 \end{bmatrix};$$

or by multiplying each entry in the first row by a suitably selected factor. Thus, we can change

$$\begin{bmatrix} 3 & 6 & \vdots & 2 \\ 2 & -3 & \vdots & 4 \end{bmatrix} \quad \text{into} \quad \begin{bmatrix} 1 & 2 & \vdots & \frac{2}{3} \\ 2 & -3 & \vdots & 4 \end{bmatrix}$$

by multiplying each entry in the first row by $\frac{1}{3}$.

In the following examples, we use notation such as

$$(2 \times C\text{③}) + C\text{①}$$

to mean "add two times each entry in the third row of matrix C to the corresponding entry in the first row of matrix C."

EXAMPLE 4   Use row operations on matrices to solve

$$\begin{aligned} x - 2y + z &= 0 \\ -2x + y - 4z &= -4 \\ -4x + 3y - 3z &= -11. \end{aligned}$$

SOLUTION   The augmented matrix A of the given system is:

$$A: \begin{bmatrix} 1 & -2 & 1 & \vdots & 0 \\ -2 & 1 & -4 & \vdots & -4 \\ -4 & 3 & -3 & \vdots & -11 \end{bmatrix}$$

Applying row operations as indicated, we obtain:

$$B: \begin{bmatrix} 1 & -2 & 1 & \vdots & 0 \\ 0 & -3 & -2 & \vdots & -4 \\ -4 & 3 & -3 & \vdots & -11 \end{bmatrix} \quad (2 \times A\text{①}) + A\text{②}$$

$$C: \begin{bmatrix} 1 & -2 & 1 & \vdots & 0 \\ 0 & -3 & -2 & \vdots & -4 \\ 0 & -5 & 1 & \vdots & -11 \end{bmatrix} \quad (4 \times B\text{①}) + B\text{③}$$

(Continued on next page)

(As you will note below, *more than one* row transformation may be applied at each step.)

$$D: \begin{bmatrix} 1 & 0 & \frac{7}{3} & \vdots & \frac{8}{3} \\ 0 & 1 & \frac{2}{3} & \vdots & \frac{4}{3} \\ 0 & 0 & \frac{13}{3} & \vdots & -\frac{13}{3} \end{bmatrix} \quad \begin{array}{l} \left(-\frac{2}{3} \times C\,②\right)+C\,① \\ \left(-\frac{1}{3} \times C\,②\right) \\ \left(-\frac{5}{3} \times C\,②\right)+C\,③ \end{array}$$

$$E: \begin{bmatrix} 3 & 0 & 7 & \vdots & 8 \\ 0 & 3 & 2 & \vdots & 4 \\ 0 & 0 & 1 & \vdots & -1 \end{bmatrix} \quad \begin{array}{l} (3 \times D\,①) \\ (3 \times D\,②) \\ \left(\frac{3}{13} \times D\,③\right) \end{array}$$

$$F: \begin{bmatrix} 3 & 0 & 0 & \vdots & 15 \\ 0 & 3 & 0 & \vdots & 6 \\ 0 & 0 & 1 & \vdots & -1 \end{bmatrix} \quad \begin{array}{l} (-7 \times E\,③)+E\,① \\ (-2 \times E\,③)+E\,② \end{array}$$

$$G: \begin{bmatrix} 1 & 0 & 0 & \vdots & 5 \\ 0 & 1 & 0 & \vdots & 2 \\ 0 & 0 & 1 & \vdots & -1 \end{bmatrix} \quad \begin{array}{l} \left(\frac{1}{3} \times F\,①\right) \\ \left(\frac{1}{3} \times F\,②\right) \end{array}$$

Matrix G corresponds to the system

$$\begin{aligned} x + 0y + 0z &= 5 \\ 0x + y + 0z &= 2 \\ 0x + 0y + z &= -1, \end{aligned}$$

from which we see that the solution set of the given (equivalent) system is $\{(5, 2, -1)\}$.

As you transform an augmented matrix, watch for a row of zeros, except for the last entry. The appearance of such a pattern indicates an *inconsistent* system. (See also Section 5.1, Example 5.)

EXAMPLE 5   Use row operations on matrices to solve

$$\begin{aligned} x - 2y + 5z &= -3 \\ x - 5y - 14z &= 11 \\ 2x - 7y - 9z &= 5. \end{aligned}$$

SOLUTION   Starting with the augmented matrix A, we have

$$A: \begin{bmatrix} 1 & -2 & 5 & \vdots & -3 \\ 1 & -5 & -14 & \vdots & 11 \\ 2 & -7 & -9 & \vdots & 5 \end{bmatrix}$$

$$B: \begin{bmatrix} 1 & -2 & 5 & \vdots & -3 \\ 0 & -3 & -19 & \vdots & 14 \\ 0 & -3 & -19 & \vdots & 11 \end{bmatrix} \quad \begin{array}{l} (-1 \times A\,①)+A\,② \\ (-2 \times A\,①)+A\,③ \end{array}$$

$$C: \begin{bmatrix} 1 & -2 & 5 & \vdots & -3 \\ 0 & -3 & -19 & \vdots & 14 \\ 0 & 0 & 0 & \vdots & -3 \end{bmatrix} \quad (-1 \times B\,②)+B\,③$$

Matrix C corresponds to the system D,

$$D\begin{cases} \text{①} & x-2y+\ 5z=-3 \\ \text{②} & 0x-3y-19z=14 \\ \text{③} & 0x+0y+\ 0z=-3 \end{cases}$$

which system is equivalent to the given system. But equation $D\text{③}$ has no solutions. Hence, system D is inconsistent, from which it follows that the given system is inconsistent and its solution set is $\emptyset$.

Exercise Set 5.2

(A) *Use elementary transformations to solve each system. (See Example 1.)*

1. $3x+2y=\ 5$
   $x-4y=11$

2. $5x+3y=4$
   $3x+2y=1$

3. $4x+\ y=3$
   $8x-3y=-6$

4. $6x-10y=-1$
   $9x+\ 4y=8$

Write the coefficient matrix and the augmented matrix for each system. (See Example 2.)

5. $x+y=8$
   $x-y=4$

6. $4x-6y=11$
   $x+9y=-4$

7. $\ \ x+\ y-3z=3$
   $3x+2y+\ z=7$
   $-2x+\ y-\ z=0$

8. $x+y+\ z=1$
   $x-y+2z=7$
   $x+y-\ z=-3$

9. $x+\ y=2$
   $y+\ z=1$
   $x-2z=-2$

10. $\ \ 3x-\ z=16$
    $x-\ y+5z=0$
    $4y+8z=-5$

For each augmented matrix, write the corresponding linear system, using variables x, y and z. (See Example 3.)

11. $\begin{bmatrix} 1 & 2 & \vdots & 3 \\ 0 & 3 & \vdots & 4 \end{bmatrix}$

12. $\begin{bmatrix} 1 & 7 & \vdots & 0 \\ 0 & 1 & \vdots & \frac{1}{2} \end{bmatrix}$

13. $\begin{bmatrix} 1 & 3 & 0 & \vdots & 4 \\ 2 & -6 & 1 & \vdots & -3 \\ 0 & -2 & 1 & \vdots & 0 \end{bmatrix}$

14. $\begin{bmatrix} -2 & 4 & -1 & \vdots & 6 \\ 3 & 9 & 11 & \vdots & 14 \\ 0 & 0 & 1 & \vdots & -8 \end{bmatrix}$

15. $\begin{bmatrix} 1 & 0 & 0 & \vdots & 2 \\ 0 & 1 & 0 & \vdots & \frac{1}{3} \\ 0 & 0 & 1 & \vdots & -\frac{1}{2} \end{bmatrix}$

16. $\begin{bmatrix} 1 & 0 & 0 & \vdots & -11 \\ 0 & 1 & 0 & \vdots & 0 \\ 0 & 0 & 1 & \vdots & -\frac{4}{5} \end{bmatrix}$

Use row operations on matrices to solve each system. (See Example 4.)

17. $\ \ 2x+3y=3$
    $-6x+6y=1$

18. $3x-\ y=-12$
    $3x+2y=\ -2$

19. $\ \ x+\ y-3z=3$
    $3x+2y+\ z=7$
    $-2x+\ y-\ z=0$

20. $x+y+\ z=\ 1$
    $x-y+2z=\ 7$
    $x+y-\ z=-3$

21. $x+\ y=2$
    $y+\ z=1$
    $x-2z=-2$

22. $\ \ 3y-\ z=18$
    $x-6y-5z=0$
    $x+2z=-9$

23. $3x+2y+4z=5$
$2x+ y+ z=1$
$x-2y-3z=1$

24. $2x+ y+ z=1$
$x+2y+ z=0$
$x+ y+2z=0$

Use row operations on matrices to show that each system is inconsistent. (See Example 5.)

25. $x+2y=6$
$3x+6y=8$

26. $-5x+ y=-2$
$2x-\frac{2}{5}y=1$

27. $3x+4y+2z=0$
$x+ y+ z=2$
$4x+5y+3z=-7$

28. $x+2y+4z=13$
$3x+5y+6z=2$
$2x+3y+2z=7$

29. $x+3y+2z=7$
$x+5y-4z=6$
$2x+8y-2z=-2$

30. $x- y+ 2z=2$
$y+ 3z=-1$
$x+2y+11z=-3$

(B) *Use row operations to solve each system.*

31. $x+y =2$
$y+z =0$
$z+w =0$
$x+z =-2$

32. $y+ z+ w=0$
$x+2z- w=4$
$2x- z+2w=3$
$-2x+2z- w=-2$

5.3 SOLVING $m \times n$ LINEAR SYSTEMS

So far we have considered systems in which the number of equations is the same as the number of variables. Row operations on matrices can also be used to solve linear systems in which the number of equations differs from the number of variables.

EXAMPLE 1 Use matrices to solve the system

$$x+2y-5z=4$$
$$-3x-7y+ z=0.$$

SOLUTION Starting with the augmented matrix of the system, we have

$$A: \begin{bmatrix} 1 & 2 & -5 & \vdots & 4 \\ -3 & -7 & 1 & \vdots & 0 \end{bmatrix}$$

$$B: \begin{bmatrix} 1 & 2 & -5 & \vdots & 4 \\ 0 & -1 & -14 & \vdots & 12 \end{bmatrix} \quad (3 \times A\,①)+A\,②$$

$$C: \begin{bmatrix} 1 & 0 & -33 & \vdots & 28 \\ 0 & 1 & 14 & \vdots & -12 \end{bmatrix} \quad \begin{array}{l}(2 \times B\,②)+B\,① \\ (-1 \times B\,②)\end{array}$$

Matrix C corresponds to the system

$$E \begin{cases} x+0y-33z=28 \\ 0x+ y+14z=-12 \end{cases} \leftrightarrow F \begin{cases} x=33z+28 \\ y=-14z-12. \end{cases}$$

In system F, note that unique values for x and for y are determined for each real number replacement of z. Hence, the solution set of the given system is the infinite set of ordered triples

$$\{(x, y, z): \; x = 33z + 28, \; y = -14z - 12, \; z \in R\}.$$

Particular solutions can be obtained by substituting any real number for z. Thus, if $z = -1$, then $(-5, 2, -1)$ is one of the solutions of the given system.

EXAMPLE 2   Use matrices to solve the system

$$\begin{aligned}
x_1 + x_2 - 4x_3 - 2x_4 &= -6 \\
3x_1 + 2x_2 - x_3 + x_4 &= 8 \\
3x_1 + x_2 + 10x_3 + 8x_4 &= 3
\end{aligned}$$

SOLUTION   Starting with the augmented matrix of the system, we have

$$A: \begin{bmatrix} 1 & 1 & -4 & -2 & \vdots & -6 \\ 3 & 2 & -1 & 1 & \vdots & 8 \\ 3 & 1 & 10 & 8 & \vdots & 3 \end{bmatrix}$$

$$B: \begin{bmatrix} 1 & 1 & -4 & -2 & \vdots & -6 \\ 0 & -1 & 11 & 7 & \vdots & 26 \\ 0 & -2 & 22 & 14 & \vdots & 21 \end{bmatrix} \quad \begin{array}{l} \left(-3 \times A \textcircled{1}\right) + A \textcircled{2} \\ \left(-3 \times A \textcircled{1}\right) + A \textcircled{3} \end{array}$$

$$C: \begin{bmatrix} 1 & 1 & -4 & -2 & \vdots & -6 \\ 0 & -1 & 11 & 7 & \vdots & 26 \\ 0 & 0 & 0 & 0 & \vdots & -31 \end{bmatrix} \quad \left(-2 \times B \textcircled{2}\right) + B \textcircled{3}$$

The last row of matrix C corresponds to the linear equation

$$0x_1 + 0x_2 + 0x_3 + 0x_4 = -31,$$

for which there are no solutions. Hence, the given system is inconsistent; its solution set is $\emptyset$.

In the final example of this section we shall not specify the particular row operations used. You will find it instructive to *verify each step* as you read. The arrows indicate the sequence of row-equivalent matrices.

EXAMPLE 3   Use matrices to solve the system

$$\begin{aligned}
x - y + z &= 0 \\
4x + y - 2z &= 4 \\
3x - 5y - 6z &= -13 \\
8x - 13z &= -5
\end{aligned}$$

(Continued on next page)

SOLUTION Starting with the augmented matrix, we have

$$\left[\begin{array}{rrr:r} 1 & -1 & 1 & 0 \\ 4 & 1 & -2 & 4 \\ 3 & -5 & -6 & -13 \\ 8 & 0 & -13 & -5 \end{array}\right] \rightarrow \left[\begin{array}{rrr:r} 1 & -1 & 1 & 0 \\ 0 & 5 & -6 & 4 \\ 0 & -2 & -9 & -13 \\ 0 & 8 & -21 & -5 \end{array}\right] \rightarrow$$

$$\left[\begin{array}{rrr:r} 1 & -1 & 1 & 0 \\ 0 & 1 & -\frac{6}{5} & \frac{4}{5} \\ 0 & -2 & -9 & -13 \\ 0 & 8 & -21 & -5 \end{array}\right] \rightarrow \left[\begin{array}{rrr:r} 1 & -1 & 1 & 0 \\ 0 & 1 & -\frac{6}{5} & \frac{4}{5} \\ 0 & 0 & -\frac{57}{5} & -\frac{57}{5} \\ 0 & 0 & -\frac{57}{5} & -\frac{57}{5} \end{array}\right] \rightarrow$$

$$\left[\begin{array}{rrr:r} 1 & -1 & 1 & 0 \\ 0 & 1 & -\frac{6}{5} & \frac{4}{5} \\ 0 & 0 & 1 & 1 \\ 0 & 0 & 1 & 1 \end{array}\right] \rightarrow \left[\begin{array}{rrr:r} 1 & 0 & -\frac{1}{5} & \frac{4}{5} \\ 0 & 1 & 0 & 2 \\ 0 & 0 & 1 & 1 \\ 0 & 0 & 0 & 0 \end{array}\right] \rightarrow \left[\begin{array}{rrr:r} 1 & 0 & 0 & 1 \\ 0 & 1 & 0 & 2 \\ 0 & 0 & 1 & 1 \\ 0 & 0 & 0 & 0 \end{array}\right]$$

From the final matrix we see that $x=1$, $y=2$ and $z=1$. Hence, the solution set is $\{(1, 2, 1)\}$.

The significance of the row of zero entries in the last matrix of Example 3 can be seen as follows. First, note that the entries correspond to

$$0x + 0y + 0z = 0, \qquad\qquad \langle 1 \rangle$$

an equation satisfied by *any* ordered triple of numbers. In particular, *all* solutions of the 3×3 system corresponding to the top three rows of entries in the matrix are solutions of $\langle 1 \rangle$. Second, the system corresponding to the last matrix is equivalent to the given system. Hence, the solution set of the system corresponding to the top three rows of the last matrix is equal to the solution set of the given system, and the row of zeros plays no further role in finding the solution set.

Exercise Set 5.3

(A) *Use matrices to solve each system of equations. If there is more than one solution, give at least three particular solutions.*

1. $\begin{aligned} x+3y- z&=5 \\ 2x-2y+4z&=-1 \end{aligned}$

2. $\begin{aligned} 3x- y-3z&=0 \\ x-2y+ z&=2 \end{aligned}$

3. $\begin{aligned} 2x- y+ z&=-4 \\ -6x+3y-3z&=1 \end{aligned}$

4. $\begin{aligned} 6x+8y+4z&=2 \\ 3x+4y+2z&=1 \end{aligned}$

5. $\begin{aligned} x_1+2x_2+3x_3-9x_4&=5 \\ -x_1- x_2+2x_3+2x_4&=-3 \\ 4x_1-3x_2+2x_3-4x_4&=-2 \end{aligned}$

6. $\begin{aligned} x_1+2x_2+3x_3- x_4&=8 \\ -x_1- x_2+2x_3+5x_4&=3 \\ 4x_1-3x_2+2x_3-3x_4&=1 \end{aligned}$

7. $\begin{aligned} x_1- x_2- 6x_3+ 2x_4&=-1 \\ 2x_1-2x_2-15x_3+12x_4&=-6 \\ 3x_1-3x_2-12x_3-10x_4&=5 \end{aligned}$

8. $\begin{aligned} 3x_1+7x_2- 5x_3+5x_4&=16 \\ 2x_1+ x_2-18x_3-4x_4&=7 \\ x_1+2x_2- 3x_3+ x_4&=5 \end{aligned}$

9. $x_1 - 2x_2 + x_4 = 0$
   $2x_1 + 3x_3 = 0$
   $x_2 - x_3 = 0$

10. $-2x_1 + 3x_3 + 5x_4 = 0$
    $x_1 + 2x_2 - x_4 = 0$
    $x_2 - 4x_3 + x_4 = 0$

11. $3x - y - 4z = 5$
    $x - 2z = 0$
    $2y - 3z = 1$
    $-x + 4y = 0$

12. $x + y - z = 1$
    $2x + 2y + 3z = 4$
    $-4x - y + 5z = 2$
    $-x + 7y - 2z = 8$

(B)

13. $x_1 + 2x_2 - x_3 + 2x_4 = 0$
    $5x_1 + 3x_2 - 4x_3 + 3x_4 = 2$

14. $x_1 + 4x_2 + 4x_3 - 5x_4 = 12$
    $5x_1 - x_2 - x_3 - 4x_4 = 6$

The graph of each of the equations of the 2×3 linear system

$$a_1 x + b_1 y + c_1 z = d_1$$
$$a_2 x + b_2 y + c_2 z = d_2$$

<2>

is a **plane** in space. From geometry, we know that if two distinct planes intersect, then they intersect in exactly one line. Hence, <2> might determine a line in space. A particular solution of the system is an ordered triple (x, y, z) that gives the coordinates of a point in that line. Find three points in the line of intersection of each of the following pairs of planes.

15. $x + y + z = 0$
    $x - 2y + 4z = 9$

16. $x + 2y + 3z = 4$
    $2x + 5y - 2z = 6$

17. $3x - y - z = 9$
    $2x + 4y - 3z = 6$

18. $x + y + z = 0$
    $2x + 2y + 2z = 7$

19. A collection of pennies, nickels and dimes consists of 50 coins totaling $2.85. How many of each kind of coin is in the collection? Is there more than one answer?

20. A poultry ranch pays $1 for a chicken, $4 for a turkey and $10 for a pheasant. If the ranch has $520 to spend, enough space for 175 birds, and if it must stock at least 100 chickens and 49 turkeys, how many birds of each kind should be bought?

5.4   SYSTEMS OF NONLINEAR EQUATIONS

In this section we consider examples of some methods useful for solving systems of two equations, where one or both of the equations is nonlinear. A rigorous study of the theory underlying the general problem of solving such nonlinear systems would carry us too far afield. Instead, we settle for some representative techniques that enable us to solve a significant number of the kinds of systems most frequently encountered. A word of caution—the techniques commonly used for solving nonlinear systems do not always generate equivalent systems. Therefore, it is essential to *check* each "possible" solution in order to verify that it really does satisfy both equations of the system. "Possible" solutions that do not satisfy both equations are called *extraneous solutions*.

Although each of the examples in this section highlights a particular approach, you should understand that there is no rigid rule about which approach to use on a given type of system. In fact, you may find it instructive to try other approaches on any or all of the examples. The first example illustrates the use of substitution.

EXAMPLE 1   Solve, and then sketch the graph of, the system:

$$① \quad 3x - y = 4$$
$$② \quad 9x^2 - y^2 = 32.$$

SOLUTION   Solving equation ① for y in terms of x yields

$$y = 3x - 4.$$

Substituting $3x - 4$ for y in equation ②, we have

$$9x^2 - (3x - 4)^2 = 32$$
$$24x = 48, \quad \text{or} \quad x = 2.$$

Substituting 2 for x in equation ①, we obtain

$$3 \cdot 2 - y = 4 \quad \leftrightarrow \quad y = 2,$$

and $(2, 2)$ is the only solution of ①. Substituting 2 for x and 2 for y in equation ②, we find that $(2, 2)$ is also a solution of ②. Hence, the solution set of the given system is $\{(2, 2)\}$. Figure 5.4−1 shows the line and the hyperbola that are the respective graphs of equations ① and ②. Observe that the line intersects the hyperbola only at the point $(2, 2)$.

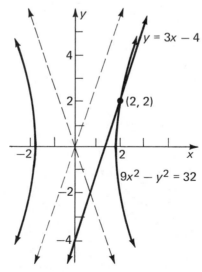

Figure 5.4-1

The next example illustrates the use of substitution together with factoring.

EXAMPLE 2   Solve for ordered pairs of *real* numbers; graph the system.

$$\text{①} \quad x^2 - 3y = 0$$
$$\text{②} \quad y^2 - 3x = 0$$

SOLUTION   Solving equation ① for y, we obtain $y = \frac{1}{3}x^2$. Substituting $\frac{1}{3}x^2$ for y in ② and simplifying, we have

$$\text{③} \quad x^4 - 27x = 0.$$

Factoring the left member of equation ③ yields

$$x(x - 3)(x^2 + 3x + 9) = 0,$$

from which we obtain $x = 0$ or $x = 3$ as the only *real* solutions for x. Substituting 0 for x in equation ②, we find that $y = 0$, and $(0, 0)$ is a possible solution. Substituting 3 for x in ② we obtain $y = \pm 3$. Hence, $(3, 3)$ and $(3, -3)$ are possible solutions. Checking each of the ordered pairs $(0, 0)$, $(3, 3)$ and $(3, -3)$ by appropriate substitutions in equation ①, we find that $(0, 0)$ and $(3, 3)$ are solutions but $(3, -3)$ is not a solution. Thus, the required solution set of the given system is $\{(0, 0), (3, 3)\}$. Figure 5.4-2 shows the two parabolas that are the graph of the system.

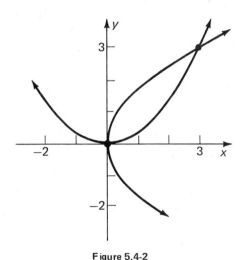

Figure 5.4-2

The method of linear combinations is applied in the next example.

EXAMPLE 3   Solve, and graph, the system

$$\text{①} \quad 9x^2 + 25y^2 = 208$$
$$\text{②} \quad x^2 + \quad y^2 = 16.$$

(Continued on next page)

SOLUTION   The variable x can be eliminated as follows. Multiply each member of equation ① by 1, each member of equation ② by -9, and add corresponding members to obtain

$$9x^2 + 25y^2 - 9x^2 - 9y^2 = 208 - 144 \quad \leftrightarrow \quad 16y^2 = 64$$
$$y^2 = 4$$
$$y = \pm 2.$$

Substituting 2 for y in equation ② we find that

$$x^2 + 4 = 16 \quad \leftrightarrow \quad x^2 = 12, \text{ or } x = \pm 2\sqrt{3}.$$

Hence, $(2\sqrt{3}, 2)$ and $(-2\sqrt{3}, 2)$ are possible solutions. Substituting -2 for y in ②, we obtain $x = \pm 2\sqrt{3}$. Hence, $(2\sqrt{3}, -2)$ and $(-2\sqrt{3}, -2)$ are possible solutions. You can check to verify that the solution set of the given system is

$$\{(2\sqrt{3}, 2), (-2\sqrt{3}, 2), (2\sqrt{3}, -2), (-2\sqrt{3}, -2)\}.$$

Figure 5.4–3 shows the ellipse and the circle that are the graph of the system.

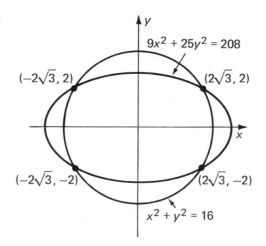

Figure 5.4-3

If at least one of the equations of a quadratic system includes an xy-term, it is often helpful to use linear combinations to *eliminate the constant term.*

EXAMPLE 4   Solve the system

$$① \quad 3x^2 + 4xy + 3y^2 = 67$$
$$② \qquad x^2 + y^2 = 17.$$

SOLUTION   Multiplying each member of ① by 17 and each member of ② by -67, we obtain

$$③ \quad 51x^2 + 68xy + 51y^2 = 1139$$
$$④ \quad -67x^2 \qquad\quad - 67y^2 = -1139.$$

Adding corresponding members of ③ and ④ gives

$$-16x^2+68xy-16y^2=0,$$

or

⑤ $\quad 4x^2-17xy+4y^2=0.$

Equation ⑤ can be written in factored form as

$$(4x-y)(x-4y)=0, \qquad\qquad <1>$$

from which it follows that

$$y=4x \qquad \text{or} \qquad y=\tfrac{1}{4}x. \qquad\qquad <2>$$

Each of equations <2> provides a substitution for y into either equation ① or ②. We choose ② because it is relatively simpler, and then solve *each* of the systems

$$\begin{array}{ccc} x^2+y^2=17 & & x^2+y^2=17 \\ y=4x; & & y=\tfrac{1}{4}x; \end{array} \qquad\qquad <3>$$

(details are not shown). As you can verify, the solution set of the given system is $\{(4,1),(-4,-1),(1,4),(-1,-4)\}$.

Steps <1>, <2> and <3> of Example 4 suggest a method for solving nonlinear systems that involves *factoring*. This method should be considered particularly when *zero* is the only constant in at least one of the equations of the system.

EXAMPLE 5 Solve the system

① $\quad 3x^2-3xy+y^2=49$
② $\quad 2x^2-xy-6y^2=0.$

SOLUTION Because zero is the only constant in ②, we factor the left member of ② and obtain the factored form equation

$$(x-2y)(2x+3y)=0.$$

Next, we solve each of the systems

$$\begin{array}{ccc} 3x^2-3xy+y^2=29 & & 3x^2-3xy+y^2=49 \\ x-2y=0 & \text{and} & 2x+3y=0 \end{array} \qquad <4>$$

as in Example 1 to obtain the solution set

$$\{(2\sqrt{7},\sqrt{7}),(-2\sqrt{7},-\sqrt{7}),(3,-2),(-3,2)\}.$$

You will find it instructive to solve both systems of <4> and check all of the solutions.

The next example is another illustration of the use of factoring.

EXAMPLE 6 Solve the system

① $\quad 4x^2+3xy-36x=0$
② $\quad 3y^2+2xy-24y=0.$

SOLUTION   Factoring each equation yields

$$\text{(3)} \quad x(4x + 3y - 36) = 0$$
$$\text{(4)} \quad y(3y + 2x - 24) = 0.$$

From equation ③ we have

$$x = 0 \quad \text{or} \quad 4x + 3y - 36 = 0.$$

From equation ④ we have

$$y = 0 \quad \text{or} \quad 3y + 2x - 24 = 0.$$

Hence, we must solve *each* of the following *four* systems:

$$
\begin{array}{ll}
x = 0 & x = 0 \\
y = 0; & 3y + 2x - 24 = 0; \\
\\
4x + 3y - 36 = 0 & 4x + 3y - 36 = 0 \\
y = 0; & 3y + 2x - 24 = 0;
\end{array}
$$

from which we obtain the solution set $\{(0, 0), (0, 8), (9, 0), (6, 4)\}$, as you can verify.

Exercise Set 5.4

*In this exercise set, solve for ordered pairs of **real numbers** only.*

(A)   *Solve and graph each system. (See Example 1.)*

1. $x^2 + y^2 = 25$
   $x + y = 1$
2. $x^2 + y^2 = 16$
   $y = 3x - 12$
3. $y = 4 - x^2$
   $y = 4 - 2x$
4. $y = x^2 - 9$
   $y = x - 7$
5. $x^2 + 16y^2 = 16$
   $x - 4y = 4$
6. $x^2 - y^2 = 16$
   $x - 3y = -4$

Solve and graph each system. (See Example 2.)

7. $y = x^2$
   $x = y^2$
8. $x^2 + y = 0$
   $y^2 + x = 0$
9. $2y = 8 - x^2$
   $y = x^2$
10. $4y = x^2 + 16$
    $y = x^2$
11. $x^2 + y^2 = 6$
    $xy = 4$
12. $3x^2 + 7y^2 = 21$
    $xy = 3$

Solve and graph each system. (See Example 3.)

13. $x^2 + y^2 = 25$
    $x^2 + 25y^2 = 25$
14. $9x^2 + 16y^2 = 160$
    $x^2 - y^2 = 15$
15. $x^2 + 4y^2 = 52$
    $x^2 + y^2 = 25$
16. $9x^2 - 4y^2 = 36$
    $x^2 + y^2 = 43$
17. $4x^2 + 9y^2 = 36$
    $x^2 + y^2 = 16$
18. $y^2 + x^2 = 4$
    $y^2 - x^2 = 9$

Solve each system. (See Example 4.)

19. $2x^2 + xy - 4y^2 = -12$
    $x^2 - 2y^2 = -4$
20. $x^2 - xy = 54$
    $xy - y^2 = 18$
21. $x^2 + 3xy - y^2 = 1$
    $2x^2 + 5xy - 9y^2 = 1$
22. $x^2 + 3xy - 2y^2 = 2$
    $2x^2 - 5xy + 6y^2 = 3$

Chapter 5   Self-Test

[5.1]   *Solve each system:*   *a.*   *By substitution or linear combinations.*   March 1

[5.2]   *b.*   *By matrix methods.*

1. $3x - 8y = 6$
   $x + 4y = 3$

2. $5x + 2y = 12$
   $7x + 3y = -4$

3. $3x + 2y - z = 11$
   $x + 2y - 4z = 4$
   $2x - y + 3z = 4$

4. $x + y + z = 2$
   $3x + 4y + 5z = -7$
   $2x + 3y + 4z = 0$

[5.3]   *Solve each system by matrix methods. If there is more than one solution, give at least three particular solutions.*

5. $x - 2y + z = 2$
   $3x - y - 3z = 0$

6. $x + 3y = -10$
   $3x - y = 10$
   $5x + 2y = 2$

[5.4]   *Solve and graph each system.*

7. $x^2 + y^2 = 16$
   $x - 3y = -12$

8. $9x^2 + 4y^2 = 36$
   $x^2 - y^2 = -9$

Solve each system.

9. $6x^2 - 5xy + 2y^2 = 3$
   $3x^2 - xy + y^2 = 3$

10. $2x^2 - 5xy + 2y^2 = 0$
    $x^2 - y^2 = 25$

Next for Exam! March 6
TUES

SUPPLEMENT F

PARTIAL FRACTIONS

In this section our concern is with rational expressions $\dfrac{N(x)}{D(x)}$ in lowest terms, with the degree of the numerator polynomial less than the degree of the denominator polynomial.

If $D(x)$ is factorable over Q into a product of polynomial factors of degree less than or equal to two, then $\dfrac{N(x)}{D(x)}$ can be "decomposed" into a sum of rational expressions called **partial fractions**, each one in lowest terms and such that the degree of the numerator is less than the degree of the denominator. For example,

$$\frac{9x - 14}{2x^2 - 5x - 12} \equiv \frac{5}{2x + 3} + \frac{2}{x - 4}. \qquad \langle 1 \rangle$$

The symbol $\equiv$ (read "is identically equal to") is used to emphasize that an equation such as $\langle 1 \rangle$ is an *identity in x* which means, as you may recall, that its solution set is equal to the replacement set for x.

The method that we use for accomplishing such a decomposition is based upon two concepts. The first concept is that of *identical polynomials*. If $P(x)$ and $Q(x)$ are polynomials such that $P(x) \equiv Q(x)$, then it can be shown that corresponding coefficients (and constant terms) must be equal. Thus, if

$$x^3 - 3x + 7 \equiv Ax^3 + Bx^2 + Cx + D,$$

then $A = 1$, $B = 0$, $C = -3$ and $D = 7$.

The second concept is a consequence of the requirement that partial fractions are to be in lowest terms, with the degree of the numerator less than the degree of the denominator. For each such partial fraction: If a denominator is linear (degree one), then the numerator is a constant polynomial or zero; if a denominator is quadratic (degree two), then the degree of the numerator is less than or equal to one. Hence, partial fractions can be represented as

$$\frac{A}{x-3}, \qquad \frac{Bx+C}{x^2+5}, \qquad \text{etc.,}$$

where A, B, C, ... are constants to be determined, as illustrated below.

Distinct Linear Factors   The first example illustrates the procedure in the case that the denominator $D(x)$ factors into distinct linear factors only.

EXAMPLE 1   Decompose into partial fractions:

$$\frac{9x - 14}{2x^2 - 5x - 12}.$$

SOLUTION $D(x)$ factors as $(2x + 3)(x - 4)$, two distinct linear factors. Hence, the numerator of each partial fraction must be a constant and we write

$$\frac{9x - 14}{(2x+3)(x-4)} \equiv \frac{A}{2x+3} + \frac{B}{x-4}. \qquad \langle 2 \rangle$$

Adding the fractions on the right and equating the numerators, we have

$$9x - 14 \equiv A(x-4) + B(2x+3)$$
$$9x - 14 \equiv (A + 2B)x + (-4A + 3B).$$

Because the coefficients of the linear terms must be equal and the constant terms must be equal, we obtain the system

$$A + 2B = 9$$
$$-4A + 3B = -14. \qquad \langle 3 \rangle$$

Solving system $\langle 3 \rangle$, we have $A = 5$ and $B = 2$. Hence,

$$\frac{9x - 14}{2x^2 - 5x - 12} \equiv \frac{5}{2x+3} + \frac{2}{x-4}.$$

Note that this agrees with equation $\langle 1 \rangle$ above.

Repeated Linear Factors   A denominator $D(x)$ may factor into linear factors, some of which may appear to powers greater than one. For example, a factor such as $(x+a)^2$ may appear. In such cases, a partial fraction corresponding to *each* of the denominators $(x+a)^2$ and $(x+a)^1$ must be included.

EXAMPLE 2   Decompose into partial fractions:

$$\frac{7x^2-21x+12}{x^3-4x^2+4x}.$$

SOLUTION $D(x)$ factors as $x(x-2)^2$. Hence, we write:

$$\frac{7x^2-21x+12}{x(x-2)^2} \equiv \frac{A}{x}+\frac{B}{(x-2)^2}+\frac{C}{x-2}$$

$$7x^2-21x+12 \equiv A(x-2)^2+Bx+Cx(x-2)$$

$$7x^2-21x+12 \equiv (A+C)x^2+(-4A+B-2C)x+4A.$$

Equating corresponding coefficients and constants, we obtain

$$A+C=7$$
$$-4A+B-2C=-21$$
$$4A=12,$$

from which $A=3$, $B=-1$ and $C=4$. Hence,

$$\frac{7x^2-21x+12}{x^3-4x^2+4x} \equiv \frac{3}{x}-\frac{1}{(x-2)^2}+\frac{4}{x-2}.$$

Quadratic Factors   The last example of this section illustrates the procedure in the case that quadratic factors appear in the factored form of $D(x)$.

EXAMPLE 3   Decompose into partial fractions:

$$\frac{9x^2-47x+8}{x^3-4x^2+2x-8}.$$

SOLUTION $D(x)$ factors as $(x^2+2)(x-4)$. Hence, we write:

$$\frac{9x^2-47x+8}{(x^2+2)(x-4)} \equiv \frac{Ax+B}{x^2+2}+\frac{C}{x-4}$$

$$9x^2-47x+8 \equiv (Ax+B)(x-4)+C(x^2+2)$$

$$9x^2-47x+8 \equiv (A+C)x^2+(-4A+B)x+(-4B+2C).$$

Equating corresponding coefficients and constants, we obtain

$$A+C=9$$
$$-4A+B=-47$$
$$-4B+2C=8,$$

from which $A=11$, $B=-3$ and $C=-2$. Hence,

$$\frac{9x^2-47x+8}{x^3-4x^2+2x-8}=\frac{11x-3}{x^2+2}-\frac{2}{x-4}.$$

Exercise Set F

Decompose into partial fractions.

1. $\dfrac{11x-3}{x^2-9}$

2. $\dfrac{11x+6}{x^2-4}$

3. $\dfrac{3x+30}{3x^2-10x-8}$

4. $\dfrac{18x+26}{8x^2+2x-3}$

5. $\dfrac{2-3x}{5x^2+19x-4}$

6. $\dfrac{8-2x}{6x^2+17x-14}$

7. $\dfrac{x^2-23x+10}{x^3-2x^2-5x+6}$

8. $\dfrac{-2x^2-21x-25}{x^3+2x^2-5x-6}$

9. $\dfrac{2x^2-11x+2}{x^3-2x^2}$

10. $\dfrac{2x^2-4x+3}{2x^3+x^2}$

11. $\dfrac{x^2+3x-1}{x^3-2x^2+x}$

12. $\dfrac{3x^2-x+2}{x^3-6x^2+4x+8}$

13. $\dfrac{x^3-3x^2+x-8}{x^4+5x^2+6}$

14. $\dfrac{3x^3-16x^2+2x-19}{3x^4-x^2-2}$

15. $\dfrac{2x^2+3x+2}{x^3+4x^2+6x+4}$

16. $\dfrac{4x^2+x+10}{x^3+x^2-2}$

17. $\dfrac{1}{x^2-a^2}$, $(x\neq a)$

18. $\dfrac{5x^3-4x}{x^4-16}$

Solve each system. (See Example 5.)

23. $x^2-xy+y^2=21$
    $x^2+2xy-8y^2=0$

24. $2x^2+3xy+3y^2=48$
    $2x^2-xy-3y^2=0$

25. $x^2-4xy-3x+12y=0$
    $x^2+4xy-2y^2=25$

26. $y^2-3xy-4y+12x=0$
    $9x^2-16y^2=144$

Solve each system. (See Example 6.)

27. $5x^2+2xy-3x=0$
    $2xy+3y^2+y=0$

28. $2x^2+3xy-3x=0$
    $6y^2-6xy-y=0$

29. $5x^3+3x^2y-4x^2=0$
    $7xy-6y^2-26y=0$

30. $6x^2-5xy-2x=0$
    $7y^3+43y^2+5xy^2=0$

(B) *Solve each system.*

31. $3y=x^3-3x^2+2x$
    $y=x^3-x^2-2x$

32. $4y=x^3-4x^2+31x$
    $y=x^3-x^2+x$

Solve and graph each system.

33. $y=x^4$
    $y=2x-x^2$

34. $y=2x^2$
    $y=x^4-2x^2$

35. $y=2\sqrt{x}$
    $y=2x-4$

36. $y=x^3-3x^2+2x$
    $y=2x-x^2$

SYSTEMS OF INEQUALITIES

6.1   LINEAR AND QUADRATIC SYSTEMS

In Section 3.4 we considered methods for graphing the solution set of a linear or a quadratic inequality. In this section we apply the same methods to obtain the graph of each of the inequalities of a given system of inequalities on the same set of axes. The *intersection* of all the graphs so obtained is the graph of the solution set of the given system or, simply the *graph of the system*. In general, the graph of a system of inequalities is a region in the plane—sometimes called the *solution space* of the system.

EXAMPLE 1   Graph the solution space of the system

$$x + 2y < 3$$
$$3x - y \geq 5.$$

SOLUTION   We sketch the graph of each of the two inequalities on the same set of axes. The solution space of the system is the intersection of the shaded regions in Figure 6.1-1, together with the solid line.

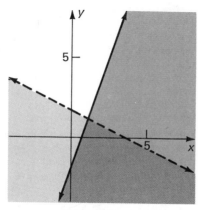

Figure 6.1-1

Recall, from geometry, that a polygonal region is a set of points that is the union of a polygon and its interior, the polygon is called the *boundary* of the region. Figure 6.1-2 shows two polygonal regions. As you

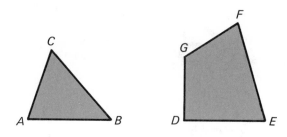

Figure 6.1-2

probably remember, points *A* through *G* are called *vertices*. A polygonal region is said to be *convex* if each two points of the region can be joined by a segment such that all points of the segment are included in the region. In Figure 6.1-3, polygonal region (a) is convex; polygonal region (b) is not convex, as you can judge by observing segment $\overline{VT}$.

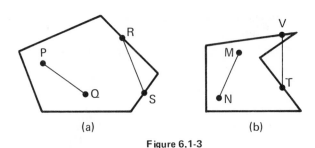

(a)   (b)

Figure 6.1-3

EXAMPLE 2   In Figure 6.1-4, the (shaded) convex triangular region ABC is the solution space of the system

$$x \leq 3$$
$$x + 2y \geq 4$$
$$2x - y \geq -2.$$

212   SYSTEMS OF INEQUALITIES

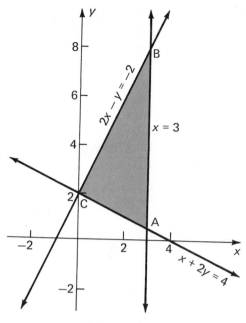

Figure 6.1-4

The vertices are determined by inspection or by solving each of the systems

$$x=3 \qquad\qquad x=3 \qquad\qquad x+2y=4$$
$$x+2y=4; \qquad 2x-y=-2; \qquad 2x-\ y=-2;$$

to obtain $A\left(3, \frac{1}{2}\right)$, $B(3, 8)$ and $C(0, 2)$.

The graph of a system involving quadratic inequalities is, in general, a region with curvilinear boundaries.

EXAMPLE 3   Graph the system

$$\textcircled{1} \quad x^2-6x+y \leq -5$$
$$\textcircled{2} \quad 9x^2+16y^2 < 144.$$

SOLUTION   The graph of each inequality is a region in the plane. To find the boundary of each of the two regions, we first sketch the graph of each of the equations

$$x^2-6x+y=-5 \qquad \text{and} \qquad 9x^2+16y^2=144,$$

or, in standard form,

$$y=-(x-3)^2+4 \qquad \text{and} \qquad \frac{x^2}{16}+\frac{y^2}{9}=1,$$

(Continued on next page)

to obtain the parabola and the ellipse shown in Figure 6.1–5. The graph indicates that (2, 0) is in the region with which we are concerned, so we choose (2, 0) as a test point. Substituting appropriately into inequalities ① and ②, we obtain the two *true* statements

$$4 - 12 + 0 \leq 5 \quad \text{and} \quad 36 + 0 < 144.$$

Hence, we conclude that the intersection of the two shaded regions is the graph of the given system.

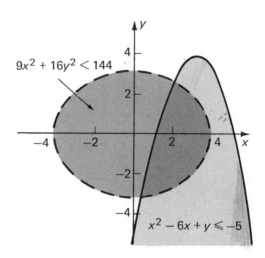

Figure 6.1-5

Exercise Set 6.1

(A) *Graph each system. (See Example 1.)*

1. $x - y < 5$
   $2x + 3y < 1$

2. $-3x + 2y > -1$
   $2x - 3y > 2$

3. $-x + 2y > 5$
   $8x - y < 4$

4. $5x + 4y < 6$
   $2x + 6y > -3$

5. $3x - 7y \geq 4$
   $2x + 5y \leq -1$

6. $6x + 5y \leq 10$
   $3x - 4y \leq -2$

Sketch the convex polygonal region that is the solution space of each system; specify the vertices. (See Example 2.)

7. $y \leq 0$
   $x \leq 0$
   $x + 2y \geq -6$

8. $x \leq 0$
   $y \geq 0$
   $2x - y \geq -6$

9. $3x + 5y \geq 15$
   $4x - y \leq 20$
   $5x - 7y \geq -21$

10. $5x - 2y \geq -10$
    $x - 8y \leq -2$
    $2x + 3y \leq 15$

214   SYSTEMS OF INEQUALITIES

11. $x - 2y + 8 \geq 0$
    $3x + 2y \leq 24$
    $x \geq 0$
    $y \geq 0$

12. $4x + 3y + 6 \geq 0$
    $2x - y - 12 \leq 0$
    $x \geq 0$
    $y \leq 0$

13. $8x + 5y \leq 72$
    $x - 2y \geq -12$
    $x + 2 \geq 0$
    $y \geq 0$

14. $x + 4y - 24 \leq 0$
    $x - y \leq 9$
    $y \geq -3$
    $x \geq 0$

Graph each system. (See Example 3.)

15. $y > x^2 - 4$
    $y - x < 2$

16. $y + 2x < -1$
    $y - x^2 > -1$

17. $x^2 + 4y^2 \geq 16$
    $4x^2 + y^2 \leq 16$

18. $x^2 - y^2 \geq 4$
    $x^2 + y^2 \leq 25$

19. $x^2 + y^2 - 6x \leq 8y$
    $xy \leq 12$

20. $x^2 + y - 10x \geq -21$
    $x^2 + y^2 + 2y \leq 10x - 1$

(B) Graph each system.

21. $|x| < 2$
    $|y| < 5$
    $y \geq x^2 - 4$

22. $|x| \leq 3$
    $|y| \geq 2$
    $x^2 + y^2 \leq 25$

23. $y \leq |x + 2|$
    $4x^2 + 9y^2 \leq 36$

24. $y \geq |x| + 1$
    $y \leq 4 - |x|$

6.2   LINEAR PROGRAMMING

A company manufactures tables and chairs, earning a profit of $8 on each table and $4 on each chair. If x tables and y chairs are sold per month, then the polynomial

$$P(x, y) = 8x + 4y$$

represents the total profit per month. Now, assuming that the company can sell all that it makes, a major concern of the management of the company is that of deciding how many tables and chairs to make each month so as to earn the greatest possible profit. That is, management must decide how to select numbers x and y so that $P(x, y)$ will be a maximum. An ideal solution to the problem would be to make as many tables each month as possible, but no chairs. In most practical situations, however, such a solution would not be acceptable because of various restrictions that affect the decision making process. For example, they may not be able to sell many tables if there are no chairs to go with them. Other restrictions arise from problems concerning use of machinery and personnel; limitations of warehouse space; supplies on hand, etc. The problem of maximizing $P(x, y)$ subject to restrictions on x and y is an example of a broad class of problems that are now being solved by methods of *linear programming*, a relatively new branch of applied mathematics. Before we can solve such problems, we need some preliminary concepts.

Maximum and Minimum of a Linear Polynomial Consider the linear polynomial function in two real variables x and y

$$P(x, y) = ax + by + c,$$

where $a, b, c \in R$. The domain of P is generally $\{(x, y): (x, y) \in R \times R\}$; its range is a subset of R. Our particular interest is with linear polynomial functions such as P where the domain is restricted to the set of all ordered pairs of real numbers that are the coordinates of the points of a convex polygonal region. For example, consider the function defined by

$$P(x, y) = 4x + 5y,$$

where the domain is the convex polygonal region S (see Figure 6.2-1) that is the solution space of the system

$$x + 2y \geq 4$$
$$3x + y \geq 6$$
$$y \leq 4$$
$$x \leq 3.$$

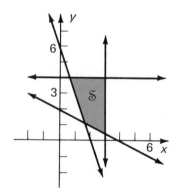

Figure 6.2-1

We say that P is defined over the convex polygonal region S, or over the solution space S. From the figure, note that $(2, 2)$, $(3, 4)$ and $(1, 3)$ are some of the ordered pairs for which P is defined; $(0, 0)$, $(-2, 1)$ and $(2, 0)$ are some ordered pairs for which P is not defined. Let us compute values of P at selected points in region S. First, as indicated in Figure 6.2-2(a), we consider some points of S in a vertical line, say $x = 2$:

$$P(2, 1) = 13; \qquad P(2, 2) = 18; \qquad P(2, 3) = 23; \qquad P(2, 4) = 28.$$

Observe that the maximum value of P in the line $x = 2$ is attained at a point of $\mathcal{S}$ where y assumes its maximum value, specifically, in segment $\overline{DC}$. The minimum value of P in the line $x = 2$ is attained at a point of $\mathcal{S}$ where y assumes its minimum value, specifically, in segment $\overline{AB}$. Next, as indicated in Figure 6.2-2(b), let us consider some points of $\mathcal{S}$ in a horizontal line, say $y = 3$:

$$P(1, 3) = 19; \qquad P(2, 3) = 23; \qquad P(3, 3) = 27.$$

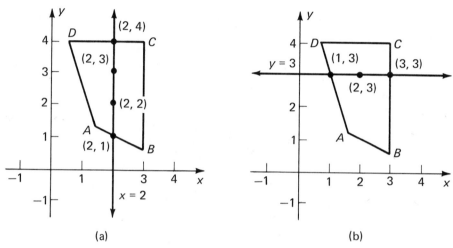

(a)   (b)

Figure 6.2-2

Note that P attains a maximum value at a point in $\overline{BC}$, and a minimum value at a point in $\overline{DA}$. These observations lead to the conjecture that a maximum or a minimum value of the function P in $\mathcal{S}$ is attained in one of the segments of the polygonal boundary of $\mathcal{S}$. In turn, because two adjacent sides of a polygon intersect at a vertex, it seems reasonable to conjecture that maximum and minimum values of functions such as P occur at vertices (sometimes called *corners*) of the polygon. That this is indeed the case is asserted by the following basic theorem of linear programming.

Theorem 6.1   If $P(x, y)$ is a linear polynomial function defined over a convex polygonal region, then maximum and/or minimum values of $P(x, y)$, if they exist, are attained at vertices of the polygonal boundary.

EXAMPLE 1   Find maximum and minimum values of $P(x, y) = 2x - 3y$ defined over the solution space of the system

$$x + 3y \geq 7$$
$$5x - y \leq 19$$
$$x - 3y + 13 \geq 0$$
$$3x - y - 1 \geq 0.$$

(Continued on next page)

SOLUTION   The shaded region in Figure 6.2–3 is the graph of the given system; the solution space is a convex polygon and we may apply Theorem 6.1. By inspection (or by solving each pair of associated equations as in Section 6.1), the vertices are $A(1, 2)$, $B(4, 1)$, $C(5, 6)$ and $D(2, 5)$. Computing $P(x, y)$ at each vertex, we obtain

$$P(1, 2) = -4; \qquad P(4, 1) = 5; \qquad P(5, 6) = -8; \qquad P(2, 5) = -11;$$

from which the maximum value of $P(x, y)$ is seen to be 5; the minimum value is -11.

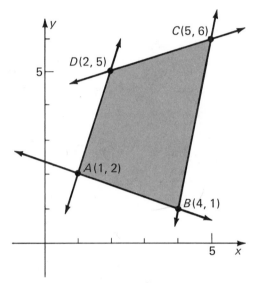

Figure 6.2-3

Maximum-Type Problems   Let us apply the preceding concepts to the problem of maximizing a linear polynomial function.

EXAMPLE 2   The ABC Truck Rebuilding Company is in the business of modifying truck bodies for several nearby trucking firms. The paint department of the ABC company is paid $40 to paint a panel truck, a job that requires 2 man-hours, and $30 to paint a pickup truck, a job that requires 1 man-hour. Because of labor costs and a lack of space, no more than 10 man-hours are allocated to the paint department each day, and at most 8 trucks can be accommodated in the paint shop. Assuming that there are always trucks waiting to be painted, how many trucks of each kind should the paint department work on in one day in order to earn a maximum amount of money? How much is the maximum amount earned?

SOLUTION   To begin, we state the conditions of the problem algebraically. Thus, on a daily basis:

Let x be the number of panel trucks painted;

Let y be the number of pickup trucks painted;

Let $P(x, y)$ be the total amount of money earned.

Then, in dollars, $40x$ is the amount earned for painting panel trucks; $30y$ is the amount earned for painting pickup trucks;

$$P(x, y) = 40x + 30y.$$

Next, we express the restrictions stated in the problem. The number of trucks of each kind is nonnegative, hence

$$x \geq 0 \quad \text{and} \quad y \geq 0.$$

The restriction that there is space for at most 8 trucks is expressed by

$$x + y \leq 8.$$

The number of man-hours required for painting x panel trucks at 2 man-hours each is $2x$; the number of man-hours for painting y pickup trucks at 1 man-hour each is y; the limitation that no more than 10 man-hours are available each day is expressed by

$$2x + y \leq 10.$$

Our problem is to select x and y so as to maximize $P(x, y) = 40x + 30y$ over the solution space of the system

$$x \geq 0$$
$$y \geq 0$$
$$x + y \leq 8 \qquad \langle 1 \rangle$$
$$2x + y \leq 10.$$

Figure 6.2-4 shows $OABC$, the solution space of system $\langle 1 \rangle$, a convex

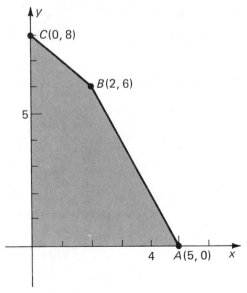

Figure 6.2-4

(Continued on next page)

polygonal region with vertices $O(0, 0)$, $A(5, 0)$, $B(2, 6)$ and $C(0, 8)$. Evaluating $P(x, y)$ at each vertex, we have:

$$P(0, 0) = 0; \qquad P(5, 0) = 200; \qquad P(2, 6) = 260; \qquad P(0, 8) = 240;$$

from which we conclude that the company should paint 2 panel trucks and 6 pickup trucks each day to earn a maximum return of $260.

Minimum-Type Problems   Problems that involve concepts such as keeping expenses as low as possible call for *minimizing* a linear polynomial function.

EXAMPLE 3   A man takes two kinds of dried food on a hiking trip. One ounce of food A provides 75 calories and 80 units of vitamins; one ounce of food B provides 125 calories and 40 units of vitamins. Each day the man requires at least 10 ounces of food A; at least 3000 calories; at least 1800 units of vitamins; not more than 50 ounces of food. If food A costs 6.8 cents per ounce and food B costs 5.2 cents per ounce, how many ounces of each kind of food should be selected so as to minimize the daily cost? How much is the least daily cost of food?

SOLUTION   On a daily basis:

Let x be the number of ounces of food A;

Let y be the number of ounces of food B;

Let $C(x, y)$ be the total cost of the food.

We want to minimize $C(x, y)$, where

$$C(x, y) = 6.8x + 5.2y,$$

over the solution space of the system:

$y \geq 0$	(the amount of food B cannot be negative)
$x \geq 10$	(at least 10 ounces of food A)
$75x + 125y \geq 3000$	(at least 3000 calories daily)
$80x + 40y \geq 1800$	(at least 1800 units of vitamins daily)
$x + y \leq 50$	(not more than 50 ounces of food daily).

This system can be replaced by the equivalent system:

$$y \geq 0$$
$$x \geq 10$$
$$3x + 5y \geq 120 \qquad \langle 2 \rangle$$
$$2x + y \geq 45$$
$$x + y \leq 50.$$

Figure 6.2–5 shows $PQRST$, the solution space of system $\langle 2 \rangle$, a convex polygonal region with vertices at $P(40, 0)$, $Q(50, 0)$, $R(10, 40)$, $S(10, 25)$ and $T(15, 15)$. Evaluating $C(x, y)$ at each vertex, we have:

$$C(40, 0) = 272; \qquad C(50, 0) = 340; \qquad C(10, 40) = 276;$$
$$C(10, 25) = 198; \qquad C(15, 15) = 180;$$

from which we conclude that 15 ounces of food A and 15 ounces of food B, at a minimum daily cost of $1.80, should be chosen.

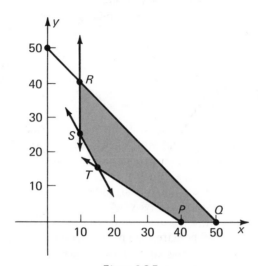

Figure 6.2-5

Exercise Set 6.2

(A) *Find maximum and minimum values of $P(x, y)$ defined over the solution space of the given system. (See Example 1)*

1. $P(x, y) = 2x + y$
   $x + 2y \le 7$
   $x - y \le 1$
   $x + y \ge -1$
   $2x - y \ge -1$

2. $P(x, y) = 5x - 4y$
   $x - 5y \le 1$
   $3x - y \le 17$
   $x + 5y \le 27$
   $5x - y \ge 5$

3. $P(x, y) = 3x - 2y$
   $x - 2y \ge -4$
   $2x + y \le 7$
   $x - 2y \le 1$
   $2x + y \ge 2$

4. $P(x, y) = x + 3y$
   $y \ge 0$
   $3x - y \ge 0$
   $x - 3y \ge -8$
   $2x - y \le 4$

For Exercises 5 through 14, see Examples 2 and 3.

5. A company makes two types of radios, A and B. Processes I and II are required to produce the radios, according to the schedule

	Process I	Process II
Type A	2 hours	1 hour
Type B	1 hour	4 hours

Because of maintenance problems, process I is available at most 10 hours per day; process II at most 12 hours per day. The profit on type A is $5 per unit; $7 per unit on type B.

a. If x type A radios and y type B radios are made daily, then $2x + y \leq 10$ asserts that at most 10 hours of process I are available daily. Write a similar inequality with respect to process II.

b. How many units of each type of radio should be made each day to maximize the profit?

6. A company makes two products, P and Q. Three machines, A, B and C are used to make each product, according to the schedule

	Product P	Product Q
Machine A	6 hours	8 hours
Machine B	5 hours	5 hours
Machine C	8 hours	4 hours

Each machine is available for no more than 40 hours per week. The profit on product P is $25 per unit; $30 per unit on product Q. How many units of each product should be made per week to maximize the profit?

7. An investor decides to buy no more than 200 shares of two stocks, A and B. Stock B involves the greater risk, so the investor decides to buy at most 90 shares of B and at least 80 shares of A. The dividend on B is $1.50 per share; $1.00 per share on A. How many shares of each stock should the investor buy to maximize his stock dividends?

8. An ice cream vendor has $100 with which to buy ice cream to be sold at a state fair. Ice cream bars cost 8 cents each and sell for 15 cents. Sundae cups cost 12 cents each and sell for 20 cents. The vendor's truck has room for no more than 800 bars and 500 sundaes. He expects to sell at most 900 pieces of ice cream, and at least twice as many bars as sundaes. How many bars and sundaes should he buy to maximize his profit?

9. A gasoline station can store at most 2000 gallons of gasoline. Each day the station can expect to sell at least 800 gallons of regular gasoline, at least 300 gallons of premium gasoline, and at least three times as much regular as premium gasoline. The profit on regular is 2 cents per gallon; 3 cents per gallon on premium. How many gallons of each kind of gasoline should the station store to maximize its profit?

10. Each capsule of a certain medication is required to contain at least 10 grains of L, 15 grains of M and 24 grains of N. A pharmaceutical company formulates the capsules from two chemicals, W and V; the quantities of L, M and N in each unit of W and V are

	L	M	N
Chemical W	2 grains	1 grain	2 grains
Chemical V	1 grain	3 grains	3 grains

The cost of W is 10 cents per unit; the cost of V is 20 cents per unit. How many units of W and of V should each capsule of the medication contain to minimize the cost per capsule?

In Exercises 11 and 12, you may find it helpful to construct tables such as those in Exercises 5, 6 and 10.

11. The nutritionist at a boys camp determines that the average weekly meat diet per boy shall include at least 3 pounds of lean meat and at least 2 pounds of fat meat. Beef at $1.00 per pound provides 80% lean and 20% fat; pork at 70 cents per pound provides 40% lean and 60% fat. The camp has 200 boys and storage space for at most 1200 pounds of meat. How many pounds of beef and pork should be purchased per week to keep meat costs at a minimum?

12. A certain diet is designed to provide a minimum weekly requirement of 18 units of fat, 24 units of carbohydrates and 16 units of protein. One pound of food A, at $1.60 per pound, provides 2 units of fat, 12 units of carbohydrates and 4 units of protein. One pound of food B, at $1.00 per pound, provides 6 units of fat, 2 units of carbohydrates and 2 units of protein. How many pounds of each food are required to meet the minimum weekly diet requirements at minimum cost?

(B)

13. An appliance store chain has two warehouses that stock a certain refrigerator. Warehouse 1 has 100 of the refrigerators in stock and warehouse 2 has 60 of the refrigerators. Store A requires 75 refrigerators and store B requires 50 refrigerators. Given the following unit shipping cost schedule, how many refrigerators should be shipped from each warehouse so as to fill the orders of stores A and B and minimize shipping costs? (Hint: Let x be the number of refrigerators from warehouse 1 to store A. Then, $75 - x$ is the number of refrigerators from warehouse 2 to store A.)

	Store A	Store B
Warehouse 1	$3.00	$2.50
Warehouse 2	$2.00	$2.00

14. A certain company sells its products through the mail and with the aid of door-to-door salesmen. The average direct sale is $200 at a selling cost of $5, whereas the average by-mail sale is $80 at a selling cost of $3. To maintain a desirable profit margin, the company projects that its total yearly sales must be at least $70,000, at a yearly selling cost of no more than $3,000. Since the company depends on repeat sales, it feels that to operate profitably in a given area it must have at least 500 customers. Because the company has difficulty finding employees to do the selling jobs it requires, it attempts to minimize the time its salesmen spend with customers. If direct sales require 2 hours per customer and correspondence sales require 1 hour per customer, how many customers should be contacted through each method of selling? (Hint: Minimize the time spent with custom-

ers. That is, minimize $T=2x+y$, where x is the number of direct sales customers and y is the number of mail order customers.)

Chapter 6   Self-Test

[6.1] *Graph each system*

1. $5x+6y \geq 10$
   $4x-3y \geq -2$

2. $4x+5y > 6$
   $y \geq 3$

Sketch the solution space of each system; specify the vertices.

3. $x \geq 0$
   $y \leq 10$
   $5y+7x-35 \geq 0$
   $5x-2y-25 \leq 0$

4. $x-5y \leq 2$
   $2x+y \leq 15$
   $x+4y-32 \leq 0$
   $x+y \geq 2$

Graph each system.

5. $x-y < 2$
   $x > y^2 - 4$

6. $x^2+y^2 \leq 25$
   $9x^2+25y^2 \geq 225$

[6.2]

7. Find maximum and minimum values of $P(x, y)=3x-5y$ over the solution space of Exercise 3.

8. Find maximum and minimum values of $P(x, y)=y-10x$ over the solution space of Exercise 4.

9. A newly organized company makes canister and tank type vacuum cleaners. A canister type requires $\frac{1}{2}$ hour of welding time and 1 hour of assembly time; a tank type requires $\frac{1}{2}$ hour each of welding and assembly time. There are 4 hours of welding time and 6 hours of assembly time available per day. How many of each type of vacuum cleaner should be made to maximize daily profits if the profit on canister and tank types is: $5 and $3, respectively; $3 and $5 respectively?

10. A paint company mixes paint that must have at least 18, 24 and 16 units each of pigment, oil and thinner. For convenience, the company stocks two pre-mixed kinds of paint, A and B, with ingredients per can as follows:

	pigment	oil	thinner
paint A	2	12	4
paint B	6	2	2

If the price per can of paint A and paint B is $1.60 and $.90 respectively, how many cans of each kind of paint should the company mix so as to minimize the cost of their paint?

MATRICES AND DETERMINANTS

7.0   INTRODUCTION; NOTATION

In a sense, the history of matrices in mathematics parallels the history of electricity in the physical sciences. First introduced in the mid-nineteenth century, the theoretical properties of matrices, as well as of electricity, were intensively studied, while little attention was given to the possibility of finding practical applications for either. Beginning in the 1920's, applications of matrices as mathematical models in such varied disciplines as the physical and behavioral sciences, management decision making, and economics, were found; new applications are still being found. The advent of high-speed computers extended our ability to use matrices as mathematical models in even more complicated problems involving great numbers of variables. Today, matrix theory plays a role that is significant enough to warrant its study as a basic preparation for many different fields. In this chapter we introduce some elementary properties of matrices.

As stated in Section 5.2, a **matrix** is a rectangular array of numbers called the **entries** or **elements** of the matrix. (In this book *all entries in matrices are understood to be real numbers*, unless specified otherwise.) A matrix with *m rows* and *n columns* is referred to as an *$m \times n$* **matrix**; $m \times n$ is called the **order** of the matrix. Note that we specify the number of rows first, the number of columns second. If

$$A = \begin{bmatrix} 2 & 4 & 0 \\ 1 & 5 & 3 \end{bmatrix}; \quad B = \begin{bmatrix} -1 & 2 & 4 \end{bmatrix}; \quad C = \begin{bmatrix} a_1 & a_2 \\ b_1 & b_2 \\ c_1 & c_2 \end{bmatrix}; \quad X = \begin{bmatrix} x_1 \\ x_2 \\ x_3 \end{bmatrix};$$

then A is a 2×3 matrix; B is a 1×3 matrix; C is a 3×2 matrix; D is a 3×1 matrix. A matrix such as B, consisting of a single row, is sometimes called a *row matrix* or a *row vector*; a matrix such as X consisting of a single column is sometimes called a *column matrix* or a *column vector*. Symbols such as $A_{2\times3}$, $B_{1\times3}$, etc., are used to indicate that A is of order 2×3, B is of order 1×3, etc.

A much used notation for naming entries in a matrix involves *double subscripts*. A single letter such as a is selected and two subscripts are appended to it; the first subscript specifies the row and the second subscript specifies the column in which a given entry appears. For example, consider

$$A_{3\times3} = \begin{bmatrix} a_{11} & a_{12} & a_{13} \\ a_{21} & a_{22} & a_{23} \\ a_{31} & a_{32} & a_{33} \end{bmatrix}.$$

Note that a_{11} is the entry in the first row and first column, a_{12} is the entry in the first row and second column, etc. More generally, a_{ij} specifies the entry in the ith row and j^{th} column of matrix A. When considering more than one matrix, we customarily use a_{ij} for entries in matrix A, b_{ij} for entries in matrix B, and so forth.

7.1   MATRIX OPERATIONS

Matrices A and B are said to be **equal matrices**, and we write $A = B$, if and only if both matrices are of the same order and $a_{ij} = b_{ij}$ for each i, j. Thus,

$$\begin{bmatrix} \sqrt{4} & 4 & \frac{6}{2} \\ 0 & -\sqrt{9} & 1 \end{bmatrix} = \begin{bmatrix} 2 & \frac{8}{2} & 3 \\ 0 & -3 & 1 \end{bmatrix},$$

because $\sqrt{4} = 2$, $4 = \frac{8}{2}$, $\frac{6}{2} = 3$, etc. Also, if

$$\begin{bmatrix} x \\ y \\ z \end{bmatrix} = \begin{bmatrix} -2 \\ 3 \\ 0 \end{bmatrix},$$

then $x = -2$, $y = 3$ and $z = 0$.

Matrix Sums   The Western Citrus Company grows citrus fruits at each of three ranches. Each year a table is made that lists the number of boxes of each kind of fruit harvested at each ranch. A typical yearly table is:

	Ranch 1	Ranch 2	Ranch 3
Oranges	700	600	800
Lemons	400	100	600
Grapefruit	300	400	200

Such a table can be represented by a yearly *production matrix*. For example, production matrices for two successive years are:

$$\begin{bmatrix} 700 & 600 & 800 \\ 400 & 100 & 600 \\ 300 & 400 & 200 \end{bmatrix} \quad \text{and} \quad \begin{bmatrix} 750 & 600 & 800 \\ 525 & 250 & 500 \\ 300 & 500 & 250 \end{bmatrix}. \qquad <1>$$

To determine the total number of boxes of each fruit harvested at each ranch in two (or more) years, we need only *add corresponding entries* in two (or more) production matrices. Thus, from ⟨1⟩ we obtain the two-year production matrix

$$\begin{bmatrix} 700+750 & 600+600 & 800+800 \\ 400+525 & 100+250 & 600+500 \\ 300+300 & 400+500 & 200+250 \end{bmatrix} = \begin{bmatrix} 1450 & 1200 & 1600 \\ 925 & 350 & 1100 \\ 600 & 900 & 450 \end{bmatrix}.$$

Results such as these suggest the concept of *matrix addition*.

Definition 7.1   The **matrix sum** of matrices $A_{m \times n}$ and $B_{m \times n}$, denoted by $A + B$, is the matrix $C_{m \times n}$ where $c_{ij} = a_{ij} + b_{ij}$ for $i = 1, 2, 3, \ldots, m$ and $j = 1, 2, 3, \ldots, n$.

Note that a matrix sum is defined only for matrices of the same order—such matrices are said to be **conformable** with respect to addition.

EXAMPLE 1

$$\begin{bmatrix} 1 & -3 & 0 \\ 2 & 4 & -8 \end{bmatrix} + \begin{bmatrix} 6 & 5 & 7 \\ 11 & -2 & 9 \end{bmatrix} = \begin{bmatrix} 1+6 & -3+5 & 0+7 \\ 2+11 & 4-2 & -8+9 \end{bmatrix}$$

$$= \begin{bmatrix} 7 & 2 & 7 \\ 13 & 2 & 1 \end{bmatrix}.$$

Consider now the matrix sum $A + A$. By Definition 7.1,

$$\begin{bmatrix} a_{11} & a_{12} & a_{13} \\ a_{21} & a_{22} & a_{23} \end{bmatrix} + \begin{bmatrix} a_{11} & a_{12} & a_{13} \\ a_{21} & a_{22} & a_{23} \end{bmatrix} = \begin{bmatrix} 2a_{11} & 2a_{12} & 2a_{13} \\ 2a_{21} & 2a_{22} & 2a_{23} \end{bmatrix}. \quad \text{⟨2⟩}$$

Equation ⟨2⟩ suggests that the matrix sum $A + A$ can be obtained by multiplying each entry in A by 2, and that we may write $2A$ in place of $A + A$. That is, $2A$ may be considered as "2 times matrix A." Such a multiplication of a matrix by a real number is called a *scalar multiplication*.

Definition 7.2   The **product** of a real number r and an $m \times n$ matrix A, denoted by rA, is the $m \times n$ matrix that is formed by multiplying each entry in A by r.

EXAMPLE 2   If $A = \begin{bmatrix} 1 & 2 \\ 3 & 4 \end{bmatrix}$, then

a. $3A = 3\begin{bmatrix} 1 & 2 \\ 3 & 4 \end{bmatrix} = \begin{bmatrix} 3 \cdot 1 & 3 \cdot 2 \\ 3 \cdot 3 & 3 \cdot 4 \end{bmatrix} = \begin{bmatrix} 3 & 6 \\ 9 & 12 \end{bmatrix}.$

b. $-1A = -1\begin{bmatrix} 1 & 2 \\ 3 & 4 \end{bmatrix} = \begin{bmatrix} -1 \cdot 1 & -1 \cdot 2 \\ -1 \cdot 3 & -1 \cdot 4 \end{bmatrix} = \begin{bmatrix} -1 & -2 \\ -3 & -4 \end{bmatrix}.$

Let us reconsider Example 2(b). Observe that each entry in $-1A$ is the additive inverse (negative) of the corresponding entry in A. As with real numbers, we shall write $-A$ instead of $-1A$, and refer to $-A$ as the

additive inverse matrix of A. We do this because the sum of the two matrices, $A + (-A)$,

$$\begin{bmatrix} 1 & 2 \\ 3 & 4 \end{bmatrix} + \begin{bmatrix} -1 & -2 \\ -3 & -4 \end{bmatrix} = \begin{bmatrix} 0 & 0 \\ 0 & 0 \end{bmatrix}, \qquad \langle 3 \rangle$$

is a matrix in which each entry is zero. We call such a matrix a **zero matrix**. In general, a zero matrix may be of any order and is denoted by $0_{m \times n}$. Thus, the right-hand matrix in $\langle 3 \rangle$ may be named $0_{2 \times 2}$.

Now, if A and B are matrices of the same order, then $A - B$, the *difference between* A and B, can be found by the matrix sum $A + (-B)$. Thus, we have *matrix subtraction*.

EXAMPLE 3

$$\begin{bmatrix} 2 & 3 \\ 4 & -2 \end{bmatrix} - \begin{bmatrix} -2 & 0 \\ 1 & 4 \end{bmatrix} = \begin{bmatrix} 2 & 3 \\ 4 & -2 \end{bmatrix} + \begin{bmatrix} 2 & 0 \\ -1 & -4 \end{bmatrix} = \begin{bmatrix} 4 & 3 \\ 3 & -6 \end{bmatrix}.$$

Matrix Equations   Matrix operations such as addition, subtraction and scalar multiplication can be used to solve certain matrix equations by techniques that parallel equation solving techniques developed for real and complex numbers.

EXAMPLE 4   Find X by solving the matrix equation $2X + 3A = B$, where

$$A = \begin{bmatrix} 3 & -4 \\ 6 & -5 \end{bmatrix}; \quad B = \begin{bmatrix} 1 & 0 \\ 0 & 1 \end{bmatrix} \quad \text{and} \quad X = \begin{bmatrix} x_1 & x_2 \\ y_1 & y_2 \end{bmatrix}.$$

SOLUTION

$$2\begin{bmatrix} x_1 & x_2 \\ y_1 & y_2 \end{bmatrix} + 3\begin{bmatrix} 3 & -4 \\ 6 & -5 \end{bmatrix} = \begin{bmatrix} 1 & 0 \\ 0 & 1 \end{bmatrix}$$

$$\begin{bmatrix} 2x_1 + 9 & 2x_2 - 12 \\ 2y_1 + 18 & 2y_2 - 15 \end{bmatrix} = \begin{bmatrix} 1 & 0 \\ 0 & 1 \end{bmatrix}.$$

Equating corresponding entries in both matrices, we have

$$2x_1 + 9 = 1 \leftrightarrow x_1 = -4; \qquad 2x_2 - 12 = 0 \leftrightarrow x_2 = 6;$$

$$2y_1 + 18 = 0 \leftrightarrow y_1 = -9; \qquad 2y_2 - 15 = 1 \leftrightarrow y_2 = 8.$$

Thus,

$$X = \begin{bmatrix} -4 & 6 \\ -9 & 8 \end{bmatrix}.$$

Matrix Multiplication   One of the three ranches of the Western Citrus Company has 80 boxes of oranges, 40 boxes of lemons and 60 boxes of grapefruit ready to be shipped to a buyer. To save transportation costs, all the fruit is to be sent to the same city. Market conditions are such that the offered price in dollars per box for each kind of fruit in four western cities are as given in the table:

	Oranges	Lemons	Grapefruit
Los Angeles	3	1	3
San Francisco	4	2	3
Portland	4	3	2
Phoenix	3	2	3

$\langle 4 \rangle$

To which city should the fruit be sent in order to obtain a maximum amount of money from the sale of the fruit? The total amount of money obtainable in each city can be computed as follows:

Los Angeles: $\quad$ $\$3 \times 80 + \$1 \times 40 + \$3 \times 60 = \$460$
San Francisco: $\quad$ $4 \times 80 + 2 \times 40 + 3 \times 60 = 580$
Portland: $\quad$ $4 \times 80 + 3 \times 40 + 2 \times 60 = 560$
Phoenix: $\quad$ $3 \times 80 + 2 \times 40 + 3 \times 60 = 500.$

$\langle 5 \rangle$

From $\langle 5 \rangle$, it is clear that shipping to San Francisco will bring a maximum return of $580. If such computations have to be performed often, or for several ranches rather than for just one ranch, then matrix methods provide an efficient way of handling the computations. To see how this can be done, consider (below) the 4×3 *price matrix A* which summarizes the price per box of fruit in each city as listed in Table $\langle 4 \rangle$ above; the 3×1 *produce matrix B* which lists the number of boxes of each kind of fruit; and let us use the computations shown in equations $\langle 5 \rangle$ to obtain the 4×1 matrix C as follows:

$$\begin{matrix} A \\ \begin{bmatrix} 3 & 1 & 3 \\ 4 & 2 & 3 \\ 4 & 3 & 2 \\ 3 & 2 & 3 \end{bmatrix} \end{matrix} \begin{matrix} B \\ \begin{bmatrix} 80 \\ 40 \\ 60 \end{bmatrix} \end{matrix} = \begin{bmatrix} 3 \cdot 80 + 1 \cdot 40 + 3 \cdot 60 \\ 4 \cdot 80 + 2 \cdot 40 + 3 \cdot 60 \\ 4 \cdot 80 + 3 \cdot 40 + 2 \cdot 60 \\ 3 \cdot 80 + 2 \cdot 40 + 3 \cdot 60 \end{bmatrix} = \begin{matrix} C \\ \begin{bmatrix} 460 \\ 580 \\ 560 \\ 500 \end{bmatrix} \end{matrix} .$$

Using a_{ij}, b_{ij} and c_{ij} for the entries in A, B and C, respectively, we may informally describe c_{11} as the sum of the products of the entries in the *first row* of A with corresponding entries in the *first column* of B. That is,

$$c_{11} = a_{11}b_{11} + a_{12}b_{21} + a_{13}b_{31}.$$

Similarly,

$$c_{21} = a_{21}b_{11} + a_{22}b_{21} + a_{23}b_{31},$$
$$c_{31} = a_{31}b_{11} + a_{32}b_{21} + a_{33}b_{31},$$
$$c_{41} = a_{41}b_{11} + a_{42}b_{21} + a_{43}b_{31}.$$

We call C the *matrix product* of A and B and write $AB = C$. This procedure is formally stated next.

Definition 7.3 $\quad$ The **matrix product** of the matrices $A_{m \times p}$ and $B_{p \times n}$, denoted by AB, is the matrix $C_{m \times n}$ with entries given by

$$c_{ij} = a_{i1}b_{1j} + a_{i2}b_{2j} + \cdots + a_{ip}b_{pj}$$

for $i = 1, 2, 3, \ldots, m$ and $j = 1, 2, 3, \ldots, n$.

Note that the matrix product AB exists only when the number of *columns* of A is equal to the number of *rows* of B. When this condition is satisfied, we say that A and B are *conformable* for multiplication. We do not define a matrix product for non-conformable matrices.

EXAMPLE 5   If $A = \begin{bmatrix} 1 & 2 \\ 3 & 4 \end{bmatrix}$, $B = \begin{bmatrix} 1 & 3 \\ 2 & 4 \end{bmatrix}$ and $C = \begin{bmatrix} 1 & 2 & 3 \\ 4 & 5 & 6 \end{bmatrix}$, find AB; BA; CA.

SOLUTION

a. $AB = \begin{bmatrix} 1 & 2 \\ 3 & 4 \end{bmatrix}\begin{bmatrix} 1 & 3 \\ 2 & 4 \end{bmatrix} = \begin{bmatrix} 1\cdot1+2\cdot2 & 1\cdot3+2\cdot4 \\ 3\cdot1+4\cdot2 & 3\cdot3+4\cdot4 \end{bmatrix} = \begin{bmatrix} 5 & 11 \\ 11 & 25 \end{bmatrix}$.

b. $BA = \begin{bmatrix} 1 & 3 \\ 2 & 4 \end{bmatrix}\begin{bmatrix} 1 & 2 \\ 3 & 4 \end{bmatrix} = \begin{bmatrix} 1\cdot1+3\cdot3 & 1\cdot2+3\cdot4 \\ 2\cdot1+4\cdot3 & 2\cdot2+4\cdot4 \end{bmatrix} = \begin{bmatrix} 10 & 14 \\ 14 & 20 \end{bmatrix}$.

c. C and A are nonconformable because C has three columns and A has only two rows.

In Example 5(a) and (b), observe that AB is not equal to BA. In general, matrix multiplication is not commutative. With respect to AB, we say that A is *right-multiplied* by B or that B is *left-multiplied* by A. From Example 5(c), observe that C and A, *in that order*, are noncomformable, but A and C are conformable for multiplication. That is, we can only left-multiply C by A or right-multiply A by C. You may verify that AC is a 2×3 matrix.

EXAMPLE 6   The following table, and the associated produce matrix P, lists the number of boxes of each kind of fruit waiting to be shipped from Ranch 2 and Ranch 3 of the Western Citrus Company:

	Ranch 2	Ranch 3
Oranges	90	60
Lemons	50	80
Grapefruit	60	75

$$P = \begin{bmatrix} 90 & 60 \\ 50 & 80 \\ 60 & 75 \end{bmatrix}.$$

If the price per box of fruit in each of four western cities is given by the 4×3 price matrix A (taken from Table $\langle 4 \rangle$, page 229) to which city should each ranch ship its fruit in order to obtain a maximum amount of money from the sale of the fruit?

SOLUTION   We require $AP = D$, where D lists the total amounts of money for each ranch from each city. (You should verify the computations.)

$$\begin{array}{ccc} A & P & D \end{array}$$

$$\begin{bmatrix} 3 & 1 & 3 \\ 4 & 2 & 3 \\ 4 & 3 & 2 \\ 3 & 2 & 3 \end{bmatrix}\begin{bmatrix} 90 & 60 \\ 50 & 80 \\ 60 & 75 \end{bmatrix} = \begin{bmatrix} 500 & 485 \\ 640 & 625 \\ 630 & 630 \\ 550 & 565 \end{bmatrix} \begin{matrix} \text{(Los Angeles)} \\ \text{(San Francisco)} \\ \text{(Portland)} \\ \text{(Phoenix)} \end{matrix}$$

From the entries in D it becomes apparent that Ranch 2 obtains a maximum of \$640 in San Francisco; Ranch 3 obtains a maximum of \$630 in Portland.

Exercise Set 7.1

In Exercises 1 through 30, use the following matrices:

$$A = \begin{bmatrix} 3 & 1 \\ 2 & -4 \end{bmatrix} \qquad B = \begin{bmatrix} 6 & -5 \\ 0 & -3 \end{bmatrix} \qquad C = \begin{bmatrix} 6 & 2 & 4 \\ 3 & -1 & 5 \end{bmatrix}$$

$$D = \begin{bmatrix} -1 & -2 & -3 \\ 4 & 5 & 6 \end{bmatrix} \qquad E = \begin{bmatrix} 4 & 1 \\ -2 & -3 \\ 6 & 5 \end{bmatrix} \qquad F = \begin{bmatrix} 2 & 8 \\ -2 & 8 \\ 3 & 4 \end{bmatrix}$$

$$X_{2\times2} = \begin{bmatrix} x_1 & x_2 \\ y_1 & y_2 \end{bmatrix} \qquad X_{2\times3} = \begin{bmatrix} x_1 & x_2 & x_3 \\ y_1 & y_2 & y_3 \end{bmatrix} \qquad X_{3\times2} = \begin{bmatrix} x_1 & x_2 \\ y_1 & y_2 \\ z_1 & z_2 \end{bmatrix}$$

(A) *Write each sum or difference as a single matrix, if possible. (See Examples 1 and 3.)*

1. $A + B$ 2. $C + D$ 3. $E + F$ 4. $D + C$
5. $D + E$ 6. $B + C$ 7. $C - D$ 8. $E - F$

Write each product. (See Example 2.)

9. $-2D$ 10. $-3C$ 11. $4E$ 12. $5F$

Solve each matrix equation. Choose the appropriate matrix $X_{m\times n}$ from above. (See Example 4.)

13. $4A + 3B = X$ 14. $-2B + 6A = X$ 15. $3C - X = D$
16. $5X - D = 2C$ 17. $2E + 5X = 3F$ 18. $X - 2F = 3E$

Write each matrix product as a single matrix, if possible. (See Example 5.)

19. AB 20. BA 21. CD 22. DC
23. EF 24. FE 25. BC 26. CB
27. DE 28. ED 29. CF 30. FC

Write each product as a single matrix.

31. $\begin{bmatrix} \frac{1}{2} & \frac{3}{2} \\ \frac{5}{2} & -\frac{1}{2} \end{bmatrix} \begin{bmatrix} 2 \\ -4 \end{bmatrix}$

32. $\begin{bmatrix} \frac{3}{7} & \frac{1}{7} \\ \frac{1}{7} & -\frac{2}{7} \end{bmatrix} \begin{bmatrix} 4 \\ -5 \end{bmatrix}$

33. $\begin{bmatrix} \frac{13}{9} & -\frac{5}{9} & -\frac{7}{9} \\ \frac{2}{9} & \frac{2}{9} & \frac{1}{9} \\ -\frac{8}{9} & \frac{1}{9} & \frac{5}{9} \end{bmatrix} \begin{bmatrix} 1 \\ 1 \\ 5 \end{bmatrix}$

34. $\begin{bmatrix} -\frac{1}{2} & \frac{1}{2} & \frac{1}{2} \\ 1 & 0 & -1 \\ \frac{3}{2} & -\frac{1}{2} & -\frac{1}{2} \end{bmatrix} \begin{bmatrix} 6 \\ 7 \\ 9 \end{bmatrix}$

35. In Example 6, page 230, suppose that the produce matrix is

$$P = \begin{bmatrix} 60 & 40 \\ 80 & 70 \\ 65 & 85 \end{bmatrix}.$$

Using the same 4×3 price matrix A, determine to which city each ranch should ship its fruit for a maximum return of money.

36. Same as Exercise 35 but with

$$P = \begin{bmatrix} 85 & 70 & 80 \\ 65 & 50 & 60 \\ 70 & 80 & 75 \end{bmatrix}$$

7.2 SQUARE MATRICES; INVERSE MATRICES

If the number of rows of a matrix is the same as the number of columns, the resulting $n \times n$ matrix is called a **square matrix of order** n and is denoted by $A_{n \times n}$, $B_{n \times n}$, etc.

Notation and Terminology   In the $n \times n$ square matrix

$$A_{n \times n} = \begin{bmatrix} a_{11} & a_{12} & \cdots & a_{1n} \\ a_{21} & a_{22} & \cdots & a_{2n} \\ \vdots & \vdots & & \vdots \\ a_{n1} & a_{n2} & \cdots & a_{nn} \end{bmatrix},$$

the entries $a_{11}, a_{22}, \cdots, a_{nn}$ extending from the upper left corner to the lower right corner constitute the **principal diagonal** of the matrix. A **diagonal matrix** is a square matrix in which all of the entries that are *not* in the principal diagonal are zeros.

EXAMPLE 1   Consider the following square matrices of order 3.

$$A = \begin{bmatrix} 2 & 4 & 6 \\ -3 & 8 & 1 \\ 9 & 7 & 3 \end{bmatrix}, \qquad B = \begin{bmatrix} 3 & 0 & 0 \\ 0 & -2 & 0 \\ 0 & 0 & 0 \end{bmatrix}, \qquad I = \begin{bmatrix} 1 & 0 & 0 \\ 0 & 1 & 0 \\ 0 & 0 & 1 \end{bmatrix}.$$

a) The principal diagonal of A includes 2, 8 and 3.

b) B is a diagonal matrix.

c) I is a diagonal matrix in which each entry in the principal diagonal is 1.

Diagonal matrices (such as I in Example 1) in which each entry in the principal diagonal is 1 play a special role in matrix algebra. We reserve the symbol I, or $I_{n \times n}$, to name only such matrices. That is, it is to be understood that

$$I_{2 \times 2} = \begin{bmatrix} 1 & 0 \\ 0 & 1 \end{bmatrix}, \qquad I_{3 \times 3} = \begin{bmatrix} 1 & 0 & 0 \\ 0 & 1 & 0 \\ 0 & 0 & 1 \end{bmatrix}, \qquad \text{etc.}$$

Properties of Square Matrices   For each $n \geq 2$, from Definitions 7.1 and 7.3 it follows that $n \times n$ square matrices are always conformable with respect to both matrix addition and matrix multiplication. That is, for $n \times n$ matrices A and B, the sum $A + B$ and the products AB and BA always exist. Another property of interest is *associativity* of matrix multiplication of square matrices.

Theorem 7.1   If A, B and C are square matrices of order n, then

$$(AB)C = A(BC).$$

From Theorem 7.1 it follows that products of three (or more) square matrices of order n may be written without parentheses as ABC. It should also be noted that the associative property applies to all *conformable* matrices, even though Theorem 7.1 states this result for square matrices only.

Consider the matrix products:

$$AI = \begin{bmatrix} 6 & 7 \\ 4 & -3 \end{bmatrix}\begin{bmatrix} 1 & 0 \\ 0 & 1 \end{bmatrix} = \begin{bmatrix} 6 & 7 \\ 4 & -3 \end{bmatrix} = A;$$

$$IA = \begin{bmatrix} 1 & 0 \\ 0 & 1 \end{bmatrix}\begin{bmatrix} 6 & 7 \\ 4 & -3 \end{bmatrix} = \begin{bmatrix} 6 & 7 \\ 4 & -3 \end{bmatrix} = A.$$

Note first that matrix I behaves as an identity element for matrix multiplication and, second, that $AI = IA$, suggesting that multiplication involving the matrix I is commutative. These results are generalized next.

Theorem 7.2   For every $n \times n$ matrix A and the matrix $I_{n \times n}$,
$$AI = IA = A.$$

Moreover, it can be shown (see Exercise 7.2–25) that if B is a square matrix of order n such that
$$AB = BA = A$$
for *all* square matrices $A_{n \times n}$, then B must be equal to I. This concept, together with Theorem 7.2, tells us that for each $n \geq 2$ there is exactly one **identity matrix of order** n for matrix multiplication, and it is $I_{n \times n}$. (We shall use the simpler symbol I whenever the order $n \times n$ is clear from the context.)

Inverses of Square Matrices   Recall that if a and b are real numbers such that $ab = 1$, then a and b are said to be multiplicative inverses of each other, and that we write $b = a^{-1}$. Recall also that there is no multiplicative inverse of zero. Similar concepts exist with respect to square matrices. Consider

$$AB = \begin{bmatrix} 5 & 7 \\ 2 & 3 \end{bmatrix}\begin{bmatrix} 3 & -7 \\ -2 & 5 \end{bmatrix} = \begin{bmatrix} 1 & 0 \\ 0 & 1 \end{bmatrix} = I,$$

$$BA = \begin{bmatrix} 3 & -7 \\ -2 & 5 \end{bmatrix}\begin{bmatrix} 5 & 7 \\ 2 & 3 \end{bmatrix} = \begin{bmatrix} 1 & 0 \\ 0 & 1 \end{bmatrix} = I,$$

and note that each of the products AB and BA is equal to the identity matrix I. Matrices A and B are said to be *multiplicative inverse matrices* or, more simply, *inverse matrices*. The symbol A^{-1} is often used to name the inverse matrix of A.

Definition 7.4   For a given square matrix A of order n and multiplicative identity matrix I of order n, if there exists a square matrix A^{-1} of order n such that
$$AA^{-1} = I \quad \text{and} \quad A^{-1}A = I,$$
then A^{-1} is the **multiplicative inverse matrix** of A.

Just as not every real number has a multiplicative inverse (e.g. zero), not every square matrix has an inverse matrix. This can readily be shown by exhibiting at least one matrix for which no inverse exists. Consider

$$B = \begin{bmatrix} 1 & 1 \\ 1 & 1 \end{bmatrix} \quad \text{and} \quad X = \begin{bmatrix} x_1 & x_2 \\ y_1 & y_2 \end{bmatrix}.$$

If there exist numbers x_1, x_2, y_1 and y_2 such that $BX = I$ and $XB = I$, then X is the inverse matrix of B; if there are no such numbers then B has no inverse matrix. Now, if BX is equal to I, then

$$BX = \begin{bmatrix} 1 & 1 \\ 1 & 1 \end{bmatrix} \begin{bmatrix} x_1 & x_2 \\ y_1 & y_2 \end{bmatrix} = \begin{bmatrix} x_1 + y_1 & x_2 + y_2 \\ x_1 + y_1 & x_2 + y_2 \end{bmatrix} = \begin{bmatrix} 1 & 0 \\ 0 & 1 \end{bmatrix}. \quad \langle 1 \rangle$$

Equations $\langle 1 \rangle$ imply that

$$\begin{array}{cc} x_1 + y_1 = 1 \\ x_1 + y_1 = 0 \end{array} \quad \text{and} \quad \begin{array}{cc} x_2 + y_2 = 0 \\ x_2 + y_2 = 1. \end{array} \quad \langle 2 \rangle$$

But there are no solutions to either of the 2×2 systems in $\langle 2 \rangle$, as you can verify. It follows that B has no inverse matrix.

A square matrix for which no inverse matrix exists is said to be **singular**. If a square matrix A has an inverse matrix A^{-1}, then A is said to be **nonsingular**, or **invertible**. To *invert a matrix A* means to *find the inverse matrix A^{-1}*.

Inverting Nonsingular Matrices   From Section 5.2 (see page 195), recall that a matrix A is said to be *row-equivalent* to matrix B if A can be transformed into B by applying one or more of the three types of elementary row operations. The concept of row-equivalence, together with the next theorem, furnishes a method for obtaining the inverse of a nonsingular square matrix.

Theorem 7.3   A square matrix A is row-equivalent to I if and only if A is nonsingular.

Given an invertible square matrix A, Theorem 7.3 implies that a finite sequence of elementary row operations that will transform A into I can always be found. It can be shown (see Supplement I) that *the same sequence of row operations* that transforms A into I will also transform I into A^{-1}. Thus, in order to find A^{-1}, we apply the same row operations to both A and I. As an efficient procedure, we work on A and I at the same time by transforming the augmented matrix $[A \vdots I]$ into $[I \vdots A^{-1}]$, as illustrated by the next example.

EXAMPLE 2   Find A^{-1}, the inverse matrix of

$$A = \begin{bmatrix} 1 & 1 & 1 \\ 0 & 1 & 2 \\ 2 & 1 & 1 \end{bmatrix}.$$

SOLUTION   We apply elementary row operations to the augmented matrix $B = [A \mid I]$ until A is transformed into I.

$$B = \begin{bmatrix} 1 & 1 & 1 & \vdots & 1 & 0 & 0 \\ 0 & 1 & 2 & \vdots & 0 & 1 & 0 \\ 2 & 1 & 1 & \vdots & 0 & 0 & 1 \end{bmatrix}$$

$$C = \begin{bmatrix} 1 & 1 & 1 & \vdots & 1 & 0 & 0 \\ 0 & 1 & 2 & \vdots & 0 & 1 & 0 \\ 0 & -1 & -1 & \vdots & -2 & 0 & 1 \end{bmatrix} \quad (-2 \times B\textcircled{1}) + B\textcircled{3}*$$

$$D = \begin{bmatrix} 1 & 0 & -1 & \vdots & 1 & -1 & 0 \\ 0 & 1 & 2 & \vdots & 0 & 1 & 0 \\ 0 & 0 & 1 & \vdots & -2 & 1 & 1 \end{bmatrix} \quad \begin{array}{l} (-1 \times C\textcircled{2}) + C\textcircled{1} \\[4pt] \\[4pt] C\textcircled{2} + C\textcircled{3} \end{array}$$

$$E = \begin{bmatrix} 1 & 0 & 0 & \vdots & -1 & 0 & 1 \\ 0 & 1 & 0 & \vdots & 4 & -1 & -2 \\ 0 & 0 & 1 & \vdots & -2 & 1 & 1 \end{bmatrix} \quad \begin{array}{l} D\textcircled{3} + D\textcircled{1} \\[4pt] (-2 \times D\textcircled{3}) + D\textcircled{2} \end{array}$$

From $E = [I \mid A^{-1}]$, we have that

$$A^{-1} = \begin{bmatrix} -1 & 0 & 1 \\ 4 & -1 & -2 \\ -2 & 1 & 1 \end{bmatrix}.$$

You should verify that $AA^{-1} = A^{-1}A = I$.

During the process of inversion, singular matrices can be identified by a row of zero entries in the left portion of the augmented matrix.

EXAMPLE 3   Show that A is a singular matrix:

$$A = \begin{bmatrix} 1 & 0 & 2 \\ 2 & 1 & -2 \\ 3 & 1 & 0 \end{bmatrix}.$$

SOLUTION   Starting with the augmented matrix $[A \mid I]$, we have

$$B = \begin{bmatrix} 1 & 0 & 2 & \vdots & 1 & 0 & 0 \\ 2 & 1 & -2 & \vdots & 0 & 1 & 0 \\ 3 & 1 & 0 & \vdots & 0 & 0 & 1 \end{bmatrix}$$

$$C = \begin{bmatrix} 1 & 0 & 2 & \vdots & 1 & 0 & 0 \\ 0 & 1 & -6 & \vdots & -2 & 1 & 0 \\ 0 & 1 & -6 & \vdots & -3 & 0 & 1 \end{bmatrix} \quad \begin{array}{l} (-2 \times B\textcircled{1}) + B\textcircled{2} \\ (-3 \times B\textcircled{1}) + B\textcircled{3} \end{array}$$

$$D = \begin{bmatrix} 1 & 0 & 2 & \vdots & 1 & 0 & 0 \\ 0 & 1 & -6 & \vdots & -2 & 1 & 0 \\ 0 & 0 & 0 & \vdots & -1 & -1 & 1 \end{bmatrix} \quad (-1 \times C\textcircled{2}) + C\textcircled{3}$$

---

*See page 195.   *(Continued on next page)*

The row of three zero entries in the A portion of augmented matrix D prevents us from transforming A into I and we conclude that A is singular; A^{-1} does not exist. In general, any square matrix A is singular if and only if A is row-equivalent to a matrix with a zero row.

Matrix Form of an $n \times n$ System   An inverse matrix method can be used to solve $n \times n$ linear systems after such systems are written in *matrix form*. Consider the matrix product:

$$\begin{bmatrix} a_{11} & a_{12} \\ a_{21} & a_{22} \end{bmatrix}\begin{bmatrix} x \\ y \end{bmatrix} = \begin{bmatrix} a_{11}x + a_{12}y \\ a_{21}x + a_{22}y \end{bmatrix}.$$

This result suggests that the 2×2 linear system

$$a_{11}x + a_{12}y = c_1$$
$$a_{21}x + a_{22}y = c_2$$

can be written in matrix form as

$$\begin{bmatrix} a_{11} & a_{12} \\ a_{21} & a_{22} \end{bmatrix}\begin{bmatrix} x \\ y \end{bmatrix} = \begin{bmatrix} c_1 \\ c_2 \end{bmatrix}$$

and denoted by the matrix equation $AX = C$. Note that A is the coefficient matrix of the system.

EXAMPLE 4   The 3×3 system on the left is given in matrix form on the right:

$$\begin{array}{rcl} 2x + y - 3z = -5 \\ 3y + z = -3 \\ -3x + 9z = 18, \end{array} \quad \leftrightarrow \quad \begin{bmatrix} 2 & 1 & -3 \\ 0 & 3 & 1 \\ -3 & 0 & 9 \end{bmatrix}\begin{bmatrix} x \\ y \\ z \end{bmatrix} = \begin{bmatrix} -5 \\ -3 \\ 18 \end{bmatrix}.$$

Given an $n \times n$ linear system, if

$$AX = C \qquad\qquad\qquad \langle 3 \rangle$$

is the matrix form of the system and if A is nonsingular, then we can left-multiply each member of $\langle 3 \rangle$ by A^{-1} to obtain

$$A^{-1}(AX) = A^{-1}C$$
$$(A^{-1}A)X = A^{-1}C$$
$$IX = A^{-1}C$$
$$X = A^{-1}C.$$

Thus, $A^{-1}C$ is the matrix X that is the solution of $AX = C$. The inverse matrix method is applied in the next example.

EXAMPLE 5   A chemical company stocks three commercially prepared fertilizers: Types I, II and III. Each pound of fertilizer contains ingredients r, s and t in the amounts listed in the table:

	I	II	III
Units of r	1	1	1
Units of s	0	1	2
Units of t	2	1	1

How many pounds of each type of fertilizer should be combined to fill an order for a mixture that will contain 12 units of r, 10 units of s and 16 units of t?

SOLUTION Suppose that the company mixes x pounds of Type I, y pounds of Type II and z pounds of Type III. Then

$$1x + 1y + 1z, \qquad 0x + 1y + 2z, \qquad 2x + 1y + 1z$$

represent the number of units of ingredients r, s and t, respectively, in the mixture. To fill the given order requires that

$$\begin{aligned} 1x + 1y + 1z &= 12 \\ 0x + 1y + 2z &= 10 \\ 2x + 1y + 1z &= 16. \end{aligned} \qquad \langle 4 \rangle$$

In matrix form, system $\langle 4 \rangle$ becomes

$$\overset{A}{\begin{bmatrix} 1 & 1 & 1 \\ 0 & 1 & 2 \\ 2 & 1 & 1 \end{bmatrix}} \overset{X}{\begin{bmatrix} x \\ y \\ z \end{bmatrix}} = \overset{C}{\begin{bmatrix} 12 \\ 10 \\ 16 \end{bmatrix}}. \qquad \langle 5 \rangle$$

Now, the matrix A^{-1} computed in Example 2 is the inverse matrix to matrix A in equation $\langle 5 \rangle$. Hence, left-multiplying each member of $\langle 5 \rangle$ by A^{-1} (verify the details), we obtain

$$\overset{X}{\begin{bmatrix} x \\ y \\ z \end{bmatrix}} = \overset{A^{-1}}{\begin{bmatrix} -1 & 0 & 1 \\ 4 & -1 & -2 \\ -2 & 1 & 1 \end{bmatrix}} \overset{C}{\begin{bmatrix} 12 \\ 10 \\ 16 \end{bmatrix}} = \begin{bmatrix} 4 \\ 6 \\ 2 \end{bmatrix},$$

from which we have that $x = 4$, $y = 6$ and $z = 2$. That is, 4 pounds of Type I, 6 pounds of Type II and 2 pounds of Type III fertilizer are required.

It may occur to you to ask "Why do we need yet another method for solving linear systems?" One answer is that the inverse matrix method furnishes an advantage that other methods do not offer. After the inverse matrix has been obtained, the result can be used repeatedly. For instance, the chemical company of Example 5 can replace matrix C appropriately in the right member of the matrix equation $X = A^{-1}C$ to determine the mixtures needed for many *different* orders. Thus, suppose an order calls for a mixture that requires 8 units of s, 3 units of r and 14 units of t. The company needs only to compute

$$A^{-1}C = \begin{bmatrix} -1 & 0 & 1 \\ 4 & -1 & -2 \\ -2 & 1 & 1 \end{bmatrix} \begin{bmatrix} 8 \\ 3 \\ 14 \end{bmatrix} = \begin{bmatrix} 6 \\ 1 \\ 1 \end{bmatrix}$$

to conclude that 6 pounds of Type I, 1 pound of Type II and 1 pound of Type III should be mixed to fill the order. Furthermore, the same procedure can be used to determine that a particular order *cannot* be filled from the stock on hand (see Exercise 7.2-20).

Exercise Set 7.2

(A) *For each equation, verify Theorem 7.1 by computing the products in each member as indicated by the parentheses.*

1. $\left(\begin{bmatrix} 2 & 3 \\ -1 & 5 \end{bmatrix} \begin{bmatrix} 1 & -3 \\ 4 & 7 \end{bmatrix} \right) \begin{bmatrix} 2 & 5 \\ -2 & 3 \end{bmatrix} = \begin{bmatrix} 2 & 3 \\ -1 & 5 \end{bmatrix} \left(\begin{bmatrix} 1 & -3 \\ 4 & 7 \end{bmatrix} \begin{bmatrix} 2 & 5 \\ -2 & 3 \end{bmatrix} \right)$

2. $\left(\begin{bmatrix} 1 & 0 & 3 \\ 2 & 1 & 1 \\ 0 & 1 & 2 \end{bmatrix} \begin{bmatrix} 2 & 1 & 0 \\ 1 & 0 & 2 \\ 3 & 1 & 1 \end{bmatrix} \right) \begin{bmatrix} 1 & 2 & 3 \\ 0 & 0 & 1 \\ 1 & 0 & 1 \end{bmatrix} = \begin{bmatrix} 1 & 0 & 3 \\ 2 & 1 & 1 \\ 0 & 1 & 2 \end{bmatrix} \left(\begin{bmatrix} 2 & 1 & 0 \\ 1 & 0 & 2 \\ 3 & 1 & 1 \end{bmatrix} \begin{bmatrix} 1 & 2 & 3 \\ 0 & 0 & 1 \\ 1 & 0 & 1 \end{bmatrix} \right)$

For each pair of matrices A and B, show that $AB = BA = I$ and therefore that A and B are inverse matrices.

3. $A = \begin{bmatrix} 1 & 2 \\ 3 & 4 \end{bmatrix}$, $B = \begin{bmatrix} -2 & 1 \\ \frac{3}{2} & -\frac{1}{2} \end{bmatrix}$

4. $A = \begin{bmatrix} 2 & 3 \\ 3 & 5 \end{bmatrix}$, $B = \begin{bmatrix} 5 & -3 \\ -3 & 2 \end{bmatrix}$

5. $A = \begin{bmatrix} 3 & -2 & -1 \\ -4 & 1 & -1 \\ 2 & 0 & 1 \end{bmatrix}$, $B = \begin{bmatrix} 1 & 2 & 3 \\ 2 & 5 & 7 \\ -2 & -4 & -5 \end{bmatrix}$

6. $A = \begin{bmatrix} 1 & 1 & 1 \\ 2 & 3 & 2 \\ 3 & 3 & 4 \end{bmatrix}$, $B = \begin{bmatrix} 6 & -1 & -1 \\ -2 & 1 & 0 \\ -3 & 0 & 1 \end{bmatrix}$

Obtain the inverse matrix, if it exists. (See Examples 2 and 3.)

7. $\begin{bmatrix} 1 & 3 \\ 1 & 2 \end{bmatrix}$ 8. $\begin{bmatrix} 4 & 2 \\ 6 & 3 \end{bmatrix}$ 9. $\begin{bmatrix} 5 & 10 \\ 1 & 2 \end{bmatrix}$ 10. $\begin{bmatrix} 3 & 2 \\ 4 & -1 \end{bmatrix}$

11. $\begin{bmatrix} 2 & 1 & 4 \\ 1 & 0 & 2 \\ 1 & 2 & 2 \end{bmatrix}$ 12. $\begin{bmatrix} 1 & 2 & 0 \\ -1 & 1 & 0 \\ 2 & 3 & 2 \end{bmatrix}$ 13. $\begin{bmatrix} 2 & 3 & 2 \\ -1 & 0 & 2 \\ 1 & 1 & 1 \end{bmatrix}$ 14. $\begin{bmatrix} 2 & 1 & -3 \\ 4 & -2 & 1 \\ 6 & -1 & -2 \end{bmatrix}$

Solve each system by the inverse matrix method. (See Example 5.)

15. $2x - 3y = 6$
    $x + 3y = 3$

16. $3x + 2y = 5$
    $6x + 5y = 5$

17. $x + 2z = 5$
    $y + z = -1$
    $3x + y = 7$

18. $x - y + z = -3$
    $2x + y = -7$
    $x + z = -6$

For Exercises 19 and 20, refer to Example 5.

19. The chemical company receives orders for the following mixtures:
    a. 6 units of r, 5 units of s, 7 units of t;
    b. 2 units of r, 3 units of s, 2 units of t.
    Specify how many pounds of each type of fertilizer are needed to fill each order.

20. Show that each of the following orders cannot be filled:
    a. 2 units of r, 4 units of s, 7 units of t;
    b. 3 units of r, 1 unit of s, 4 units of t.

(B) 21. A mixture of nuts comes in two different size cartons: carton I contains 3 pounds of almonds and 5 pounds of cashews; Carton II contains 6 pounds of almonds and 4 pounds of cashews. If only unopened cartons can be shipped, how many of each type of carton should be ordered to obtain 57 pounds of almonds and 53 pounds of cashews?

22. Use the information in Exercise 21 to fill (if possible) the following orders:
    a. 87 pounds of almonds and 85 pounds of cashews.
    b. 124 pounds of almonds and 112 pounds of cashews.

23. A company combines three mixtures of vitamins—types I, II and III. Each gram of each type contains units of vitamins A, B and C as follows:

	I	II	III
Units of A	1	3	3
Units of B	1	4	3
Units of C	1	3	4

How many grams of each type of mixture should be combined to make capsules that will contain:
    a. 25 units of A, 29 units of B and 27 units of C?
    b. 30 units of A, 34 units of B and 35 units of C?

24. Use the information in Exercise 23 to decide which one of the following capsules cannot be made by using mixtures of I, II and III.
    a. 28 units of A, 30 units of B and 34 units of C.
    b. 29 units of A, 34 units of B and 35 units of C.
    c. 27 units of A, 31 units of B and 30 units of C.

25. Prove: If B is a square matrix of order n such that $AB=BA=A$ for *every* square matrix A of order n, then $B=I$. (Hint: If the equations are valid for *every* A, they must be valid when A is equal to I.)

7.3 DETERMINANT FUNCTIONS

In Sections 7.1 and 7.2, certain matrix operations are introduced—addition, subtraction, multiplication by a scalar and matrix multiplication. As you now know, each of these operations yields a *matrix*. In this section we consider an operation on square matrices with real number entries that yields *real numbers* and enables us to construct a type of function called a *determinant function*, symbolized by *det*.

To each square matrix A we associate a square array of entries between two vertical bars; the resulting symbol is called a **determinant of order** n. Thus,

$$\text{if } A = \begin{bmatrix} a_{11} & a_{12} \\ a_{21} & a_{22} \end{bmatrix}, \text{ then } det\ A = \begin{vmatrix} a_{11} & a_{12} \\ a_{21} & a_{22} \end{vmatrix};$$

$$\text{if } B = \begin{bmatrix} b_{11} & b_{12} & b_{13} \\ b_{21} & b_{22} & b_{23} \\ b_{31} & b_{32} & b_{33} \end{bmatrix}, \text{ then } det\ B = \begin{vmatrix} b_{11} & b_{12} & b_{13} \\ b_{21} & b_{22} & b_{23} \\ b_{31} & b_{32} & b_{33} \end{vmatrix}.$$

Det A and *det B* are second and third order determinants, respectively. As you will see in the discussion below, each such determinant is actually a numeral for a real number. It follows that each square matrix *A* can be paired with a real number. This pairing is the basis for generating a **determinant function**, where the domain is a set of $n \times n$ matrices; the range is a set of real numbers.

The strong resemblance between determinants and square matrices sometimes causes confusion. It must be remembered that *a determinant is a numeral* for a number whereas a matrix, as we use it, is a rectangular array of numbers and not a numeral at all.

Second and Third Order Determinants   To *evaluate a determinant* means to find the real number named by the determinant. We call this real number the *value* of the determinant. Methods of evaluating determinants are based upon the next three definitions.

Definition 7.5   For all 2×2 matrices A,

$$det\ A = \begin{vmatrix} a_{11} & a_{12} \\ a_{21} & a_{22} \end{vmatrix} = a_{11}a_{22} - a_{21}a_{12}.$$

This definition is sometimes remembered as "the product of the entries in the principal diagonal minus the product of the entries in the other diagonal."

EXAMPLE 1   By Definition 7.5,

$$\begin{vmatrix} 2 & -3 \\ 7 & -5 \end{vmatrix} = 2(-5) - 7(-3) = 11.$$

Definition 7.6   For all 3×3 matrices A,

$$det\ A = \begin{vmatrix} a_{11} & a_{12} & a_{13} \\ a_{21} & a_{22} & a_{23} \\ a_{31} & a_{32} & a_{33} \end{vmatrix} = a_{11}a_{22}a_{33} - a_{11}a_{23}a_{32} + a_{12}a_{23}a_{31}$$
$$- a_{12}a_{21}a_{33} + a_{13}a_{21}a_{32} - a_{13}a_{22}a_{31}.$$

This definition is much too cumbersome for routine use. Instead, we use it, together with the next definition, as a basis for developing a general method for evaluating determinants.

Definition 7.7   The **minor** A_{ij} of the entry a_{ij} in a given nth order determinant is the determinant of order $n-1$ obtained by deleting the ith row and the jth column of the given determinant.

For example, minors A_{11} and A_{32} are obtained as follows:

$$\begin{vmatrix} a_{11} & a_{12} & a_{13} \\ a_{21} & a_{22} & a_{23} \\ a_{31} & a_{32} & a_{33} \end{vmatrix}; \qquad A_{11} = \begin{vmatrix} a_{22} & a_{23} \\ a_{32} & a_{33} \end{vmatrix}$$

$$\begin{vmatrix} a_{11} & a_{12} & a_{13} \\ a_{21} & a_{22} & a_{23} \\ a_{31} & a_{32} & a_{33} \end{vmatrix}; \qquad A_{32} = \begin{vmatrix} a_{11} & a_{13} \\ a_{21} & a_{23} \end{vmatrix}.$$

EXAMPLE 2   Evaluate minors B_{12} and B_{21} of

$$\det B = \begin{vmatrix} 2 & 1 & 5 \\ -3 & 4 & 0 \\ 6 & -2 & 3 \end{vmatrix}.$$

SOLUTION

$$B_{12} = \begin{vmatrix} -3 & 0 \\ 6 & 3 \end{vmatrix} = -9 - 0 = -9; \qquad B_{21} = \begin{vmatrix} 1 & 5 \\ -2 & 3 \end{vmatrix} = 3 - (-10) = 13.$$

Expansion by Minors   If A is the 3×3 matrix

$$A = \begin{bmatrix} a_{11} & a_{12} & a_{13} \\ a_{21} & a_{22} & a_{23} \\ a_{31} & a_{32} & a_{33} \end{bmatrix}, \quad \text{then} \quad \det A = \begin{vmatrix} a_{11} & a_{12} & a_{13} \\ a_{21} & a_{22} & a_{23} \\ a_{31} & a_{32} & a_{33} \end{vmatrix}.$$

By Definition 7.6,

$$\det A = a_{11}a_{22}a_{33} - a_{11}a_{23}a_{32} + a_{12}a_{23}a_{31} - a_{12}a_{21}a_{33} \\ + a_{13}a_{21}a_{32} - a_{13}a_{22}a_{31}. \qquad \langle 1 \rangle$$

By regrouping and partially factoring, equation $\langle 1 \rangle$ can be written as

$$\det A = (a_{11}a_{22}a_{23} - a_{11}a_{23}a_{32}) - (a_{12}a_{21}a_{33} - a_{13}a_{21}a_{32}) \\ + (a_{12}a_{23}a_{31} - a_{13}a_{22}a_{31})$$

$$\det A = a_{11}(a_{22}a_{33} - a_{23}a_{32}) - a_{21}(a_{12}a_{33} - a_{13}a_{32}) \\ + a_{31}(a_{12}a_{23} - a_{13}a_{22}). \qquad \langle 2 \rangle$$

By Definitions 7.5 and 7.7, equation $\langle 2 \rangle$ can be written as

$$\det A = a_{11}\begin{vmatrix} a_{22} & a_{23} \\ a_{32} & a_{33} \end{vmatrix} - a_{21}\begin{vmatrix} a_{12} & a_{13} \\ a_{32} & a_{33} \end{vmatrix} + a_{31}\begin{vmatrix} a_{12} & a_{13} \\ a_{22} & a_{23} \end{vmatrix}$$

$$\det A = a_{11}A_{11} - a_{21}A_{21} + a_{31}A_{31}. \qquad \langle 3 \rangle$$

The right member of $\langle 3 \rangle$ is called the *expansion of det A about the first column*. Observe that each term of the expansion is the product of one entry and its minor.

Equation $\langle 1 \rangle$ can be transformed in other ways. For example, in Exercise 7.3–39 you are asked to verify that *det A* can be *expanded about the second row* to obtain

$$\det A = -a_{21}A_{21} + a_{22}A_{22} - a_{23}A_{23}. \qquad \langle 4 \rangle$$

Expansions $\langle 3 \rangle$ and $\langle 4 \rangle$ illustrate the fact that a third order determinant can be evaluated by expansion about *any row* or *any column*, provided that the sign ($+$ or $-$) preceding each term of the expansion is selected correctly. One method for selecting the correct sign is as follows. For the minor A_{ij}, add the row number i to the column number j. If $i+j$ is an *even* integer, prefix a plus ($+$) sign to the term $a_{ij}A_{ij}$; if $i+j$ is an *odd* integer, prefix a minus ($-$) sign to $a_{ij}A_{ij}$.

EXAMPLE 3   Given the expansion about the second column of the following determinant. (a) Replace each question mark by the correct $+$ or $-$ sign. (b) Evaluate the determinant.

$$det\ A = \begin{vmatrix} 2 & 7 & 5 \\ -3 & 4 & 0 \\ 6 & -2 & 3 \end{vmatrix} = ?\ 7 \begin{vmatrix} -3 & 0 \\ 6 & 3 \end{vmatrix} ?\ 4 \begin{vmatrix} 2 & 5 \\ 6 & 3 \end{vmatrix} ?(-2) \begin{vmatrix} 2 & 5 \\ -3 & 0 \end{vmatrix}$$

SOLUTION   The entry 7 is in the first row ($i=1$) and second column ($j=2$); $i+j=1+2=3$, an odd integer. Thus, a minus sign precedes the first term. The entry 4 is in the second row and second column; $i+j=2+2=4$, an even integer. Thus, a plus sign precedes the second term. You can verify that a minus sign precedes the third term and so

$$det\ A = -7 \begin{vmatrix} -3 & 0 \\ 6 & 3 \end{vmatrix} + 4 \begin{vmatrix} 2 & 5 \\ 6 & 3 \end{vmatrix} - (-2) \begin{vmatrix} 2 & 5 \\ -3 & 0 \end{vmatrix}.$$

$$det\ A = -7(-9) + 4(-24) + 2(15) = -3.$$

EXAMPLE 4   Let us evaluate $det\ (A)$ of Example 3 by expansion about the third row:

$$det\ A = +6 \begin{vmatrix} 7 & 5 \\ 4 & 0 \end{vmatrix} - (-2) \begin{vmatrix} 2 & 5 \\ -3 & 0 \end{vmatrix} + 3 \begin{vmatrix} 2 & 7 \\ -3 & 4 \end{vmatrix}$$

$$det\ A = 6(-20) + 2(15) + 3(29) = -3.$$

Applications   We conclude this section with a look at two of the many applications of determinants. Further applications of determinants are considered in Exercise Set 7.3 and in Supplement H.

Let (x_1, y_1) and (x_2, y_2) be two points in a line, and let (x, y) be the coordinates of any other point in the line. By Definition 2.3, the slope of the line is

$$\frac{y - y_1}{x - x_1} \quad \text{or} \quad \frac{y_2 - y_1}{x_2 - x_1}.$$

Hence, an equation of the line is:

$$\frac{y - y_1}{x - x_1} = \frac{y_2 - y_1}{x_2 - x_1} \quad \leftrightarrow \quad \frac{y - y_1}{x - x_1} - \frac{y_2 - y_1}{x_2 - x_1} = 0. \qquad \langle 3 \rangle$$

Transforming the right-hand equation in $<3>$, we have

$$(y-y_1)(x_2-x_1)-(y_2-y_1)(x-x_1)=0$$
$$x_2y-x_1y-x_2y_1+x_1y_1-xy_2+x_1y_2+xy_1-x_1y_1=0$$
$$(x_1y_2-x_2y_1)+(-xy_2+x_2y)+(xy_1-x_1y)=0$$
$$(x_1y_2-x_2y_1)-(xy_2-x_2y)+(xy_1-x_1y)=0$$
$$\begin{vmatrix} x_1 & y_1 \\ x_2 & y_2 \end{vmatrix} - \begin{vmatrix} x & y \\ x_2 & y_2 \end{vmatrix} + \begin{vmatrix} x & y \\ x_1 & y_1 \end{vmatrix} = 0. \qquad <4>$$

Equation $<4>$, viewed as an expansion by minors, can be written as

$$\begin{vmatrix} x & y & 1 \\ x_1 & y_1 & 1 \\ x_2 & y_2 & 1 \end{vmatrix} = 0. \qquad <5>$$

Equation $<5>$ is a *determinantal equation* of the line determined by two distinct points. Admittedly, the derivation of $<5>$ may be described as "tricky," but our major concern is with the result itself rather than with the means of obtaining the result.

EXAMPLE 5 Find a standard form equation of the line determined by points (2, 3) and (−5, 6).

SOLUTION Substituting appropriately into equation $<5>$, we have

$$\begin{vmatrix} x & y & 1 \\ 2 & 3 & 1 \\ -5 & 6 & 1 \end{vmatrix} = 0.$$

Expanding the determinant about the third column, we have

$$\begin{vmatrix} 2 & 3 \\ -5 & 6 \end{vmatrix} - \begin{vmatrix} x & y \\ -5 & 6 \end{vmatrix} + \begin{vmatrix} x & y \\ 2 & 3 \end{vmatrix} = 0$$
$$27-(6x+5y)+3x-2y=0$$
$$3x+7y-27=0. \qquad <6>$$

Equation $<6>$ is the required equation of the line.

Recall, from geometry, that three points are *collinear* if they are included in the same line. Equation $<5>$ can be used to show that three points are, or are not, collinear.

EXAMPLE 6 Are the points (0, 0), (2, 3) and (−5, 6) collinear?

SOLUTION Recall that a point is on the graph of an equation if and only if the coordinates of the point satisfy the equation. Thus, point (0, 0) is on the line determined by (2, 3) and (−5, 6) if and only if (0, 0) is a solution of the equation

$$\begin{vmatrix} x & y & 1 \\ 2 & 3 & 1 \\ -5 & 6 & 1 \end{vmatrix} = 0. \qquad <7>$$

(Continued on next page)

Substituting 0 for x, 0 for y, and evaluating the resulting determinant, we obtain

$$\begin{vmatrix} 0 & 0 & 1 \\ 2 & 3 & 1 \\ -5 & 6 & 1 \end{vmatrix} = 27 \neq 0.$$

Hence, $(0, 0)$ is *not* a solution of equation $\langle 7 \rangle$ and we conclude that the three points are *not* collinear.

Exercise Set 7.3

(A)   *Evaluate each determinant. (See Example 1.)*

1. $\begin{vmatrix} 1 & 2 \\ 3 & 4 \end{vmatrix}$ 2. $\begin{vmatrix} -3 & -5 \\ 2 & 0 \end{vmatrix}$

3. $\begin{vmatrix} 2 & 4 \\ 4 & 8 \end{vmatrix}$ 4. $\begin{vmatrix} -3 & 6 \\ -9 & 18 \end{vmatrix}$

Evaluate each of the following minors of det B of Example 2.

5. B_{11} 6. B_{22} 7. B_{23} 8. B_{32}

Evaluate each determinant twice: first by expanding about any row; second by expanding about any column. (See Examples 3 and 4.)

9. $\begin{vmatrix} 2 & 0 & 3 \\ -4 & 1 & 0 \\ -1 & -3 & 4 \end{vmatrix}$ 10. $\begin{vmatrix} 5 & 0 & -1 \\ 6 & 4 & 1 \\ -2 & 3 & -2 \end{vmatrix}$

11. $\begin{vmatrix} 0 & 0 & 5 \\ -1 & 2 & 4 \\ 6 & -8 & 3 \end{vmatrix}$ 12. $\begin{vmatrix} 11 & -2 & 0 \\ 9 & 6 & 0 \\ 14 & 1 & 5 \end{vmatrix}$

Evaluate each determinant.

13. $\begin{vmatrix} 3 & 1 & 5 \\ 0 & -1 & 0 \\ 4 & 2 & 6 \end{vmatrix}$ 14. $\begin{vmatrix} 7 & 11 & 1 \\ 8 & -4 & 0 \\ -2 & 5 & 1 \end{vmatrix}$

15. $\begin{vmatrix} 2 & 3 & 7 \\ 0 & 1 & 8 \\ 4 & 1 & -2 \end{vmatrix}$ 16. $\begin{vmatrix} 8 & 3 & 7 \\ 2 & -2 & 0 \\ 8 & 1 & -12 \end{vmatrix}$

17. $\begin{vmatrix} 3 & 4 & -7 & 8 \\ 0 & 4 & 0 & 0 \\ -1 & 2 & 2 & -3 \\ -2 & 1 & 0 & -2 \end{vmatrix}$ 18. $\begin{vmatrix} 1 & 2 & 0 & -1 \\ 1 & 0 & -1 & 2 \\ 0 & 1 & 1 & 1 \\ 2 & -1 & 0 & 1 \end{vmatrix}$

Use equation $\langle 5 \rangle$ to find a standard form equation of the line determined by each pair of points. (See Example 5.)

19. $(2, 5)$, $(-2, -9)$ 20. $(5, 3)$, $(-1, 1)$

21. $(11, 5)$, $(-10, -7)$ 22. $(-6, -10)$, $(9, 17)$

For each three points, determine whether they are, or are not, collinear. (See Example 6.)

23. $(0, -6)$, $(3, 6)$, $(-5, -26)$ 24. $(1, -2)$, $(4, 10)$, $(-3, -12)$

25. $(2, 8)$, $(5, 12)$, $(-2, -4)$ 26. $(-4, -10)$, $(0, 2)$, $(4, 14)$

Solve each equation.

27. $\begin{vmatrix} 3 & 5 & 0 \\ 2 & x & -4 \\ 0 & 1 & 2 \end{vmatrix} = 10$ 28. $\begin{vmatrix} 2 & -4 & 3 \\ x & 0 & 1 \\ 0 & 5 & -2 \end{vmatrix} = 4$

29. $\begin{vmatrix} x & 1 & -3 \\ -1 & x & 2 \\ 2 & 0 & 1 \end{vmatrix} = 0$ 30. $\begin{vmatrix} -2 & x^2 & -1 \\ 1 & 3x & 0 \\ 5 & 1 & 2 \end{vmatrix} = 0$

The area of the triangle determined by three points (x_1, y_1), (x_2, y_2) and (x_3, y_3) is given by the absolute value of

$$\frac{1}{2} \begin{vmatrix} x_1 & y_1 & 1 \\ x_2 & y_2 & 1 \\ x_3 & y_3 & 1 \end{vmatrix}.$$

Find the area of the triangle determined by each three points.

31. $(0, 0)$, $(12, 0)$, $(0, 18)$ 32. $(0, 0)$, $(-12, 0)$, $(0, -18)$

33. $(1, 2)$, $(5, 9)$, $(-3, -7)$ 34. $(3, 5)$, $(-6, 0)$, $(-8, -1)$

(B) *In three dimensions, a point is given by an **ordered triple** of numbers (x, y, z); the graph of the standard form equation*

$$ax + by + cz + d = 0$$

*is a **plane**. A determinantal equation of the plane determined by three points (x_1, y_1, z_1), (x_2, y_2, z_2) and (x_3, y_3, z_3) is*

$$\begin{vmatrix} x & y & z & 1 \\ x_1 & y_1 & z_1 & 1 \\ x_2 & y_2 & z_2 & 1 \\ x_3 & y_3 & z_3 & 1 \end{vmatrix} = 0.$$

Find a standard-form equation of the plane determined by each three points.

35. $(0, 0, 0)$, $(1, 2, 3)$, $(-2, 1, 5)$ 36. $(0, 0, 0)$, $(4, 0, 2)$, $(-2, -3, -4)$

37. $(2, 0, 0)$, $(0, 5, 0)$, $(0, 0, -7)$ 38. $(-5, 0, 0)$, $(0, -6, 0)$, $(0, 0, 9)$

See det A of Definition 7.6 for Exercises 39 and 40.

39. Show that: $\det A = -a_{21}A_{21} + a_{22}A_{22} - a_{23}A_{23}$.

40. Show that: $\det A = a_{13}A_{13} - a_{23}A_{23} + a_{33}A_{33}$.

(C) *The BASIC computer program on page 246 can be used to evaluate a 3×3 determinant. Run the program with the test data provided. Explain why the program works. Use the program to check your answers to Exercises 13-16.*

```
10 DIM A(3,3)
20 MAT READ A
30 LET D1=A(1,1)*A(2,2)*A(3,3)
40 LET D2=A(1,2)*A(2,3)*A(3,1)
50 LET D3=A(1,3)*A(2,1)*A(3,2)
60 LET D4=A(1,3)*A(2,2)*A(3,1)
70 LET D5=A(1,2)*A(2,1)*A(3,3)
80 LET D6=A(1,1)*A(2,3)*A(3,2)
90 LET D=D1+D2+D3-D4-D5-D6
100 PRINT "THE DETERMINANT OF"
110 MAT PRINT A
120 PRINT "IS" D
130 GO TO 10
140 DATA 1,2,3,4,5,6,7,8,9,1,1,2,2,2,3,3,4,4,1,1,1,5,6,7,8,8,8
150 END
```

Chapter 7   Self-Test

[7.1] *In Exercises 1–10, use the following matrices:*

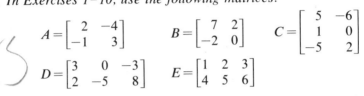

$$A = \begin{bmatrix} 2 & -4 \\ -1 & 3 \end{bmatrix} \qquad B = \begin{bmatrix} 7 & 2 \\ -2 & 0 \end{bmatrix} \qquad C = \begin{bmatrix} 5 & -6 \\ 1 & 0 \\ -5 & 2 \end{bmatrix}$$

$$D = \begin{bmatrix} 3 & 0 & -3 \\ 2 & -5 & 8 \end{bmatrix} \qquad E = \begin{bmatrix} 1 & 2 & 3 \\ 4 & 5 & 6 \end{bmatrix}$$

Write each sum, difference or matrix product as a single matrix, if possible.

  1. $B + 3A$ 2. $A + 3C$ 3. $2D - E$ 4. $4D - 3E$
  5. AD 6. CE 7. DA 8. EC

Write an appropriate matrix (with entries that are variables) for X and solve each equation.

  9. $D - 2X = E$ 10. $X + DC = B$

[7.2] *Solve each system by the inverse matrix method.*

  11. $2x + 3y = -2$ 12. $x + z = -1$
      $5x + 6y = 1$ $2x + y = 5$
                            $3y + z = 7$

  13. Show that $\begin{bmatrix} 3 & 9 \\ 4 & 12 \end{bmatrix}$ is singular (has no inverse).

  14. A homeowner finds that a nearby shop stocks only two types of 100 lb sacks of pre-mixed sand and cement, as follows:

	I	II
Sand	40 lbs	50 lbs
Cement	60 lbs	50 lbs

  How many sacks of each type should be bought if 220 lbs of sand and 280 lbs of cement are needed? If 360 lbs of sand and 440 lbs of cement are needed?

[7.3] *Evaluate each determinant.*

  15. $\begin{vmatrix} 11 & -4 & 5 \\ 1 & 0 & 1 \\ 7 & 8 & -2 \end{vmatrix}$ 16. $\begin{vmatrix} -2 & -1 & 0 & 3 \\ 1 & 2 & 2 & 4 \\ 0 & 2 & 0 & -7 \\ -2 & -3 & 0 & 8 \end{vmatrix}$

SUPPLEMENT G

THEOREMS ON DETERMINANTS

The following theorems enable us to more readily evaluate determinants, and also to transform determinants into equivalent forms for both theoretical and computational purposes.

Zero Value Theorems   It is sometimes possible to determine, almost at a glance, that the value of a given determinant is zero.

Theorem G-1   If each entry in any row (or column) of $det\ M$ is zero, then $det\ M = 0$.

Theorem G-2   If two rows (or two columns) in $det\ M$ have the same entries, then $det\ M = 0$.

Theorem G-3   If each entry in a row (or column) is equal to the same multiple of the corresponding entry in another row (or column) of $det\ M$, then $det\ M = 0$.

EXAMPLE 1   Consider each of the following three determinants.

$$\begin{vmatrix} 1 & 2 & 3 \\ 0 & 0 & 0 \\ 2 & -3 & 4 \end{vmatrix} \quad \begin{vmatrix} 3 & 1 & 3 \\ 0 & 5 & 0 \\ 2 & -3 & 2 \end{vmatrix} \quad \begin{vmatrix} 6 & 3 & -9 \\ 0 & 2 & 4 \\ 2 & 1 & -3 \end{vmatrix}$$

The value of the first determinant is zero, by Theorem G-1, because each entry in row two is zero. The value of the second determinant is zero, by Theorem G-2, because columns one and three have the same entries. The value of the third determinant is zero, by Theorem G-3, because each entry in row one is equal to three times the corresponding entry of row three.

Transformation Theorems   We sometimes change the entries in a given determinant to facilitate computations. The next theorem can be used to effect such changes.

Theorem G-4   If $det\ N$ is obtained from $det\ M$ by multiplying each entry in any row (or column) by any nonzero number k, then $det\ N = k \cdot det\ M$.

One customary use of Theorem G-4 can be expressed symbolically as:

$$\begin{vmatrix} ka_1 & b_1 \\ ka_2 & b_2 \end{vmatrix} = k \begin{vmatrix} a_1 & b_1 \\ a_2 & b_2 \end{vmatrix}.$$

EXAMPLE 2   By Theorem G-4,

$$\begin{vmatrix} 3 & 0 & 1 \\ 25 & 75 & 50 \\ -2 & 5 & 4 \end{vmatrix} = \begin{vmatrix} 3 & 0 & 1 \\ 25 \cdot 1 & 25 \cdot 3 & 25 \cdot 2 \\ -2 & 5 & 4 \end{vmatrix} = 25 \begin{vmatrix} 3 & 0 & 1 \\ 1 & 3 & 2 \\ -2 & 5 & 4 \end{vmatrix}.$$

The result in Example 2 suggests that Theorem G‑4 may be viewed as a *factoring* theorem. In the example, 25 is *factored out* of row two.

For a second use of Theorem G‑4, consider

$$\begin{vmatrix} a_1 & b_1 \\ a_2 & b_2 \end{vmatrix} = \begin{vmatrix} \frac{1}{k}\cdot ka_1 & \frac{1}{k}\cdot kb_1 \\ a_2 & b_2 \end{vmatrix} = \frac{1}{k}\begin{vmatrix} ka_1 & kb_1 \\ a_2 & b_2 \end{vmatrix}.$$

This result suggests that the value of a determinant is unchanged if we multiply each entry of any row (or column) by a nonzero number k, provided that we multiply the determinant itself by $\frac{1}{k}$, the reciprocal of k.

EXAMPLE 3

$$\begin{vmatrix} \frac{1}{2} & \frac{2}{3} & \frac{5}{6} \\ 3 & 4 & 2 \\ 1 & -3 & 7 \end{vmatrix} = \frac{1}{6}\begin{vmatrix} 6\cdot\frac{1}{2} & 6\cdot\frac{2}{3} & 6\cdot\frac{5}{6} \\ 3 & 4 & 2 \\ 1 & -3 & 7 \end{vmatrix} = \frac{1}{6}\begin{vmatrix} 3 & 4 & 5 \\ 3 & 4 & 2 \\ 1 & -3 & 7 \end{vmatrix}.$$

The following theorem can be used to reduce the *order* of a given determinant.

Theorem G‑5   If $k \neq 0$ and the product of k and each entry in any row (or column) of *det M* is added to the corresponding entry in another row (or column) of *det M*, then the resulting determinant is equal to *det M*.

EXAMPLE 4   Use Theorem G‑5 to reduce the following 3×3 determinant to a 2×2 determinant, and then evaluate.

$$\begin{vmatrix} 2 & 5 & 3 \\ 1 & -2 & 4 \\ 6 & 1 & 2 \end{vmatrix}$$

SOLUTION   First, we add the product of 2 and each entry in column one to the corresponding entries of column two. Object: to obtain a zero entry in row two, column two.

$$\begin{vmatrix} 2 & 5+2\cdot2 & 3 \\ 1 & -2+1\cdot2 & 4 \\ 6 & 1+6\cdot2 & 2 \end{vmatrix} = \begin{vmatrix} 2 & 9 & 3 \\ 1 & 0 & 4 \\ 6 & 13 & 2 \end{vmatrix}.$$

Next, we add the product of -4 and each entry in column one to the corresponding entries of column three. Object: to obtain a zero entry in row two, column three.

$$\begin{vmatrix} 2 & 9 & 3+2(-4) \\ 1 & 0 & 4+1(-4) \\ 6 & 13 & 2+6(-4) \end{vmatrix} = \begin{vmatrix} 2 & 9 & -5 \\ 1 & 0 & 0 \\ 6 & 13 & -22 \end{vmatrix}.$$

Finally, we expand about row two to obtain a 2×2 determinant, which we evaluate:

$$-1\begin{vmatrix} 9 & -5 \\ 13 & -22 \end{vmatrix} = -133.$$

Exercise Set G

Which theorem (G-1, G-2 or G-3) best justifies each statement?

1. $\begin{vmatrix} 2 & -1 & 0 \\ 0 & 0 & 0 \\ 3 & -3 & -5 \end{vmatrix} = 0$

2. $\begin{vmatrix} 4 & -3 & 0 \\ 2 & 11 & 0 \\ -7 & 8 & 0 \end{vmatrix} = 0$

3. $\begin{vmatrix} 5 & -4 & 5 \\ -11 & 0 & -11 \\ 8 & 7 & 8 \end{vmatrix} = 0$

4. $\begin{vmatrix} 7 & 3 & -15 \\ 1 & 2 & 1 \\ 1 & 2 & 1 \end{vmatrix} = 0$

5. $\begin{vmatrix} 1 & 4 & 2 \\ 5 & 20 & 10 \\ -3 & -12 & 6 \end{vmatrix} = 0$

6. $\begin{vmatrix} 3 & -1 & 0 \\ -1 & 5 & -2 \\ 2 & -10 & 4 \end{vmatrix} = 0$

Rewrite each determinant so that the entries in any row (or column) have no common factor other than 1 or −1.

7. $\begin{vmatrix} 3 & -2 & 5 \\ 1 & 0 & 7 \\ 4 & 12 & 8 \end{vmatrix}$

8. $\begin{vmatrix} 6 & -6 & 5 \\ -5 & 16 & 1 \\ 7 & -8 & 3 \end{vmatrix}$

Rewrite each determinant so that each entry is an integer.

9. $\begin{vmatrix} 1 & 3 & 0 \\ \frac{2}{3} & -\frac{1}{6} & \frac{5}{12} \\ 2 & 4 & -5 \end{vmatrix}$

10. $\begin{vmatrix} \frac{1}{2} & \frac{1}{4} & 2 \\ \frac{3}{5} & \frac{7}{10} & \frac{11}{20} \\ 3 & 2 & 1 \end{vmatrix}$

Use the theorems of this section to reduce each determinant to order 2, then evaluate.

11. $\begin{vmatrix} 1 & -1 & 4 \\ 3 & 2 & -5 \\ -2 & 0 & 7 \end{vmatrix}$

12. $\begin{vmatrix} 6 & -3 & 1 \\ -2 & 5 & 11 \\ 3 & 10 & 2 \end{vmatrix}$

13. $\begin{vmatrix} \frac{2}{3} & \frac{4}{5} & -\frac{7}{15} \\ -1 & 0 & 2 \\ 2 & -5 & 0 \end{vmatrix}$

14. $\begin{vmatrix} 10 & 24 & 6 \\ 9 & -12 & 3 \\ -2 & 48 & -1 \end{vmatrix}$

15. $\begin{vmatrix} 4 & \frac{3}{4} & -3 \\ -7 & \frac{5}{8} & 9 \\ -6 & -\frac{3}{2} & 15 \end{vmatrix}$

16. $\begin{vmatrix} 24 & -40 & 16 \\ 3 & 0 & 6 \\ -4 & -2 & 5 \end{vmatrix}$

17. $\begin{vmatrix} 1 & 0 & -2 & 4 \\ 3 & 5 & 1 & -1 \\ 0 & -4 & 1 & 2 \\ -3 & 6 & 1 & 0 \end{vmatrix}$

18. $\begin{vmatrix} -3 & 6 & 0 & 5 \\ 2 & 4 & -1 & 3 \\ 0 & 8 & 5 & 6 \\ 1 & -4 & 0 & 7 \end{vmatrix}$

SUPPLEMENT H

CRAMER'S RULE

Determinants can be used to solve $n \times n$ linear systems by a technique known as Cramer's rule*. Although we develop the rule for a 2×2 system, the method employed may be generalized to systems for $n > 2$.

Consider the 2×2 system

$$\begin{align} &① \quad a_1 x + b_1 y = c_1 \\ &② \quad a_2 x + b_2 y = c_2. \end{align} \qquad \langle 1 \rangle$$

Multiplying each member of ① by b_2 and each member of ② by $-b_1$, we obtain

$$a_1 b_2 x + b_1 b_2 y = c_1 b_2$$
$$-a_2 b_1 x - b_1 b_2 y = -c_2 b_1.$$

Adding corresponding members of both equations, we have

$$a_1 b_2 x - a_2 b_1 x = c_1 b_2 - c_2 b_1$$

$$x = \frac{c_1 b_2 - c_2 b_1}{a_1 b_2 - a_2 b_1}. \qquad \langle 2 \rangle$$

Starting again with system $\langle 1 \rangle$ and solving for y (see Exercise 17) it can be shown that

$$y = \frac{a_1 c_2 - a_2 c_1}{a_1 b_2 - a_2 b_1}. \qquad \langle 3 \rangle$$

By Definition 7.5, equations $\langle 2 \rangle$ and $\langle 3 \rangle$ can be written respectively as

$$x = \frac{\begin{vmatrix} c_1 & b_1 \\ c_2 & b_2 \end{vmatrix}}{\begin{vmatrix} a_1 & b_1 \\ a_2 & b_2 \end{vmatrix}} \quad \text{and} \quad y = \frac{\begin{vmatrix} a_1 & c_1 \\ a_2 & c_2 \end{vmatrix}}{\begin{vmatrix} a_1 & b_1 \\ a_2 & b_2 \end{vmatrix}}.$$

Let us use the symbols D, D_x and D_y, where

$$D = det \begin{bmatrix} a_1 & b_1 \\ a_2 & b_2 \end{bmatrix} = \begin{vmatrix} a_1 & b_1 \\ a_2 & b_2 \end{vmatrix};$$

$$D_x = \begin{vmatrix} c_1 & b_1 \\ c_2 & b_2 \end{vmatrix}; \qquad D_y = \begin{vmatrix} a_1 & c_1 \\ a_2 & c_2 \end{vmatrix}.$$

Then, if $D \neq 0$, system $\langle 1 \rangle$ can be solved by Cramer's rule as follows:

1. *Find D, the determinant of the coefficient matrix of the system.*
2. *Obtain D_x from D by replacing the coefficients of x (a_1 and a_2) by the constant terms (c_1 and c_2) of the system.*
3. *Obtain D_y from D by replacing the coefficients of y (b_1 and b_2) by the constant terms (c_1 and c_2) of the system.*

*Gabrial Cramer (1704–1752) of Geneva.

4. *Write the solution set:* $\left\{ \left(\dfrac{D_x}{D}, \dfrac{D_y}{D} \right) \right\}$.

EXAMPLE 1   Solve the following system by Cramer's rule.

$$3x - y = 2$$
$$x + 2y = 5$$

SOLUTION   The coefficient matrix of the system is $\begin{bmatrix} 3 & -1 \\ 1 & 2 \end{bmatrix}$. Then,

$$D = \begin{vmatrix} 3 & -1 \\ 1 & 2 \end{vmatrix} = 7; \qquad D_x = \begin{vmatrix} 2 & -1 \\ 5 & 2 \end{vmatrix} = 9; \qquad D_y = \begin{vmatrix} 3 & 2 \\ 1 & 5 \end{vmatrix} = 13.$$

Hence,

$$x = \frac{D_x}{D} = \frac{9}{7}, \qquad y = \frac{D_y}{D} = \frac{13}{7}$$

and the solution set of the system is $\left\{ \left(\dfrac{9}{7}, \dfrac{13}{7} \right) \right\}$.

Cramer's rule can be readily extended to higher order linear systems. For 3×3 systems, we need only introduce the symbol D_z, the meaning of which will be made clear in the next example.

EXAMPLE 2   Solve the following system by Cramer's rule.

$$3x - 2y - z = 1$$
$$2x + 3y - z = 4$$
$$x - y + 2z = 7$$

SOLUTION   Let D denote the determinant of the coefficient matrix of the system. Then (verify the details):

$$D = \begin{vmatrix} 3 & -2 & -1 \\ 2 & 3 & -1 \\ 1 & -1 & 2 \end{vmatrix} = 30; \qquad D_x = \begin{vmatrix} 1 & -2 & -1 \\ 4 & 3 & -1 \\ 7 & -1 & 2 \end{vmatrix} = 60;$$

$$D_y = \begin{vmatrix} 3 & 1 & -1 \\ 2 & 4 & -1 \\ 1 & 7 & 2 \end{vmatrix} = 30; \qquad D_z = \begin{vmatrix} 3 & -2 & 1 \\ 2 & 3 & 4 \\ 1 & -1 & 7 \end{vmatrix} = 90.$$

By Cramer's rule: $x = \dfrac{D_x}{D} = \dfrac{60}{30}; \ y = \dfrac{D_y}{D} = \dfrac{30}{30}; \ z = \dfrac{D_z}{D} = \dfrac{90}{30};$ and the solution set is $\{(2, 1, 3)\}$.

Note that Cramer's rule is not applicable if $D = 0$. In such cases matrix methods are effective.

Exercise Set H

Solve by Cramer's rule. If not applicable, so state.

1. $2x - y = 12$
   $3x - 6y = -6$

2. $4x + y = 3$
   $8x - 3y = -6$

3. $\frac{1}{2}x + \frac{1}{4}y = 4$

   $\frac{1}{3}x - \frac{1}{2}y = 0$

4. $\frac{2}{3}s + 3t = -3$

   $s + \frac{1}{3}t = 8$

5. $-3s + 4t = 6$
   $-12s + 16t = 24$

6. $12x - 4t = 2$
   $3s - t = 1$

7. $x + y + 2z = 0$
   $y + z = 0$
   $x - z = 4$

8. $x + y = 4$
   $y + z = 2$
   $x + z = 2$

9. $3r - 5s + t = 6$
   $4r + 3s + 3t = 0$
   $r + 8s + 2t = -3$

10. $r + s - 2t = 2$
    $3r - s + t = 5$
    $3r + 3s - 6t = -4$

11. $u + 2v - w = 2$
    $2u - 2v + w = -1$
    $6u - 4v + 3w = 5$

12. $2u + 3v - w = 4$
    $-u + v + 2w = -2$
    $3u - v + 2w = 1$

13. $x - y + 2z = 5$
    $2w + y - z = 3$
    $w + 5x - 3y = -2$
    $-4w + x - 4z = -25$

14. $w - x + y + z = 0$
    $-w + 2x - y + 2z = 3$
    $w + 5x + y - 3z = 1$
    $3w + x + 4y - z = -2$

View each system as linear in $\dfrac{1}{x}$ and $\dfrac{1}{y}$ and solve by Cramer's rule.

15. $\dfrac{3}{x} + \dfrac{5}{y} = -18$

    $\dfrac{4}{x} - \dfrac{7}{y} = 58$

16. $\dfrac{5}{x} - \dfrac{2}{y} = \dfrac{1}{3}$

    $\dfrac{8}{x} - \dfrac{4}{y} = \dfrac{1}{5}$

17. Solve system $\langle 1 \rangle$, page 250, for y.

SUPPLEMENT I

ELEMENTARY MATRICES

In this section we relate matrix multiplications to elementary row operations on matrices.

Given a matrix A, the identity matrix I and the elementary row operation "*multiply each entry in the second row by k, $k \neq 0$*," let us compare the following results. Applying the row operation directly to A:

$$A = \begin{bmatrix} a_{11} & a_{12} \\ a_{21} & a_{22} \end{bmatrix} \quad \rightarrow \quad \begin{bmatrix} a_{11} & a_{12} \\ ka_{21} & ka_{22} \end{bmatrix} = B. \qquad \langle 1 \rangle$$

Now, let us apply the same row operation to I,

$$I = \begin{bmatrix} 1 & 0 \\ 0 & 1 \end{bmatrix} \quad \rightarrow \quad \begin{bmatrix} 1 & 0 \\ 0 & k \end{bmatrix} = T, \qquad \langle 2 \rangle$$

and then left-multiply A by T:

$$TA = \begin{bmatrix} 1 & 0 \\ 0 & k \end{bmatrix} \begin{bmatrix} a_{11} & a_{12} \\ a_{21} & a_{22} \end{bmatrix} = \begin{bmatrix} a_{11} & a_{12} \\ ka_{21} & ka_{22} \end{bmatrix}. \qquad \langle 3 \rangle$$

Comparing $\langle 3 \rangle$ with $\langle 1 \rangle$, we see that left-multiplying by T transforms A into B, the same result as that obtained by the direct application of the given row operation to A. Any matrix (such as T) that left-multiplies a given matrix A to effect an elementary row operation on A is called an **elementary matrix**. As suggested by equations $\langle 2 \rangle$, elementary matrices of order n are obtained by applying a particular elementary row operation to the identity matrix $I_{n \times n}$.

EXAMPLE 1 Applying the elementary row operation "interchange first and second rows" to I, we have

$$I = \begin{bmatrix} 1 & 0 \\ 0 & 1 \end{bmatrix} \quad \rightarrow \quad \begin{bmatrix} 0 & 1 \\ 1 & 0 \end{bmatrix} = T.$$

Elementary matrix T has the property that a left-multiplication of A by T will interchange first and second rows:

$$TA = \begin{bmatrix} 0 & 1 \\ 1 & 0 \end{bmatrix} \begin{bmatrix} a_{11} & a_{12} \\ a_{21} & a_{22} \end{bmatrix} = \begin{bmatrix} a_{21} & a_{22} \\ a_{11} & a_{12} \end{bmatrix}.$$

As indicated by the above discussion, we see that if a given elementary row operation transforms matrix A into B, then there is an associated elementary matrix T such that $TA = B$. This association of elementary row operations with matrices enables us to use the methods of matrix algebra to analyze the effects of elementary row operations. For example, Theorem 7.3 (page 234) implies that every invertible matrix

$A_{n \times n}$ is row-equivalent to the identity matrix $I_{n \times n}$. It follows that there exists a finite sequence of associated elementary matrices $T_1, T_2, \ldots,$ T_m such that

$$T_m \cdots T_2 T_1 A = I. \tag{4}$$

On the basis of equation $\langle 4 \rangle$, we can now justify our method of inverting a nonsingular square matrix. Essentially, the method depends upon the rather surprising result that any sequence of elementary matrices that transforms $A_{n \times n}$ into $I_{n \times n}$ will also transform $I_{n \times n}$ into $A_{n \times n}^{-1}$, the inverse of A. To see how this comes about, consider the following argument. If A is nonsingular, then there exists a matrix A^{-1} such that

$$AA^{-1} = I. \tag{5}$$

Now, left-multiplying each member of $\langle 5 \rangle$ by the product $T_m \cdots T_2 T_1$ of matrices that converts A to I, we have

$$T_m \cdots T_2 T_1 A A^{-1} = T_m \cdots T_2 T_1 I$$

from which, by the associativity of matrix multiplication, it follows that

$$(T_m \cdots T_2 T_1 A) A^{-1} = T_m \cdots T_2 T_1 I.$$

Substituting from equation $\langle 4 \rangle$, we have

$$IA^{-1} = T_m \cdots T_2 T_1 I \quad \leftrightarrow \quad A^{-1} = T_m \cdots T_2 T_1 I,$$

the desired result.

Exercise Set I
Find an elementary matrix of order two associated with each of the following elementary row operations. Verify by a left-multiplication of $A_{2 \times 2}$.
1. Multiply each entry in the first row by $\frac{1}{3}$.
2. Add 2 times each entry in the first row to the corresponding entry in the second row.

Find an elementary matrix of order three associated with each of the following elementary row operations. Verify by a left-multiplication of $A_{3 \times 3}$.
3. Interchange rows two and three.
4. Multiply each entry of row one by $\frac{1}{2}$.
5. Add -3 times each entry in row one to the corresponding entry in row three.

If: $A = \begin{bmatrix} 3 & 4 \\ 1 & 2 \end{bmatrix}$; $T_1 = \begin{bmatrix} 0 & 1 \\ 1 & 0 \end{bmatrix}$; $T_2 = \begin{bmatrix} 1 & 0 \\ -3 & 1 \end{bmatrix}$; $T_3 = \begin{bmatrix} 1 & 0 \\ 0 & -\frac{1}{2} \end{bmatrix}$;

$$T_4 = \begin{bmatrix} 1 & -2 \\ 0 & 1 \end{bmatrix};$$

6. Show that $T_4 T_3 T_2 T_1 A = I$.

7. Show that $T_4 T_3 T_2 T_1 I = A^{-1}$.

8. Given $B = \begin{bmatrix} 1 & 5 \\ -2 & -6 \end{bmatrix}$, find B^{-1} by the method of Chapter 7.

9. Find a sequence of elementary matrices $T_m, \cdots, T_2, T_1$ corresponding to the sequence of transformations in Exercise 8.

10. For the sequence of elementary matrices obtained in Exercise 9, show that $T_m \cdots T_2 T_1 I = B^{-1}$, where B^{-1} is the inverse matrix of Exercise 8.

11. Given $C = \begin{bmatrix} a & b \\ c & d \end{bmatrix}$, find C^{-1} by the method of Chapter 7. Specify a condition under which C^{-1} fails to exist.

12. Find a sequence of elementary matrices corresponding to the sequence of transformations of Exercise 11.

If TA denotes a left-multiplication of A by T, then AT denotes a right-multiplication.

13. Write three different elementary matrices of order two. Investigate and describe the effect of right-multiplying matrix $A_{2 \times 2}$ by each of your elementary matrices.

14. Same as Exercise 13 for order three, and matrix $A_{3 \times 3}$.

EXPONENTIAL AND LOGARITHMIC FUNCTIONS

8.1 EXPONENTIAL FUNCTIONS

If b is a positive number and x is a rational number, then b^x is a unique real number. Hence, for all *rational* numbers x and for each $b > 0$, the equation

$$y = b^x \qquad \langle 1 \rangle$$

defines a function. In this section we introduce functions defined by equations such as $\langle 1 \rangle$ where the replacement set for x is the set R of real numbers rather than the set Q of rational numbers.

Increasing and Decreasing Functions   Let $y = f(x)$ define a function and let x_1 and x_2 be *any* two numbers in the domain of f. If function values "increase" as x "increases," that is, if $f(x_1) < f(x_2)$ whenever $x_1 < x_2$, then f is called an **increasing function**. If function values "decrease" as x "increases," that is, if $f(x_1) > f(x_2)$ whenever $x_1 < x_2$, then f is called a **decreasing function**.

EXAMPLE 1   Figures 8.1–1(a) and (b) show the respective graphs of

$$y = 2^x \qquad \text{and} \qquad y = \left(\tfrac{1}{2}\right)^x,$$

for $\{x: -3 \le x \le 3,\ x \in J\}$. By inspection, we see that if $x_1 < x_2$, then

$$2^{x_1} < 2^{x_2} \qquad \text{and} \qquad \left(\tfrac{1}{2}\right)^{x_1} > \left(\tfrac{1}{2}\right)^{x_2}.$$

Hence, over the specified domain: $y = 2^x$ defines an increasing function; $y = \left(\tfrac{1}{2}\right)^x$ defines a decreasing function.

The results of Example 1 are particular cases of the following theorem.

Theorem 8.1   For $x \in Q$:

   I   If $b > 1$, then $y = b^x$ defines an increasing function;

  II   If $0 < b < 1$, then $y = b^x$ defines a decreasing function.

Theorem 8.1 omits the case that $b = 1$ because, if $b = 1$, then $y = b^x$ becomes simply $y = 1$, a constant function.

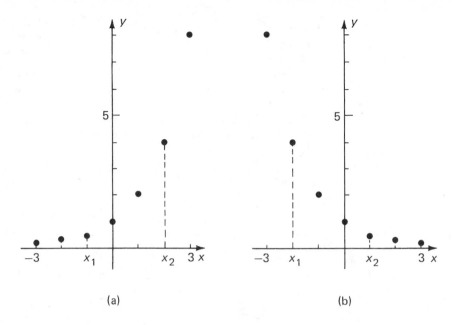

Figure 8.1-1

Exponential Functions   A graph of $y = 2^x$, $x \in Q$, is shown in Figure 8.1–2(a). Because the domain of the function is the set of *rational* numbers, there are no points in the graph corresponding to any *irrational* x-coordinates. Hence, the graph appears as a set of disconnected points

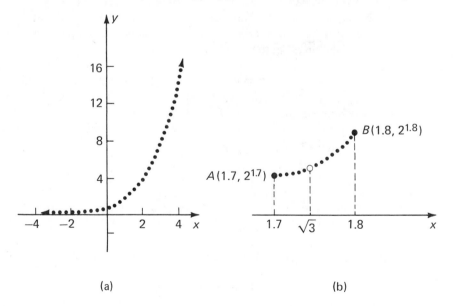

Figure 8.1-2

rather than a continuous curve. For instance, as indicated in Figure 8.1–2(b), the graph includes points $A(1.7, 2^{1.7})$ and $B(1.8, 2^{1.8})$, but it does not include a point with the irrational x-coordinate $\sqrt{3}$ even though $1.7 < \sqrt{3} < 1.8$. Thus, at $x = \sqrt{3}$, we have a "hole" in the graph. The holes corresponding to all irrational replacements for x can be "filled" if it can be shown that, for $b > 0$, b^x names a unique real number for each irrational replacement for x, but such a proof requires mathematics not available at this level. However, we can consider a plausibility argument, based upon Theorem 8.1–I, to the effect that the symbol $2^{\sqrt{3}}$ names a unique real number. First, note that rational approximations to $\sqrt{3}$ can be found to any desired degree of accuracy. For example,

$$1.7 < 1.73 < 1.732 < \cdots < \sqrt{3} < \cdots < 1.733 < 1.74 < 1.8. \quad \langle 2 \rangle$$

Second, by Theorem 8.1–I it follows that

IF	THEN
$1.7 < x < 1.8$	$2^{1.7} < 2^x < 2^{1.8}$
$1.73 < x < 1.74$	$2^{1.73} < 2^x < 2^{1.74}$
$1.732 < x < 1.733$	$2^{1.732} < 2^x < 2^{1.733}$
$\vdots$	$\vdots$

Furthermore (again by Theorem 8.1–I),

$$2^{1.7} < 2^{1.73} < 2^{1.732} < \cdots < 2^x < \cdots < 2^{1.733} < 2^{1.74} < 2^{1.8}. \quad \langle 3 \rangle$$

A comparison of inequalities $\langle 2 \rangle$ and $\langle 3 \rangle$ suggests that the exponent x in 2^x should be replaced by $\sqrt{3}$ to obtain $2^{\sqrt{3}}$. Last, because of the one-to-one correspondence between the real numbers and the points in a number line, we may conclude that $2^{\sqrt{3}}$ names a unique real number. Hence, the "hole" in the graph of Figure 8.1–2 can be filled by the point with coordinates $\left(\sqrt{3}, 2^{\sqrt{3}}\right)$. More generally, it can be shown that, if $b > 0$ and x is any irrational number ($x \in H$), then b^x names a unique real number. Thus, 2^x names a unique real number for any $x \in H$. It follows that the holes in the graph of Figure 8.1–2(a) can be filled with the set of points with coordinates $(x, 2^x)$, where $x \in H$, to obtain the graph of the function defined by

$$y = 2^x, \quad x \in R$$

as in Figure 8.1–3. This function is an example of an *exponential function*. In general, for each $b > 0$, $b \neq 1$, the function defined by

$$y = b^x, \quad x \in R,$$

[or by $f(x) = b^x$], is called the **exponential function to the base** b.

A consequence of the preceding discussion is that Theorem 0.22 (Properties of Rational Exponents, page 25) and Theorem 8.1 are both valid for *all real number exponents*. Henceforth, if Theorems 0.22 or 8.1 are cited, it is to be understood that R is the replacement set for x.

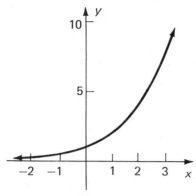

Figure 8.1-3

Properties of Exponential Functions   The **domain** of an exponential function is the set R of real numbers; the **range** is the set R^+ of positive numbers. By Theorem 8.1, an exponential function to the base b is an increasing function when $b > 1$ and a decreasing function when $0 < b < 1$. Each exponential function is continuous over the domain of real numbers; its graph is **asymptotic** to the x-axis.

EXAMPLE 2   Graph $y = 5^x$; $y = \left(\frac{1}{5}\right)^x$.

SOLUTION   First we compute the coordinates of a few of the points in the graph and plot them as in Figure 8.1–4. Next, we sketch a continuous curve through these points in such a manner as to indicate that the curve approaches the x-axis as an asymptote. (Note the use of different scales on the axes.)

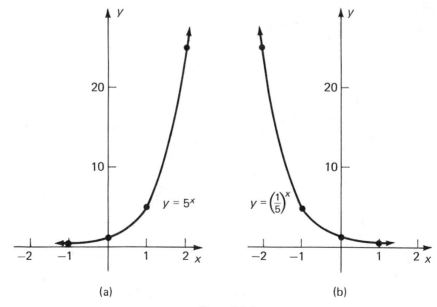

Figure 8.1-4

Base e   Mathematical models for the study of physical phenomena frequently lead to exponential functions. Often, the study of situations concerning either decay or growth (see Section 8.5) involves some form of the **exponential function to the base** e, defined by

$$y = e^x,$$

where e names a certain irrational number between 2 and 3 ($e \approx 2.71828$). Table 3, page 345 can be used to obtain coordinates of points in the graph of $y = e^x$, as shown in Figure 8.1–5.

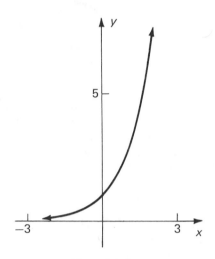

Figure 8.1-5

Exercise Set 8.1

(A) *Specify whether each graph is the graph of an increasing function; a decreasing function; neither.*

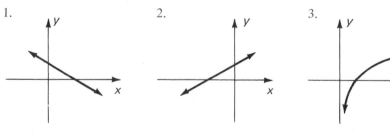

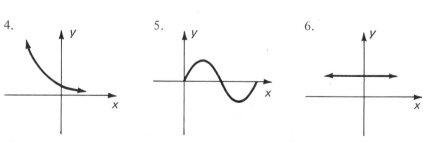

Graph each function. (See Example 2.)

7. $y = 3^x$ 8. $y = \left(\frac{1}{3}\right)^x$ 9. $y = \left(\frac{1}{4}\right)^x$

10. $y = 4^x$ 11. $y = 10^x$ 12. $y = \left(\frac{1}{10}\right)^x$

Graph both functions on the same coordinate axes.

13. $y = 2^x$; $y = \left(\frac{1}{2}\right)^x$ 14. $y = 6^x$; $y = \left(\frac{1}{6}\right)^x$

Graph each function. (Use Table 00.)

15. $y = 2e^x$ 16. $y = e^{-x}$ 17. $y = 2e^{-x}$

18. $y = e^{(1/2)x}$ 19. $y = e^{(1/3)x}$ 20. $y = e^{(-1/2)x}$

21. Graph $y = 3^{-x}$ and $y = -3^x$ on the same coordinate axes.

22. Graph $y = |3^x|$ and $y = 3^{|x|}$ on the same coordinate axes.

(B) 23. Prove: If $b > 1$, then $b^x > 0$ for all $x \in R$. [Hint: Consider the three cases $x = 0$, $x > 0$, $x < 0$. For $x > 0$ (Case 2), apply Theorem 8.1–I to show that $b^x > b^0$. For $x < 0$ (Case 3), let $x = -y$, where $y > 0$. Consider b^{-y} and use the result of Case 2.]

24. Prove: If $0 < b < 1$, then $b^x > 0$ for all $x \in R$.

For Exercises 25 through 32, you may wish to consider Supplement D. Use the graph of $y = 2^x$ to construct the graph of each of the following equations.

25. $y = -2^x$ 26. $y = 2(2^x)$ 27. $y - 1 = 2^x$

28. $y = 2^{x-3}$ 29. $y - 1 = 2^{x-3}$ 30. $y = \dfrac{1}{2^{x-3}}$

Use the addition of ordinates method to graph each equation.

31. $y = 2^x + x$ 32. $y = 2^x - x$

8.2   LOGARITHMIC FUNCTIONS

In Figure 8.2–1 on page 262, observe that a horizontal line test (see page 139) indicates that each exponential function is one-to-one, from which it follows that each exponential function has an inverse function.

Inverse of an Exponential Function   Let g be the inverse function of the exponential function defined by $y = b^x$, where $b > 0$, $b \neq 1$. Interchanging the variables x and y, we have

$$g = \{(x, y): x = b^y, \ b > 0, \ b \neq 1\}.$$

Now, following the procedure for constructing inverses (see Section 3.5), when we try to obtain a defining equation for g in the form $y = g(x)$, we discover that there are no algebraic methods for solving the equation

$$x = b^y \qquad\qquad \langle 1 \rangle$$

explicitly for y in terms of x. Hence, we introduce the symbol $\log_b x$ (read

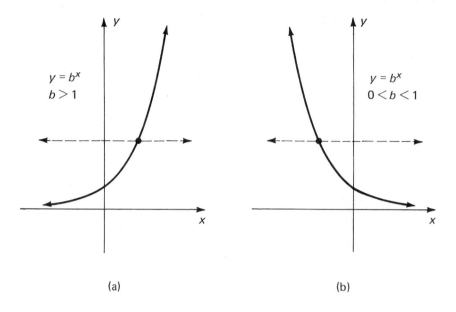

Figure 8.2-1

"logarithm to the base b of x" or "log of x to the base b") to name the unique real number that g associates with each real number in the domain of g, and we write

$$y = log_b x. \qquad \langle 2 \rangle$$

For each $b > 0$, $b \neq 1$, the function defined by equation $\langle 2 \rangle$ is called the **logarithmic function to the base** b. Equations $\langle 1 \rangle$ and $\langle 2 \rangle$ are equivalent equations because any ordered pair (x, y) that satisfies one equation also satisfies the other. That is, they define the same function, just as $x = y - 3$ and $y = x + 3$ define the same function. Hence, we have that

$$y = log_b x \qquad \text{if and only if} \qquad x = b^y. \qquad \langle 3 \rangle$$

EXAMPLE 1 Express each exponential equation in logarithmic form.

a. $6^3 = 216$ b. $10^0 = 1$

SOLUTION In each case we identify x, b and y, and use the equivalent equations:

$$x = b^y \ \leftrightarrow \ y = log_b x.$$

a. $x = 216$, $b = 6$ and $y = 3$. Hence,

$$6^3 = 216 \ \leftrightarrow \ 3 = log_6 216.$$

b. $x = 1$, $b = 10$ and $y = 0$. Hence,

$$10^0 = 1 \ \leftrightarrow \ 0 = log_{10} 1.$$

EXAMPLE 2 Evaluate each logarithm by changing it to exponential form.

a. $log_3 81$ b. $log_{1/10} 100$

SOLUTION   In each case we use: $y = \log_b x \leftrightarrow x = b^y$.

a. $y = \log_3 81 \leftrightarrow 81 = 3^y$. But we know that $81 = 3^4$. Hence, $y = 4$ and $\log_3 81 = 4$.

b. $y = \log_{1/10} 100 \leftrightarrow 100 = \left(\frac{1}{10}\right)^y$. But we know that $100 = \left(\frac{1}{10}\right)^{-2}$. Hence, $y = -2$ and $\log_{1/10} 100 = -2$.

EXAMPLE 3   Solve each equation for x or b.

a. $\log_4 x = 2$ b. $\log_b 36 = 2$.

SOLUTION   In each case we use: $y = \log_b x \leftrightarrow x = b^y$.

a. $\log_4 x = 2 \leftrightarrow x = 4^2$, or $x = 16$. Hence, the solution set is $\{16\}$.

b. $\log_b 36 = 2 \leftrightarrow 36 = b^2$, or $b = \pm 6$. But the base b must be *positive*. Hence, we conclude that the solution set is $\{6\}$.

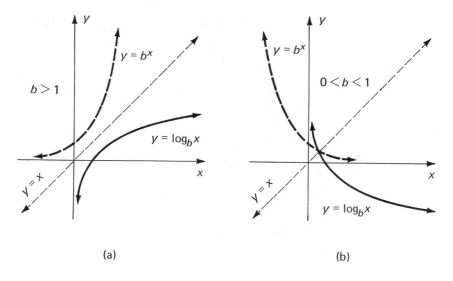

(a)   (b)

Figure 8.2-2

Graphs of Logarithmic Functions   Recall that the graphs of a function and its inverse are reflections of each other about the line $y = x$ (see page 138). We may use this concept, together with the graph of $y = b^x$, to sketch graphs of logarithmic functions, as shown in Figure 8.2-2. Alternatively, points in the graph of $y = \log_b x$ may be found by substituting values for y in the equivalent equation $x = b^y$, and then computing associated values of x.

Properties of Logarithmic Functions   Certain properties of logarithmic functions are apparent from the graphs in Figure 8.2-2.

1. A logarithmic function is an increasing function if $b > 1$.
2. A logarithmic function is a decreasing function if $0 < b < 1$.

3. The *domain* of a logarithmic function is R^+; the *range* is R.

4. The graph is asymptotic to the y-axis.

Note, from property 3, that $log_b x$ is defined only for *positive* values of x. Henceforth, in the text or exercises in this book, for any expression involving the logarithm to the base b of one or more variables, it is to be understood that:

1. $b > 0$ *and* $b \neq 1$:

2. *Replacement sets for variables are such that the expression represents the logarithm of a positive number.*

The inverse relationship between logarithmic and exponential functions leads to several further properties of logarithmic functions. For example, we can show that

$$log_b x_1 x_2 = log_b x_1 + log_b x_2$$

as follows. Because the range of a logarithmic function is R, we may let

$$m = log_b x_1 \quad \text{and} \quad n = log_b x_2,$$

where $m, n \in R$. From $y = log_b x \leftrightarrow x = b^y$ (equations $<3>$), it follows that

$$x_1 = b^m \quad \text{and} \quad x_2 = b^n,$$

from which

$$x_1 x_2 = b^m b^n = b^{m+n}.$$

Applying equations $<3>$ to $x_1 x_2 = b^{m+n}$, we obtain

$$log_b x_1 x_2 = m + n = log_b x_1 + log_b x_2,$$

and we have proved Part I of the following theorem.

Theorem 8.2   If $b > 0$, $b \neq 1$, p is a real number and $x, x_1, x_2 \in R^+$, then

I $log_b x_1 x_2 = log_b x_1 + log_b x_2$;

II $log_b \dfrac{x_1}{x_2} = log_b x_1 - log_b x_2$;

III $log_b x^p = p \, log_b x.$ **(P)**

EXAMPLE 4

a. $log_2(4 \cdot 2) = log_2 4 + log_2 2 = 2 + 1 = 3.$

b. $log_5 \dfrac{625}{5} = log_5 625 - log_5 5 = 4 - 1 = 3.$

c. $log_{10} 10^2 = 2 log_{10} 10 = 2 \cdot 1 = 2.$

EXAMPLE 5 Write $log_b\dfrac{\sqrt{x}\,y^3}{z_2}$ as an expression involving only logarithms of x, y and z.

SOLUTION
$$log_b\frac{\sqrt{x}\,y^3}{z^2}=log_b\sqrt{x}\,y^3-log_b z^2 \qquad \text{(Theorem 8.2-II)}$$
$$=log_b\sqrt{x}+log_b y^3-log_b z^2 \quad \text{(Theorem 8.2-I)}$$
$$=\tfrac{1}{2}log_b x+3log_b y-2log_b z \quad \text{(Theorem 8.2-III)}$$

EXAMPLE 6 Write $\tfrac{1}{2}(log_b x+log_b y+3\,log_b z)$ as a single logarithm with coefficient 1.

SOLUTION
$$\tfrac{1}{2}(log_b x+log_b y+3log_b z)=\tfrac{1}{2}(log_b x+log_b y+log_b z^3) \quad \text{(Theorem 8.2-III)}$$
$$=\tfrac{1}{2}(log_b xyz^3) \qquad\qquad \text{(Theorem 8.2-I)}$$
$$=log_b\sqrt{xyz^3} \qquad\qquad \text{(Theorem 8.2-III)}$$

Supplement J, page 280, provides a review of computations using logarithms to the base 10 together with a table of Common Logarithms.

Exercise Set 8.2

(A) *Express each equation in logarithmic form. (See Example 1.)*

1. $2^5=32$
2. $10^3=1000$
3. $32^{1/5}=2$
4. $1000^{1/3}=10$
5. $16^{-1/2}=\frac{1}{4}$
6. $27^{-1/3}=\frac{1}{3}$

Evaluate each logarithm by changing it to exponential form. (See Example 2.)

7. $log_2 16$
8. $log_5 125$
9. $log_{16}2$
10. $log_8 2$
11. $log_{10}10,000$
12. $log_{10}\frac{1}{100}$
13. $log_{1/10}\frac{1}{100}$
14. $log_{1/10}\frac{1}{1000}$
15. $log_7 7$
16. $log_{1/3}\frac{1}{3}$
17. $log_3\sqrt[3]{3}$
18. $log_4\sqrt[3]{4}$

Solve for x or b. (See Example 3.)

19. $log_2 x=5$
20. $log_3 x=5$
21. $log_b 49=2$
22. $log_b 32=5$
23. $log_b 1000=3$
24. $log_{1/10}x=-2$

Graph each pair of inverse functions, together with the line $y=x$, on the same set of coordinate axes.

25. $y=2^x$; $y=log_2 x$
26. $y=3^x$; $y=log_3 x$
27. $y=10^x$; $y=log_{10}x$
28. $y=(\frac{1}{3})^x$; $y=log_{1/3}x$
29.* $y=e^x$; $y=log_e x$
30. $y=(\frac{1}{2})^x$; $y=log_{1/2}x$

Graph each equation $y=log_b x$ by using the equivalent equation $x=b^y$ to compute coordinates of points of the graph.

31. $y=log_{10}x$
32. $y=log_3 x$
33. $y=log_e x$
34. $y=log_2 x$
35. $y=log_{1/2}x$
36. $y=log_{1/3}x$

Evaluate each logarithm. (See Example 4.)

37. $log_3(81\cdot 243)$
38. $log_4(64\cdot 16)$
39. $log_5(625\div 3125)$

*See Table 3, page 345.

40. $log_7(2401 \div 49)$ 41. $log_{10}(1000)^{2/5}$ 42. $log_8 8^{25}$

Evaluate each expression. (Hint: Use Theorem 8.2.)
43. $log_5 75 - log_5 3$ 44. $log_3 108 - log_3 4$
45. $log_4 32 + log_4 2$ 46. $log_6 4 + log_6 54$

Write each logarithm as an expression involving only the logarithms of the individual variables. (See Example 5.)

47. $log_b(xyz)$ 48. $log_b \dfrac{xy}{z}$

49. $log_b \dfrac{x^2}{y^{1/3}z^{1/2}}$ 50. $log_b \sqrt{(x-y)(y-z)}$

51. $log_b \sqrt[3]{\dfrac{(x+y)}{(x-y)^2}}$ 52. $log_b \left(\dfrac{xy^{2/3}}{z^2}\right)^{3/2}$

Write each expression as a single logarithm with coefficient 1. (See Example 6.)
53. $log_b x + log_b y + log_b z$
54. $log_b x - 2\ log_b y$
55. $2\ log_b \sqrt[3]{x} - log_b y^3 - 5\ log_b z$
56. $\frac{1}{3}(log_b x - 2\ log_b y + 4\ log_b z)$
57. $-3[log_b(x-y) + 2\ log_b(y-z) - log_b(z-x)]$
58. $-\frac{1}{2}[3\ log_b(x+y) - 4\ log_b(y+z)]$

(B)   59. Show that $log_4 x = \frac{1}{2}log_2 x$
       60. Show that $log_9 x = \frac{1}{2}log_3 x$

Graph each equation. (See also Supplement D.)
61. $y = |log_2 x|$ 62. $y = log_2 x^2$ 63. $y - 3 = log_2(x+3)$
64. Find the error in this proof that $2 < 1$.

$$\tfrac{1}{9} < \tfrac{1}{3} \leftrightarrow \left(\tfrac{1}{3}\right)^2 < \tfrac{1}{3}$$
$$2\ log_b \tfrac{1}{3} < log_b \tfrac{1}{3}$$
$$2 < 1.$$

(P)   I   Prove Theorem 8.2−II. (Hint: See the proof of Theorem 8.2−I, page 264.)

     II   Prove Theorem 8.2−III. (Hint: Let $log_b x^p = M$ and $log_b x = N$. Then $x^p = b^M$ and $x = b^N$. From the latter equation we have that $x^p = (b^N)^p$ from which $b^M = b^{pN}$. Continue with $log_b b^M = log_b b^{pN}$ and complete the proof.)

8.3   NATURAL LOGARITHMS

The number e (introduced in Section 8.1) is of great importance in the theory of mathematics as well as in many practical applications. Logarithms to the base e are called **natural logarithms** and are usually denoted by the symbol *ln x*, read "the natural logarithm of *x*." Table 2, a table of natural logarithms, appears on page 344.

EXAMPLE 1   From Table 2:

$$\textit{ln}\ 1.9 = 0.6419; \qquad \textit{ln}\ 19 = 2.9444; \qquad \textit{ln}\ 190 = 5.2470.$$

For entries in the table corresponding to numbers less than 1, it is customary to omit printing the term -10. Thus,

$$\textit{ln}\ 0.1 = 7.6794 - \mathbf{10}, \qquad \text{or} \qquad -2.3206.$$

Note that Table 2 does not conform to the characteristic-mantissa technique for obtaining logarithms to the base 10. Rather, each entry (except for $x < 1$) is complete as is. To obtain the natural logarithm of a number not listed in the table, we consider two alternatives. The first alternative involves scientific notation, as illustrated next.

EXAMPLE 2   To find $\textit{ln}\ 380$:

$$\textit{ln}\ 380 = \textit{ln}(3.8 \times 10^2) = \textit{ln}\ 3.8 + 2\ \textit{ln}\ 10$$
$$\textit{ln}\ 380 = 1.3350 + 4.6052$$
$$\textit{ln}\ 380 = 5.9402.$$

The second alternative, based upon the following theorem, enables us to compute natural logarithms in terms of logarithms to the base 10.

Theorem 8.3   If M and N are positive numbers and $b \neq 1$, then

$$M = N \qquad \text{if and only if} \qquad log_b M = log_b N. \qquad \textbf{(P)}$$

Now, let $y = \textit{ln}\ x$ and consider the following. First,

$$y = \textit{ln}\ x \ \leftrightarrow\ x = e^y.$$

Second, we apply Theorem 8.3 to $x = e^y$, taking the logarithm *to the base 10* of each member:

$$log_{10} x = y\ log_{10} e. \qquad \langle 1 \rangle$$

Solving $\langle 1 \rangle$ for y (which we replace by $\textit{ln}\ x$), we have

$$\textit{ln}\ x = \frac{log_{10} x}{log_{10} e}. \qquad \langle 2 \rangle$$

For convenient reference, $log_{10} e = \mathbf{0.4343}$.

EXAMPLE 3   By equation $\langle 2 \rangle$:

$$\textit{ln}\ 486 = \frac{log_{10} 486}{log_{10} e} = \frac{2.6866}{0.4343} = 6.1860.$$

The procedure used to derive equation $\langle 2 \rangle$ can be generalized to derive a **change of base formula** that enables us to compute logarithms in one base, say a, in terms of logarithms to a second base, say b, as follows. Let $y = log_a x$. Then,

$$y = log_a x \ \leftrightarrow\ x = a^y.$$

Applying Theorem 8.3 and taking logarithms *to the base b*, we have

$$log_b x = y \, log_b a.$$

Solving for y (which we replace by $log_a x$), we obtain

$$log_a x = \frac{log_b x}{log_b a}. \tag{3}$$

EXAMPLE 4   Find $log_{25} 64$ in terms of logarithms to the base 10.

SOLUTION   By equation $\langle 3 \rangle$,

$$log_{25} 64 = \frac{log_{10} 64}{log_{10} 25} = \frac{1.8062}{1.3979} = 1.2921.$$

Given the logarithm of a number, for a particular problem we may need to find the number. That is, given N such that

$$N = log_b x, \tag{4}$$

we may need to find x. Equation $\langle 4 \rangle$ is sometimes given in the equivalent form

$$x = antilog_b N.$$

Table 2 can be used to find antilogarithms to the base e.

EXAMPLE 5   From Table 2:

a. $antilog_e 1.2528 = 3.5; \quad antilog_e 3.5553 = 35.$

b. $antilog_e(-0.9163) = antilog_e(-0.9163 + 10 - 10)$
$\qquad\qquad\qquad\quad = antilog_e(9.0837 - 10),$
$\qquad\qquad\qquad\quad = 0.4.$

Exercise Set 8.3

(A)   *Use Table 2 to find each logarithm. (See Example 1.)*

1. *ln* 6.3   2. *ln* 8.4   3. *ln* 0.8   4. *ln* 0.6
5. *ln* 35   6. *ln* 45   7. *ln* 130   8. *ln* 180

Find each logarithm by the method of Example 2.

9. *ln* 550   10. *ln* 850   11. *ln* 83
12. *ln* 68   13. *ln* 0.048   14. *ln* 0.0034

Find each logarithm by the method of Example 3.

15. *ln* 74   16. *ln* 28   17. *ln* 6.34
18. *ln* 8.51   19. *ln* 245   20. *ln* 398

Find each logarithm in terms of logarithms to the base 10. (See Example 4.)

21. $log_{25} 81$ 22. $log_{25} 8.1$ 23. $log_3 36$
24. $log_3 3600$ 25. $log_5 65$ 26. $log_{100} 212$

Use Table 2 to find each antilogarithm. (See Example 5.)

27. $antilog_e 4.0073$ 28. $antilog_e 0.3365$ 29. $antilog_e 2.1748$
30. $antilog_e 5.1358$ 31. $antilog_e(-1.2040)$ 32. $antilog_e(-1.6094)$

33. Apply the *method* used to derive the change of base formula $<3>$ to show that $log_b x = (log_{1/b} x)\left(log_b \frac{1}{b}\right)$.

34. Prove: $log_{b^2} x = \frac{1}{2} log_b x$.

35. Prove: $(log_a b)(log_b a) = 1$.

(P) I A proof of Theorem 8.3 calls for two parts:

1. If $M = N$, then $log_b M = log_b N$.

2. If $log_b M = log_b N$, then $M = N$.

Construct a proof of Part 1 by letting $M = b^x$, $N = b^x$ and solving each equation for x. Prove Part 2 by reversing the proof of Part 1.

8.4 LOGARITHMIC AND EXPONENTIAL EQUATIONS

In this section we consider some techniques for solving equations involving logarithmic or exponential terms. Throughout this section the notation $log\ x$, where no base is specified, shall be understood to mean $log_{10} x$. Also, recall from section 8.2 that

$$y = log_b x \leftrightarrow x = b^y. \qquad <1>$$

As you probably know, entries in logarithm and exponential tables (Tables 1, 2 and 3) are rounded-off approximations of irrational numbers. It follows that calculations based upon such tables lead only to approximate results, suggesting the use of the $\approx$ symbol. However, because our primary concern is with mathematical techniques rather than with any calculations that may be encountered, we shall use the $=$ symbol.

EXAMPLE 1 Specify the replacement set and solve the equation
$$log(80x + 40) = 3.$$

SOLUTION Because logarithmic functions are defined only for *positive* numbers, the replacement set for x is
$$\{x: 80x + 40 > 0\} = \{x: x > -\tfrac{1}{2}\}.$$
That is, only numbers greater than $-\frac{1}{2}$ can be solutions. To solve, we first use equations $<1>$ to write the given equation in equivalent exponential form, and then solve the resulting equation:
$$log(80x + 40) = 3$$
$$80x + 40 = 10^3$$
$$80x = 960$$
$$x = 12.$$
Because $12 > -\frac{1}{2}$, we conclude that $\{12\}$ is the solution set. As a check: $log(80 \cdot 12 + 40) = log\ 1000 = 3$.

EXAMPLE 2 Specify the replacement set and solve:
$$log\ x + log(2x - 1) = 1. \qquad <2>$$

SOLUTION The replacement set for x is

$$\{x: x > 0 \text{ and } 2x - 1 > 0\} = \{x: x > \tfrac{1}{2}\}.$$

Applying Theorem 8.2-I to equation $\langle 2 \rangle$, we have

$$\begin{aligned}
log\ x(2x - 1) &= 1 \\
log(2x^2 - x) &= 1 \\
2x^2 - x &= 10 \\
2x^2 - x - 10 &= 0.
\end{aligned}$$

You can verify that $\tfrac{5}{2}$ and -2 are solutions of the quadratic equation. However, -2 is *not* in the replacement set for x, hence we discard it—the solution set is $\{\tfrac{5}{2}\}$. You will find it instructive to check that $\tfrac{5}{2}$ satisfies the original equation.

Equations involving terms in which the variable appears in an exponent can often be solved by applying Theorem 8.3—taking the logarithm of each member of the equation.

EXAMPLE 3 Solve $5^x = 8$.

SOLUTION

$$\begin{aligned}
5^x &= 8 \\
log\ 5^x &= log\ 8 && \text{(Theorem 8.3)} \\
x\ log\ 5 &= log\ 8 && \text{(Theorem 8.2-III)} \\
x &= \frac{log\ 8}{log\ 5}.
\end{aligned}$$

The solution set is $\left\{\dfrac{log\ 8}{log\ 5}\right\}$. To the nearest hundredth,

$$x = \frac{log\ 8}{log\ 5} = \frac{0.9031}{0.6990} = 1.29.$$

EXAMPLE 4 Solve $14^x = 6^{3x-1}$.

SOLUTION

$$\begin{aligned}
14^x &= 6^{3x-1} \\
log\ 14^x &= log\ 6^{3x-1} && \text{(Theorem 8.3)} \\
x\ log\ 14 &= (3x - 1)\ log\ 6 && \text{(Theorem 8.2-III)} \\
x\ log\ 14 &= 3x\ log\ 6 - log\ 6 \\
log\ 6 &= 3x\ log\ 6 - x\ log\ 14 \\
log\ 6 &= x(3\ log\ 6 - log\ 14) \\
x &= \frac{log\ 6}{3\ log\ 6 - log\ 14}
\end{aligned}$$

and the solution set is $\left\{\dfrac{log\ 6}{3\ log\ 6 - log\ 14}\right\}$. You may wish to verify that, to the nearest hundredth, $x = 0.65$.

The equations of Examples 3 and 4 can also be solved by using natural logarithms (base e), but no real advantage is gained. On the other hand, the use of natural logarithms is a distinct advantage when we can use the fact that $ln\ e = 1$. In the next example, the same equation is solved first using natural logarithms, and then using common logarithms (base 10). You may decide for yourself which approach you prefer.

EXAMPLE 5   Solve for t (to the nearest hundredth): $18 = 36e^{-0.5t}$.

SOLUTION
$$18 = 36e^{-0.5t}$$
$$\tfrac{1}{2} = e^{-0.5t}$$
$$2 = e^{0.5t}$$

$ln\ 2 = (0.5t)\ ln\ e$	$log\ 2 = (0.5t)\ log\ e$
$t = \dfrac{ln\ 2}{(0.5)\ ln\ e}$	$t = \dfrac{log\ 2}{(0.5)\ log\ e}$
$t = \dfrac{0.6931}{(0.5)(1)}$	$t = \dfrac{0.3010}{(0.5)(0.4343)}$
$t = 1.39$	$t = 1.39$

Exercise Set 8.4

(A)   *Solve each equation. (See Examples 1 and 2.)*

1. $log(x+3) = 2$
2. $log(5x-40) = 3$
3. $log(2x+3) = 1$
4. $log(x-1) = 4$
5. $log(x+2) + log(x-1) = 1$
6. $log\ x + log(4x+8) = 2$
7. $log(2x-5) - log(x+3) = 3$
8. $log\ x^2 - log(x-1) = 1$
9. $log(x-1) + log(x+1) = 1 + log\ x$
10. $log(2x^2 - 10) - log(x-5) = 1 + log(x+1)$

Solve each equation. (See Examples 3 and 4.)

11. $5^{2+x} = 28$
12. $26^{x-5} = 145$
13. $2^{x^2} = 9$
14. $4^{x^2-1} = 9$
15. $25^{-x} = 0.52$
16. $(.2)^{-x} = 8$
17. $9^x = 7^{x+2}$
18. $20^{2x+1} = 30^{3x-2}$
19. $100^{x+2} = 10^{x+1}$
20. $1000^{x^2+2x} = 100^{x^2-2}$

Solve for t to the nearest hundredth. (See Example 5.)

21. $25 = 50e^{-0.012t}$
22. $75 = 300e^{-0.43t}$
23. $25 = 50e^{0.012t}$
24. $75 = 300e^{0.43t}$

Solve for t, P, A or r.

25. $460 = 175(1.07)^t$
26. $3200 = 810(1.06)^t$
27. $8000 = P(1.08)^{20}$
28. $4240 = P(1.065)^9$
29. $A = 6500(1.045)^8$
30. $A = 1500(1.05)^{12}$
31. $A = 6500(.9)^8$
32. $A = 1500(.05)^{12}$
33. $2 = \left(1 + \dfrac{r}{4}\right)^{40}$
34. $3 = \left(1 + \dfrac{r}{2}\right)^{20}$

(B)   *Solve each equation.*

35. $log\ x^2 = (log\ x)^2$
36. $log(log\ x) = 1$
37. Solve for t: $A = P\left(1 + \dfrac{r}{n}\right)^{nt}$.

8.5 APPLICATIONS

In this section we consider some of the many applications of exponential and logarithmic functions. Because the accuracy of the calculations in the examples is limited to that obtainable from the four-place tables in this book, answers are given in rounded-off form. (For greater accuracy, more extensive tables are available.)

Interest   A sum of money that is borrowed or lent is usually referred to as the *principal*, *P*. Money that is paid or received for the use of the principal is called the *interest*, *I*. Interest is computed at a *rate*, *r*, most often specified as a percent per year. The sum of the principal and the interest is called the *amount*, *A*. If the interest is computed one or more times on the original principal, it is called **simple interest** and is given by the familiar formula $I = Prt$. If the interest is computed and then added to the principal to create a new principal on which future interest is to be computed, it is called **compound interest**.

EXAMPLE 1   Given $P = \$1900$, $r = 7\%$ and $t = 2$ years. The simple interest on P is

$$I = 1900(.07)(2) = 266;$$

the amount after two years is $\$1900 + \$266 = \$2166$. The compound interest on P, compounded annually, can be computed as follows. After one year:

$$I = 1900(.07)(1) = 133;$$

the amount is $\$1900 + \$133 = \$2033$. This amount is the new principal for the second year. After two years:

$$I = 2033(.07)(1) = 142.31;$$

the amount is $\$2033 + \$142.31 = \$2175.31$.

If the compound interest in Example 1 had to be computed over a period of many years, the computations would be repetitive and laborious. Instead, a formula for finding the amount A of a given principal P compounded annually at a yearly rate r is readily obtained. Consider (and verify the factoring process):

AFTER	AMOUNT
1 year	$A = P + Pr = P(1 + r)$
2 years	$A = P(1 + r) + P(1 + r)(r) = P(1 + r)^2$
3 years	$A = P(1 + r)^2 + P(1 + r)^2(r) = P(1 + r)^3$
$\vdots$	$\vdots$

The preceding pattern suggests that, after t years,

$$A = P(1 + r)^t. \qquad \qquad <1>$$

EXAMPLE 2   If $1900 is compounded annually for 4 years at 7% then, by equation $\langle 1 \rangle$ (to the nearest dollar),

$$A = 1900(1 + .07)^4 = 1900(1.07)^4 = \$2491.$$

In most cases, interest is compounded more frequently than just annually. If the interest is compounded n times per year for t years, then the interest is calculated nt times, the effective rate of interest each time is $\frac{r}{n}$, and the compound interest formula $\langle 1 \rangle$ takes the form

$$A = P\left(1 + \frac{r}{n}\right)^{nt}.$$ $\langle 2 \rangle$

EXAMPLE 3   The interest on a $1000 two-year time certificate is compounded quarterly at 6%. For the amount after two years, we have: $P = 1000$, $n = 4$, $r = .06$ and $t = 2$. From equation $\langle 2 \rangle$,

$$A = 1000\left(1 + \frac{.06}{4}\right)^{4 \cdot 2} = 1000(1.015)^8 = \$1127.$$

Continuous Compounding   If $1 is invested for 1 year at the unlikely rate of 100%, then equation $\langle 2 \rangle$ takes the form

$$A = \left(1 + \frac{1}{n}\right)^n.$$

Let us consider the effect upon the amount A of increasing n, the number of times per year that the interest is compounded. An examination of Table 8.1 clearly suggests that A increase as n increases.

Annually	$(n=1)$:	$A = \left(1 + \frac{1}{1}\right)^1$	$= 2.00$;
Semiannually	$(n=2)$:	$A = \left(1 + \frac{1}{2}\right)^2$	$= 2.25$;
Quarterly	$(n=4)$:	$A = \left(1 + \frac{1}{4}\right)^4$	$= 2.44$;
Monthly	$(n=12)$:	$A = \left(1 + \frac{1}{12}\right)^{12}$	$= 2.61$;
Daily	$(n=360)$:	$A = \left(1 + \frac{1}{360}\right)^{360}$	$= 2.72$.

TABLE 8.1

This might further suggest that if the interest is compounded more often— say every hour, or every minute, or even every second—then the amount will increase beyond our wildest imaginings. That this will not happen is indicated by a closer look at Table 8.1. Note that from $n=1$ to $n=2$, A increases by 25 cents, whereas from $n=12$ to $n=360$ (a much greater "jump"), A increases by only 11 cents. These figures imply that the *rate* of increase of A is "slowing down." The remarkable fact is that if the interest is compounded as often and as rapidly as the most advanced electronic computer can perform, then $1 will grow to about e dollars (about $2.72). Mathematicians may describe this situation by a statement such as "If n increases without bound, then the difference between $\left(1+\dfrac{1}{n}\right)^n$ and e can be made as small as we please." As a consequence, it can be shown that if a principal P is compounded "continuously" at a rate r per year for t years, then the amount A is given by

$$A = Pe^{rt}. \qquad\qquad \langle 3 \rangle$$

Computations based upon equation $\langle 3 \rangle$ can be performed rapidly by electronic computers. Furthermore, because equation $\langle 3 \rangle$ is virtually equivalent to interest compounded daily, many banks with computer facilities are now able to offer "daily interest" and "interest from day of deposit to day of withdrawal" on the savings of their customers.

EXAMPLE 4   Find the amount of $2000 at 6% compounded continuously for 2 years.

SOLUTION   From equation $\langle 3 \rangle$ and Table 3 on page 345, we have:

$$A = 2000e^{(.06)(2)} = 2000e^{0.12} = \$2255.$$

Hydrogen Potential   In the field of chemistry, the hydrogen potential of a solution, denoted by the symbol **pH**, is defined by

$$\text{pH} = log_{10}\frac{1}{[\text{H}^+]}, \qquad\qquad \langle 4 \rangle$$

where $[\text{H}^+]$ denotes a numerical value for the concentration of hydrogen ions in aqueous solution in moles per liter. The pH is usually given to the nearest tenth of a unit. For computational purposes, equation $\langle 4 \rangle$ can be adapted as follows:

$$\text{pH} = log_{10}\frac{1}{[\text{H}^+]} = log_{10}1 - log_{10}[\text{H}^+] = -log_{10}[\text{H}^+]. \qquad \langle 5 \rangle$$

EXAMPLE 5   Find the pH of a solution in which $[\text{H}^+]$, the hydrogen-ion concentration, is 5.0×10^{-4}.

SOLUTION   Substituting in $\text{pH} = -log[\text{H}^+]$:

$$\text{pH} = -log_{10}(5.0 \times 10^{-4}) = -(log_{10}5.0 - 4) = 3.3.$$

From the pH of a solution we can compute the hydrogen-ion concentration $[\text{H}^+]$.

EXAMPLE 6   Find $[\text{H}^+]$ of a solution with a pH of 5.8.

SOLUTION   From equation $\langle 4 \rangle$:

$$log_{10}\frac{1}{[\text{H}^+]} = 5.8000$$

$$\frac{1}{[\text{H}^+]} = 6.31 \times 10^5$$

$$[\text{H}^+] = 0.16 \times 10^{-5} \text{ or } 1.6 \times 10^{-6}.$$

Decay; Growth   Exponential functions play a particularly significant role in the study of processes involving decay or growth.

Radioactive materials are known to decay spontaneously at a rate that depends upon the amount M present at any given time t. From this property it can be deduced that

$$M = M_0 e^{-kt}, \qquad k > 0,$$

where M_0 is the initial amount (at time $t=0$) and k is a constant that depends upon the "half-life" of the particular material. The *half-life* is the time needed for an amount M_0 to decay to the amount $\frac{1}{2}M_0$. Any material that decays according to an equation such as $<4>$ is said to *decay exponentially*.

EXAMPLE 7   Strontium-90, a by-product of a hydrogen bomb explosion, decays exponentially with a half-life of 28 years. Find an expression for the amount M of strontium-90 at any time t if $M_0 = 10.0$ grams. Find M after 72 years.

SOLUTION   At $t=0$, $M=M_0=10.0$; at $t=28$, $M=\frac{1}{2}(10.0)=5.0$. Hence,

$$M = M_0 e^{-kt} \quad \leftrightarrow \quad 5.0 = 10.0 e^{-28k}$$
$$2 = e^{28k} \qquad\qquad <6>$$

From equation $<6>$, $k = 0.025$ to the nearest thousandth, and

$$M = 10.0 e^{-0.025t}.$$

At $t=72$ years,

$$M = 10.0 e^{-0.025(72)} = 10.0 e^{-1.8}.$$

From Table 3 page 341, $e^{-1.8} = 0.1653$ and, to the nearest tenth of a gram,

$$M = 10.0(0.1653) = 1.7 \text{ grams.}$$

EXAMPLE 8   The value of a certain machine depreciates (decreases in value) 10% each year. Find an expression for the value A of the machine at any time t in years, if the initial cost is A_0.

SOLUTION   If the machine depreciates by 10% each year, then the value of the machine at the end of each year is 90% (or .9) of the value during the preceding year. Hence, at the end of

1 year:	$A = A_0(.9)$
2 years:	$A = A_0(.9)(.9) = A_0(.9)^2$
3 years:	$A = A_0(.9)^2(.9) = A_0(.9)^3$, etc.

This pattern suggests that at the end of t years, $A = A_0(.9)^t$.

EXAMPLE 9   If $A_0 = \$2600$, find the value of the machine of Example 8 after 10 years.

(Continued on next page)

SOLUTION
$$A = 2600(.9)^{10}$$
$$\log A = \log 2600 + 10 \log(.9)$$
$$\log A = 3.4150 + 10(9.9542 - 10) = 3.4150 + 99.5420 - 100$$
$$\log A = 2.9570$$
$$A = \$906.$$

The next two examples illustrate *exponential growth*.

EXAMPLE 10   A biologist observes that the number of bacteria in a culture doubles each hour. Find an expression for the number N of bacteria in the culture at any time t in hours, if the initial number of bacteria is N_0.

SOLUTION   Consider an hour by hour basis. At the end of

1 hour: $N = N_0 \cdot 2$
2 hours: $N = N_0 \cdot 2 \cdot 2 = N_0 2^2$
3 hours: $N = N_0 2^2 \cdot 2 = N_0 2^3$, etc.

This pattern suggests that, at the end of t hours, $N = N_0 2^t$.

EXAMPLE 11   The market value of a certain building appreciates (increases in value) 2% each year. Find an expression for the market value A at any time t in years, if the initial value of the building is A_0.

SOLUTION   Consider a year by year basis. After

1 year: $A = A_0 + A_0(.02) = A_0(1.02)$
2 years: $A = A_0(1.02) + A_0(1.02)(.02) = A_0(1.02)^2$
3 years: $A = A_0(1.02)^2 + A_0(1.02)^2(.02) = A_0(1.02)^3$, etc.

This pattern suggests that, at the end of t years, $A = A_0(1.02)^t$.

Radiocarbon Dating   One of the methods of computing the age of ancient specimens of organic material involves carbon-14 and is known as *radiocarbon dating*, a method that appears to be accurate to as much as 20,000 years. Carbon-14 (C14) is a radioactive material that is generated in the air by the action of cosmic rays, and thereafter decays exponentially. In living organic matter, decaying C14 is replaced from the air. As a consequence, the concentration of C14 in such matter remains constant and is equal to the concentration of C14 in the atmosphere. When life terminates, the C14 decays with a half-life of about 5570 years and is no longer replaced. Hence, the concentration of C14 in a given specimen of wood, bone, charcoal, etc., can be used to compute the age of that specimen. By a procedure similar to that of Example 7, it can be shown that $k = 0.0001244$ for the decay of carbon-14. Thus, for radioactive carbon dating we have

$$M = M_0 e^{-0.0001244t}, \qquad \langle 7 \rangle$$

where M denotes the concentration of C14 in the specimen and M_0 denotes the concentration of C14 in air. It is convenient to specify M as a percent of M_0.

EXAMPLE 12   A wooden fence post dug out of an Indian village site is found to have a concentration of carbon-14 equal to 70% of the concentration of C14 in the air. How many years ago was the tree from which the post was made chopped down?

SOLUTION   Given that $M = .7M_0$, from equation $\langle 7 \rangle$ we have

$$.7M_0 = M_0 e^{-0.0001244t} \quad \leftrightarrow \quad .7 = e^{-0.0001244t}$$
$$\frac{10}{7} = e^{0.0001244t}. \qquad \langle 8 \rangle$$

Solving equation $\langle 8 \rangle$ for t, we find that, to the nearest whole year, the tree was chopped down 2867 years ago.

Exercise Set 8.5

(A)   *For Exercises 1 through 10, see equations $\langle 2 \rangle$ and $\langle 3 \rangle$.*

1. Find the amount of \$5000 invested at 8% for 5 years compounded: annually; semiannually; quarterly; monthly.

2. Show that if a principal P is compounded n times per year at a rate r, then the time needed for P to double $(A = 2P)$ is given by

$$t = \frac{\log 2}{n \log\left(1 + \dfrac{r}{n}\right)}.$$

3. In how many years and months will \$1000 double at 6% if the interest is compounded: annually; semiannually; quarterly; monthly? (See Exercise 2.)

4. To the nearest tenth of a percent, at what interest rate will \$1000 double in 10 years if the interest is compounded: annually; semiannually; quarterly; monthly; continuously?

Find the amount of each principal if interest is compounded continuously.

5. \$5000; $6\frac{1}{2}$%; 4 years
6. \$260,000; 7%; 3 years
7. \$100,000; $8\frac{1}{2}$%; 40 years
8. \$890; $6\frac{1}{2}$%; 20 years

9. If interest at 6% is compounded continuously, how much must be deposited today in order to have the amount \$6000 in 10 years?

10. At what interest rate, compounded continuously, will \$1000 increase to \$3000 in 15 years?

Compute (to the nearest tenth) the pH of a solution with each hydrogen-ion concentration. (See Example 5.)

11. $[H^+] = 4.0 \times 10^{-3}$
12. $[H^+] = 6.3 \times 10^{-5}$

Compute $[H^+]$ for each pH. (See Example 6.)

13. $pH = 3.9$
14. $pH = 6.6$

15. Actinium decays exponentially with a half-life of 22 years. Find an expression for the amount M of actinium at any time t in years if $M_0 = 8.9$ grams. Find M when $t = 100$ years.

16. Radium decays exponentially. If the amount M of radium at any time t in years is given by $M = M_0 e^{-0.000411t}$, find the half-life of radium to the nearest year.

17. A certain vehicle depreciates 15% each year. Find an expression for the value A of the vehicle at any time t in years if the initial value of the vehicle is $7000. Find the value of the vehicle at the end of 3 years.

18. Mammoth brand automobiles depreciate 20% in the first year and 10% each subsequent year. Find an expression for the value A of a $5000 Mammoth at any time t in years. Find the value of the car after 4 years.

19. The number of bacteria in a certain culture triples each hour. Find an expression for the number N of bacteria in the culture at any time t in hours if the initial number of bacteria is N_0. Find the number of bacteria at the end of 4 hours if $N_0 = 40,000$.

20. On a television quiz show, a contestant wins $1000 for the correct answer to the first question. For a correct answer to each successive question the contestant wins double the amount won on the previous question. Find an expression for the amount A to be won for answering the nth question correctly. How much is won by answering the 6th question correctly?

21. The market value of a certain building increases by 2% each year. Find an expression for the market value V of a $30,000 building after n years. Find the market value of the building after 20 years.

22. The maintenance costs of a certain automobile increase by 20% each year. Find the maintenance costs for the third year if $180 is spent the first year.

For Exercises 23 and 24, see equation $\langle 7 \rangle$.
23. A primitive wood carving is offered for sale to a museum, with the claim that it is over 1000 years old. A test sample of the wood shows a concentration of carbon-14 equal to 90% of the concentration of C14 in air. How old is the carving? Is the claim valid?

24. To the nearest tenth of a percent, the concentration of C14 in a 20,000 year old specimen is what percent of the concentration of C14 in air?

(C) I The equation $A = Pe^{RT}$ can be solved either by taking logarithms of both sides, or by using a table for e^x such as Table 3. BASIC programming language includes the built-in functions LOG, for finding natural logarithms, and EXP for finding powers of e. The following program computes $A = Pe^{RT}$ in two different ways and compares the results. Run the program with the test data given. Use the program to check your answers to Exercises 5–8.

```
10 READ P,R,T
20 LET A1=P*EXP(R*T)
30 LET A2=EXP(LOG(P)+R*T)
40 PRINT "A1          A2          PERCENT ERROR"
50 PRINT A1,A2,ABS(A1-A2)*100/A1
60 GO TO 10
70 DATA 1,1,1,1000,0.05,10,100,0.0775,15
80 END
```

Chapter 8   Self-Test

[**8.1**] *Graph each pair of equations on the same set of axes.*

1. $y = 3^x$; $y = (\frac{1}{3})^x$

2. $y = \frac{1}{2}e^x$; $y = \frac{1}{2}e^{-x}$

[**8.2**] *Write each equation in logarithmic form.*

3. $3^4 = 81$

4. $25^{-1/2} = \frac{1}{5}$

Evaluate.

5. $log_3 81$

6. $log_{81} 3$

7. $log_6(36 \cdot 216)$

8. $log_4(64)^5$

9. $log_2 96 - log_2 6$

10. $log_6 4^2 + log_6 9^2$

11. Graph $y = 2^x$ and $y = log_2 x$ on the same set of axes, together with the line $y = x$.

12. Write as an expression involving logarithms of the individual variables only:

$$log_b\left(\frac{x^{1/2} y^2}{z}\right)^{1/3}.$$

13. Write as a single logarithm with coefficient 1:

$$log_b \sqrt[3]{x} - log_b y^2 + 2 log_b z.$$

[**8.3**] *Find each logarithm or antilogarithm.*

14. *ln* 5.7

15. *ln* 0.7

16. *ln* 130

17. *ln* 0.64

18. *ln* 73

19. *ln* 345

20. *antilog*$_e$ 0.4055

21. *antilog*$_e$ 3.9120

22. *antilog*$_e$(−0.9163)

[**8.4**] *Solve each equation.*

23. $log_{10} 4x + log_{10}(3x - 10) = 2$

24. $8^{x+2} = 24^x$

25. Solve for t: $B = B_0 e^{kt}$.

[**8.5**] 26. Find the amount of $1000 at 6% for 4 years compounded: semi-annually; monthly; continuously.

27. A typewriter depreciates 10% each year. Find the value of a $185 typewriter after 10 years.

28. A certain painting appreciates 2% each year. Find the value of a painting after 200 years if it originally sold for $1500.

SUPPLEMENT J

COMPUTATIONS WITH BASE 10 LOGARITHMS

Most computations with logarithms are performed with logarithms to the base 10, called **common logarithms**. In this section the notation $log\ x$, where no base is specified, is to be understood to mean $log_{10}x$.

Finding $log_{10}x$ Any positive number x can be written in scientific notation form as

$$a \times 10^k,$$

where $1 \le a < 10$ and k is an integer. Now, by Theorem 8.2, we have

$$log(a \times 10^k) = log\ a + k\ log\ 10.$$

But $log_{10}10 = 1$, hence

$$log(a \times 10^k) = (log\ a) + k.$$

The number $log\ a$ is called the **mantissa**; it is to be found in Table 1 on pages 338 and 339. The integer k is called the **characteristic**; it is simply the exponent on the base 10 when the number x is written in scientific notation form.

EXAMPLE 1   In Table 1 we find that $log\ 8.53 = 0.9309$.* Hence:

a. $log\ 85,300 = log(8.53 \times 10^4) = (log\ 8.53) + 4$
$$= 0.9309 + 4$$
$$= 4.9309.$$

b. $log\ 85.3 = log(8.53 \times 10^1) = (log\ 8.53) + 1$
$$= 0.9309 + 1$$
$$= 1.9309.$$

In the case of a negative characteristic, it is customary to use a technique involving the addition of zero, as illustrated next.

EXAMPLE 2 $log\ 0.0853 = log(8.53 \times 10^{-2}) = (log\ 8.53) + (-2)$
$$= 0.9309 - 2$$
$$= 0.9309 - 2 + \mathbf{10 - 10}$$
$$= 0.9309 + 8 - 10$$
$$= 8.9309 - 10.$$

With practice, characteristics such as -2 in Example 2 are immediately viewed as $8 - 10$ and there is no need for intermediate steps.

---

*Technically, the "approximately equal" symbol $\approx$ should be used throughout this section because entries in Table 1 are only 4-place rational approximations to irrational mantissas. However, it is customary to use the $=$ symbol and we follow this procedure.

Antilogarithms   Given that N is the logarithm of the number x, we often need to determine x. That is, given

$$log_{10}x=N,$$ <1>

find x. Equation <1> is sometimes given in the equivalent form

$$x=antilog_{10}N,$$ <2>

where the right member of <2> is read "the antilogarithm to the base 10 of N." In this section the notation "*antilog N*" where no base is specified, is to be understood to mean $antilog_{10}N$. Table 1 can be used to find antilogarithms.

EXAMPLE 3   Given $log\ x=2.4955$, x can be obtained as follows. First, we have that $log\ x=2.4955=0.4955+2$. Hence,

$$x=antilog\ (0.4955+2)=a\times 10^2,$$

where a is the number such that $log\ a=0.4955$. Next, from Table 1, $0.4955=log\ 3.13$. Hence, $a=3.13$ and

$$x=3.13\times 10^2=313.$$

Linear Interpolation   Note that Table 1 does not immediately provide the mantissa of the logarithm of a number with more than 3 significant digits, such as 8533. However, 4-place approximations to such mantissas can be obtained from Table 1 by the method of **linear interpolation,** illustrated next.

EXAMPLE 4   To find $log\ 8533$, we first compute the mantissa, as follows.

$$
0.010
\begin{bmatrix}
0.003
\begin{bmatrix}
log\ 8.530=0.9309 \\
log\ 8.533=\ \ \ ?
\end{bmatrix} d \\
log\ 8.540=0.9315
\end{bmatrix}
0.0006
$$

$$\frac{0.003}{0.010}=\frac{d}{0.0006}, \qquad \text{from which } d=0.00018.$$

Rounding off d to 0.0002, we now have

$$log\ 8.533=0.9309+0.0002=0.9311.$$

Hence, $log\ 8533=0.9311+3=3.9311$.

EXAMPLE 5   If $x=antilog(7.9511-10)$, we can determine x as follows. First,

$$x=antilog[0.9511+(-3)]=a\times 10^{-3},$$

where a is the number such that $log\ a=0.9511$. To find a, we interpolate from Table 1:

(Continued on next page)

$$\begin{array}{c} 0.010 \left[\begin{array}{c} d \left[\begin{array}{l} \rightarrow log\ 8.930 = 0.9509 \\ \hookrightarrow log\ a\ \ \ \ = 0.9511 \end{array} \right. \\ \rightarrow log\ 8.940 = 0.9513 \end{array} \right. \end{array} \left. \begin{array}{c} 0.0002 \end{array} \right] 0.0004$$

$$\frac{d}{0.010} = \frac{0.0002}{0.0004}, \quad \text{from which } d = 0.005.$$

Thus, $a = 8.930 + 0.005 = 8.935$ and

$$x = 8.935 \times 10^{-3} = 0.008935.$$

Computations Using Logarithms   Logarithms are particularly applicable to problems involving products, quotients, powers and roots of numbers. Computations involving logarithms depend upon properties of logarithmic functions (Theorem 8.2) together with the property that two numbers are equal if and only if their logarithms are equal (Theorem 8.3).

EXAMPLE 6   Use logarithms to compute 82.5×2.93.

SOLUTION   Let $N = 82.5 \times 2.93$. By Theorem 8.3, we can "take the logarithm" of each member, and then apply Theorem 8.2–I, as follows:

$$log\ N = log(82.5 \times 2.93)$$
$$log\ N = log\ 82.5 + log\ 2.93$$
$$log\ N = 1.9165 + 0.4669$$
$$log\ N = 2.3834$$
$$N = antilog\ 2.3834$$
$$N = 2.418 \times 10^2 = 241.8$$

EXAMPLE 7   Use logarithms to compute $\dfrac{427}{(5.2)^3}$.

SOLUTION $$N = \frac{427}{(5.2)^3}$$

$$log\ N = log\ 427 - 3\ log\ 5.2$$
$$log\ N = 2.6304 - 3(0.7160)$$
$$log\ N = 0.4824$$
$$N = antilog\ 0.4824$$
$$N = 3.037 \times 10^0 = 3.037.$$

EXAMPLE 8   Use logarithms to compute $\sqrt[3]{0.538}$.

SOLUTION $$N = \sqrt[3]{0.538}$$

$$log\ N = \tfrac{1}{3} log\ 0.538 = \tfrac{1}{3}(9.7308 - 10).$$

Because 10 is not divisible by 3, we add $20 - 20$ to $9.7308 - 10$ to obtain $29.7308 - 30$. Then,

$$\log N = \tfrac{1}{3}(29.7308 - 30) = 9.9103 - 10$$
$$N = antilog(9.9103 - 10)$$
$$N = 8.134 \times 10^{-1} = 0.8134.$$

Exercise Set J

Find the logarithm of each number. (See Examples 1, 2 and 4.)

1. 300	2. 658	3. 1050	4. 73.8
5. 0.053	6. 0.000864	7. 0.02136	8. 0.5027
9. 7.528	10. 8765	11. 0.001059	12. 125,700

Find the antilogarithm of each number. (See Examples 3 and 5.)

13. 2.3909	14. 1.5717	15. $8.0414 - 10$
16. $7.9881 - 10$	17. 0.8102	18. $9.4409 - 10$
19. $6.2891 - 10$	20. $7.8701 - 10$	21. 4.9622
22. 3.0185	23. 2.2442	24. $8.4820 - 10$

Use logarithms to perform each computation.

25. 56.8×27.3

26. $\dfrac{0.0837}{52.4}$

27. $(75.3)^4$

28. $\sqrt[3]{2.52}$

29. $\dfrac{87.2 \times 0.0048}{156 \times 0.0321}$

30. $\dfrac{(0.00562)^2(793)^3}{\sqrt{0.0007}}$

31. $\dfrac{(44.1 \times 56.4)^2}{(25.3 \times 4.9)^3}$

32. $\dfrac{\sqrt{9658 \times 3.452}}{(15.21)^2}$

33. $\sqrt[4]{(27.5)(18.6)(34.7)}$

34. $\sqrt[3]{\dfrac{7.13}{0.00521 \times (65.2)^2}}$

SUPPLEMENT K

SOLVING OPEN SENTENCES GRAPHICALLY

Equations and inequalities in one variable can often be solved by sketching the graph of each member and reading the coordinates of points of intersection, if any. The degree of accuracy that can be obtained depends upon the scale chosen for the coordinate axes. In Examples 1 and 2, the axes are scaled at 10 squares per 1 unit so that coordinates can be read to the nearest tenth of a unit.

EXAMPLE 1   Solve graphically: $2^x = \frac{2}{3}x + \frac{4}{3}$.

SOLUTION   Figure K-1 shows the graphs of $y = 2^x$ and $y = \frac{2}{3}x + \frac{4}{3}$ on the same set of axes. By inspection, we note that the graphs intersect only in two points—where $x \approx 1.0$ and $x \approx -1.5$. (In fact, 1.0 is an exact solution, as you can verify.) We write the solution set as $\{1.0, -1.5\}$.

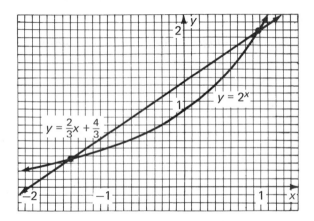

Figure K-1

EXAMPLE 2   Graphically determine the intervals over which $\sqrt{x} \geq x$ and $\sqrt{x} \leq x$.

SOLUTION   Figure K-2 shows the graphs of $y = \sqrt{x}$ and $y = x$ on the same set of axes. Now, we look for intervals on the x-axis over which the graph of $y = \sqrt{x}$ is *above* and intervals over which the graph of $y = \sqrt{x}$ is *below* the graph of $y = x$. We note that $\sqrt{x} \geq x$ over $\{x: 0 \leq x \leq 1\}$; $\sqrt{x} \leq x$ over $\{x: 1 \leq x\}$.

Even if an equation has no solutions, graphical methods may furnish other useful information. For example, Figure K-3 shows the graphs of $y = 2^x$ and $y = \ln x$, and it is clear that there are no solutions to the equation $2^x = \ln x$. Note, however, that for $x > 0$, we have the inequality

$$2^x > \ln x.$$

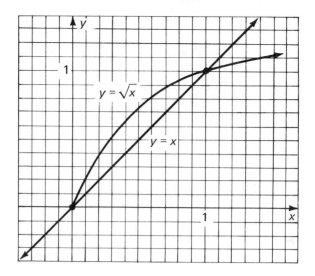

Figure K-2

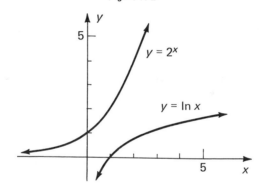

Figure K-3

Exercise Set K

All exercises in this set are to be done graphically.

Solve each equation.

1. $2^x = 3 - \frac{3}{4}x$ 2. $2^x = \frac{5}{3}x + 1$ 3. $2^x = x^2$

4. $2^x = 3 - x^2$ 5. $1 + \sqrt{x} = 3^x$ 6. $\log_{10} x = (x-2)^2 - 1$

Determine intervals over which: a. $f(x) \geq g(x)$;   b. $f(x) \leq g(x)$.

7. $f(x) = 2^x$; $g(x) = 2 - (x-1)^2$ 8. $f(x) = 3^x$; $g(x) = 3x^2$

9. $f(x) = \left(\frac{1}{2}\right)^x$; $g(x) = x^{1/2}$ 10. $f(x) = \left(\frac{1}{2}\right)^x$; $g(x) = \frac{1}{2}x$

Show that each equation has no solutions; specify an inequality suggested by your graph.

11. $x = 2^x$ 12. $x = \log_{10} x$

13. $\ln x = \sqrt{x}$ 14. $\sqrt[3]{x} = 3^x$.

15. Determine all intervals over which a real number is greater than its reciprocal.

SUPPLEMENT L

HYPERBOLIC FUNCTIONS

Certain combinations of functions involving e^x and e^{-x} appear frequently enough in both the theory and applications of mathematics to justify special attention and a special name—they are called **hyperbolic functions**. Two of the hyperbolic functions are defined as follows:

$$H_1: \quad y = \tfrac{1}{2}(e^x - e^{-x}); \qquad H_2: \quad y = \tfrac{1}{2}(e^x + e^{-x}).$$

EXAMPLE 1   Determine the intercepts of the graph of $H_1(x)$.

SOLUTION   Substituting 0 for y in the defining equation for H_1:

$$0 = \tfrac{1}{2}(e^x - e^{-x}) \quad \leftrightarrow \quad e^x - e^{-x} = 0$$
$$e^x - \frac{1}{e^x} = 0$$
$$e^{2x} - 1 = 0$$
$$e^{2x} = 1,$$

from which $x = 0$, and 0 is the x-intercept. Substituting 0 for x:

$$y = \tfrac{1}{2}(e^0 - e^{-0}) \quad \leftrightarrow \quad y = \tfrac{1}{2}(1 - 1) = 0.$$

Hence, 0 is also the y-intercept and the origin is the only point at which the graph of $H_1(x)$ intersects the axes.

EXAMPLE 2   Test the function H_2 for symmetry.

SOLUTION

a.   Substituting $-y$ for y:
$$-y = \tfrac{1}{2}(e^x + e^{-x}) \quad \leftrightarrow \quad y = -\tfrac{1}{2}(e^x + e^{-x}).$$
Hence, there is no symmetry to the x-axis.

b.   Substituting $-x$ for x:
$$y = \tfrac{1}{2}(e^{-x} + e^x) \quad \leftrightarrow \quad y = \tfrac{1}{2}(e^x + e^{-x}).$$
Hence, there is symmetry to the y-axis.

c.   Substituting $-y$ for y and $-x$ for x:
$$-y = \tfrac{1}{2}(e^{-x} + e^x) \quad \leftrightarrow \quad y = -\tfrac{1}{2}(e^x + e^{-x}).$$
Hence, there is no symmetry to the origin.

The last example illustrates an application of the quadratic formula to the study of hyperbolic functions.

EXAMPLE 3   Solve $y = \tfrac{1}{2}(e^x + e^{-x})$ for e^x.

SOLUTION

$$y = \tfrac{1}{2}(e^x + e^{-x}) \quad \leftrightarrow \quad 2y = e^x + \frac{1}{e^x}$$

$$2ye^x = e^{2x} + 1$$

$$e^{2x} - 2ye^x + 1 = 0.$$

The last equation is quadratic in e^x. Hence, by the quadratic formula we have

$$e^x = \frac{2y \pm \sqrt{4y^2 - 4}}{2}$$

$$e^x = y \pm \sqrt{y^2 - 1}.$$

Exercise Set L

In Exercises 1 and 2 find any intercepts, test for symmetry and sketch each graph. (Use Table 3, page 345, as needed.)

1. $H_1(x)$ 2. $H_2(x)$

3. Solve $y = \tfrac{1}{2}(e^x - e^{-x})$ for e^x.

4. Show that $H_1(2x) = 2H_1(x) \cdot H_2(x)$.

5. Show that $H_2(2x) = [H_2(x)]^2 + [H_1(x)]^2$.

6. Show that for each $x \in R$: $H_2(x) \geq 1$. [Hint: Consider $(e^x - 1)^2 \geq 0$.]

7. Solve a system of equations to show that the graphs of $H_1(x)$ and $H_2(x)$ do not intersect.

8. Show that the graphs of $H_1(x)$ and $y = e^x$ do not intersect.

9. Show that: $-1 < \dfrac{e^x - e^{-x}}{e^x + e^{-x}}; \quad \dfrac{e^x - e^{-x}}{e^x + e^{-x}} < 1.$

10. The hyperbolic function H_3 is defined as the quotient of H_1 and H_2.

$$H_3: \quad y = \frac{e^x - e^{-x}}{e^x + e^{-x}}.$$

Find any intercepts, test for symmetry and use the results of Exercise 9 to sketch the graph of $H_3(x)$.

SEQUENCES AND SERIES

9.1   SEQUENCES: ARITHMETIC AND GEOMETRIC

In this chapter we consider certain functions, called *sequence functions*, that play an important role in the development of such widely separated branches of mathematics as the *mathematics of finance* and *advanced calculus*.

Sequence Functions   A *sequence* of numbers is an "ordered array" of numbers. Some familiar sequences of numbers are

$$1, 2, 3, 4, \ldots \quad \text{and} \quad 2, 4, 6, 8, \ldots.$$

A **sequence function** s is a function whose domain is the set N of natural numbers. Function values of s are denoted by the symbol $s(n)$ or, more often, by s_n, where $n \in N$. Because the domain N is an infinite set, the set of all function values s_n is an infinite set. If the function values of a sequence function are arranged in the order

$$s_1, s_2, s_3, \ldots, s_n, \ldots$$

the resulting ordered array is an **infinite sequence**; s_1 is the *first term*, s_2 is the *second term*, $\ldots$, s_n is the nth *term* or *general term*. Sequence functions may be defined by equations. However, because we are usually more concerned with the resulting sequence rather than with the function itself, we also say that the sequence is defined by the equation.

EXAMPLE 1   Consider the sequence defined by

$$s_n = 5n - 1, \qquad n \in N. \qquad\qquad \langle 1 \rangle$$

We find the first four terms by successively substituting 1, 2, 3 and 4 for n to obtain:

$$s_1 = 5(1) - 1 = 4,$$
$$s_2 = 5(2) - 1 = 9,$$
$$s_3 = 5(3) - 1 = 14,$$
$$s_4 = 5(4) - 1 = 19.$$

With $5n-1$ as the nth or general term, the infinite sequence defined by $<1>$ can be expressed as

$$4, 9, 14, 19, \ldots, 5n-1, \ldots.$$

The domain of a sequence function may be restricted to a finite ordered subset of N such as $\{1, 2, 3, 4, 5\}$, in which there is a least member, a greatest member, and all the natural numbers between. The sequence of function values associated with such a sequence function is a **finite sequence**.

EXAMPLE 2   Consider the finite sequence defined by

$$s_n = (-1)^n 2^n, \qquad n \in \{1, 2, 3, 4, 5\}.$$

Successively substituting 1, 2, 3, 4 and 5 for n, we obtain

$$(-1)^1 2^1, \ (-1)^2 2^2, \ (-1)^3 2^3, \ (-1)^4 2^4, \ (-1)^5 2^5$$

or, more simply, the finite sequence: $-2, 4, -8, 16, -32$.

Unless specified otherwise: the word *sequence* is used to mean "infinite sequence;" *the replacement set for the variable n is the set N of natural numbers.*

Recursion Formulas   Another method for defining sequences is by the use of **recursion formulas**. In general, recursion formulas specify the first term of a sequence, together with an equation or other rule that tells how to obtain each term (after the first) from the term that precedes it.

EXAMPLE 3   Specify the first four terms of the sequence defined by the recursion formulas

$$s_1 = 2 \qquad \text{and} \qquad s_{n+1} = s_n + 3n.$$

SOLUTION   The first term s_1 is 2. The remaining terms are found as follows:

$$s_2 = s_1 + 3(1) = 2 + 3 = 5,$$
$$s_3 = s_2 + 3(2) = 5 + 6 = 11,$$
$$s_4 = s_3 + 3(3) = 11 + 9 = 20.$$

The first four terms of the sequence are: 2, 5, 11, 20.

Arithmetic Sequences   Consider the sequence of numbers

$$6, 13, 20, 27, \ldots$$

and verify that each term (after the first) can be obtained by adding 7 to the preceding term. Sequences that can be characterized by such a property are called *arithmetic sequences* (or *arithmetic progressions*).

Definition 9.1   A sequence $s_1, s_2, s_3, \ldots, s_n, \ldots$ is an **arithmetic sequence** if there are numbers a and d such that for each $n \in N$,

$$s_1 = a \quad \text{and} \quad s_{n+1} = s_n + d.$$

The second equation of Definition 9.1 can be written as

$$s_{n+1} - s_n = d,$$

which equation asserts that the difference between each term of an arithmetic sequence and the term that precedes it is always the same number, d. For this reason, d is called the **common difference** of the sequence.

By Definition 9.1, for each arithmetic sequence we have

$$
\begin{aligned}
s_1 &= a, \\
s_2 &= s_1 + d = a + 1d, \\
s_3 &= s_2 + d = a + 1d + d = a + 2d, \\
s_4 &= s_3 + d = a + 2d + d = a + 3d, \text{ etc.}
\end{aligned}
$$

The pattern displayed by the preceding equations suggests that the nth term s_n is given by $a + (n-1)d$. The following theorem states this result formally.

Theorem 9.1   If $a, s_2, s_3, \ldots, s_n, \ldots$ is an arithmetic sequence with common difference d, and $n \in N$, then

$$s_n = a + (n-1)d.$$

From the preceding discussion it follows that any arithmetic sequence can be written in the general form

$$a, a+d, a+2d, a+3d, \ldots, a+(n-1)d, \ldots.$$

EXAMPLE 4   For the arithmetic sequence $18, 23, 28, \ldots$: a. Find the common difference d. b. Specify the next three terms. c. Find an expression for the nth term s_n.

SOLUTION

a. $d = 23 - 18 = 5$.

b. The next three terms are: $28 + 5$, $(28 + 5) + 5$, $(28 + 5 + 5) + 5$, or, more simply, $33, 38, 43$.

c. By Theorem 9.1: $s_n = 18 + (n-1)(5) = 5n + 13$.

EXAMPLE 5   Given the arithmetic sequence $14, 6, -2, \ldots$, find s_{12}.

SOLUTION   First, we find that $d = 6 - 14 = -8$. Next, by Theorem 9.1 with $a = 14$, $n = 12$ and $d = -8$, we have

$$s_{12} = 14 + (12 - 1)(-8) = -74.$$

EXAMPLE 6   How many integers between 13 and 115 are exactly divisible by 7?

SOLUTION   Of the integers between 13 and 115 that are exactly divisible by 7, the least is 14 and the greatest is 112. Hence, we need to find n, the number of terms in the finite arithmetic sequence: 14, 21, 28, . . . , 105, 112. By Theorem 9.1,

$$s_n = a + (n-1)d$$
$$112 = 14 + (n-1)(7)$$
$$n = 15.$$

Geometric Sequences   Consider the sequence of numbers

$$3, 9, 27, 81, \ldots$$

and verify that each term (after the first) can be obtained by multiplying the preceding term by 3. Sequences that can be characterized by such a property are called *geometric sequences* (or *geometric progressions*).

Definition 9.2   A sequence $s_1, s_2, s_3, \ldots, s_n, \ldots$ is a **geometric sequence** if there are nonzero numbers a and r such that for each $n \in N$ and $r \neq 1$,

$$s_1 = a \qquad \text{and} \qquad s_{n+1} = s_n r.$$

The second equation of Definition 9.2 can be written as

$$\frac{s_{n+1}}{s_n} = r,$$

which equation asserts that the quotient of any term of a geometric sequence and the term that precedes it is always the same number, r. For this reason, r is called the **common ratio** of the sequence. Note that if $r = 1$, then $s_{n+1} = s_n$ and the resulting sequence takes the form $a, a, a, \ldots$, which is a *constant sequence* rather than a geometric sequence. This is one of the reasons for the restriction $r \neq 1$ in Definition 9.2.

By Definition 9.2, for each geometric sequence we have

$$s_1 = a,$$
$$s_2 = s_1 r = ar^1,$$
$$s_3 = s_2 r = ar \cdot r = ar^2,$$
$$s_4 = s_3 r = ar^2 \cdot r = ar^3, \text{ etc.}$$

The pattern displayed by the preceding equations suggests that the nth term is given by ar^{n-1}. The next theorem states this result formally. (See Supplement M, Exercise 14.)

Theorem 9.2   If $a \neq 0$, $r \neq 1$, $n \in N$ and $a, s_2, s_3, \ldots, s_n, \ldots$ is a geometric sequence with common difference r, then

$$s_n = ar^{n-1}.$$

From the preceding discussion it follows that any geometric sequence can be written in the general form

$$a, ar, ar^2, ar^3, \ldots, ar^{n-1}, \ldots.$$

EXAMPLE 7   For the geometric sequence $\frac{1}{2}, \frac{1}{4}, \frac{1}{8}, \ldots$: a. Find the common ratio r.   b. Specify the next three terms.   c. Find an expression for the nth term s_n.

SOLUTION

a. $r = \frac{1}{4} \div \frac{1}{2} = \frac{1}{2}$.

b.   The next three terms are: $\frac{1}{8} \cdot \frac{1}{2}, \left(\frac{1}{8} \cdot \frac{1}{2}\right)\frac{1}{2}, \left(\frac{1}{8} \cdot \frac{1}{2} \cdot \frac{1}{2}\right)\frac{1}{2}$, or simply $\frac{1}{16}, \frac{1}{32}, \frac{1}{64}$.

c.   By Theorem 9.2: $s_n = \frac{1}{2}\left(\frac{1}{2}\right)^{n-1} = \frac{1}{2^n}$.

EXAMPLE 8   Given the geometric sequence $-9, 3, -1, \ldots$, find s_7.

SOLUTION   First, $r = \frac{3}{-9} = -\frac{1}{3}$. Next, by Theorem 9.2 with $r = -\frac{1}{3}$, $a = -9$ and $n = 7$, we have

$$s_7 = (-9)\left(-\frac{1}{3}\right)^6 = -\frac{1}{81}.$$

Exercise Set 9.1

(A)   *Each equation defines a sequence. Write the sequence by specifying the first four terms and the nth term. (See Example 1.)*

1. $s_n = 2n - 1$ 2. $s_n = 2n + 1$
3. $s_n = 2^{n-1}$ 4. $s_n = 3^n$
5. $s_n = \dfrac{1}{3^n}$ 6. $s_n = \dfrac{1}{2^n}$
7. $s_n = (-1)^n(2n)^{n-1}$ 8. $s_n = (-1)^{n+1}(n)^n$

Write the finite sequence defined by each equation over the domain $\{1, 2, 3, 4, 5, 6\}$. (See Example 2.)

9. $s_n = 1 + \dfrac{1}{n}$ 10. $s_n = \dfrac{n}{2n - 1}$
11. $s_n = (-1)^n\dfrac{n-2}{n}$ 12. $s_n = \dfrac{n^2 - 3}{n + 1}$

Specify the first four terms of the sequence defined by each set of recursion formulas. (See Example 3.)

13. $s_1 = 2, \ s_{n+1} = s_n + 5$ 14. $s_1 = 3, \ s_{n+1} = 2s_n - 4$
15. $s_1 = \frac{1}{2}, \ s_{n+1} = \frac{1}{2}s_n$ 16. $s_1 = \frac{2}{3}, \ s_{n+1} = \frac{1}{3}s_n$
17. $s_1 = \frac{x}{2}, \ s_{n+1} = \frac{x^2}{2}s_n$ 18. $s_1 = -x, \ s_{n+1} = (-x^2)s_n$

For each arithmetic sequence: find the common difference; specify the next three terms; write an expression for the nth term. (See Example 4.)

19. 7, 14, 21, . . .   20. 9, 18, 27, . . .

21. 27, 18, 9, . . .   22. 21, 14, 7, . . .

23. $x+c$, $2x+3c$, $3x+5c$, . . .   24. $2y-c$, y, c, . . .

Find the indicated term of each arithmetic sequence. (See Example 5.)

25. $3, \frac{13}{5}, \frac{11}{5}, \ldots ; s_{16}$

26. $-7, -3, 1, \ldots ; s_7$

27. $x, 2x+1, 3x+2, \ldots ; s_{12}$

28. $y, 3, 6-y, \ldots ; s_8$

For each geometric sequence; find the common ratio; specify the next three terms; write an expression for the nth term. (See Example 7.)

29. 2, 4, 8, . . .   30. 3, 9, 27, . . .

31. $\frac{1}{3}, -\frac{1}{9}, \frac{1}{27}, \ldots$ 32. $-\frac{1}{2}, \frac{1}{4}, -\frac{1}{8}, \ldots$

33. $\frac{c}{x}, 1, \frac{x}{c}, \ldots$ 34. $\frac{x}{c}, \frac{x}{c^2}, \frac{x}{c^3}, \ldots$

Find the indicated term of each geometric sequence. (See Example 8.)

35. 12, 24, 48, . . . ; s_7

36. $-27, -9, -3, \ldots ; s_5$

37. $-\frac{x^2}{2}, x^5, -2x^8, \ldots ; s_6$

38. $\sqrt{2}, -4, 8\sqrt{2}, \ldots ; s_8$

39. In an arithmetic sequence: $s_3=8$ and $s_{10}=22$. Find the common difference, the first term and the 24th term.

40. In an arithmetic sequence: $s_4=-11$ and $s_{19}=-21$. Find s_{13}.

41. Find the first term of the geometric sequence in which $s_4=48$ and the common ratio is 2.

42. In the arithmetic sequence 2, 5, 8, . . . , which term is 128?

43. How many even integers are there between 11 and 91?

44. How many integral multiples of 7 are there between 8 and 110?

(B) 45. For what value (or values) of k will $k-2$, $k-6$, $2k+3$, . . . form an arithmetic sequence?

46. Same as Exercise 45, but form a geometric sequence.

*Numbers inserted between the first and last terms of a finite sequence are called **arithmetic means** if the sequence is arithmetic and **geometric means** if the sequence is geometric.*

47. Insert two arithmetic means between 3 and 24. (Hint: Find x and y so that 3, x, y, 24 forms an arithmetic sequence.)

48. Insert two arithmetic means between 7 and 25.

49. Insert one geometric mean between 2 and 54.

50. Insert two geometric means between 2 and 54.

9.2   SUMS AND SUMMATION NOTATION

It is often necessary to find the sum of n consecutive terms of a sequence. If n is relatively large, the computations can be quite time consuming. Hence, when possible, it is desirable to obtain an explicit representation of such a sum, preferably in terms of n and free of the ellipsis symbol. Although this may be a difficult problem for many sequences, for arithmetic and geometric sequences such representations can be found.

Sum of a Finite Arithmetic Sequence   We use the symbol S_n to mean the *sum* of the first n successive terms of a sequence. Thus, for the sequence

$$1, 3, 5, 7, 9, 11, \ldots, 2n+1, \ldots$$

we write

$$S_1 = 3$$
$$S_2 = 3 + 5$$
$$S_3 = 3 + 5 + 7$$
$$\vdots \qquad \vdots$$
$$S_n = 3 + 5 + 7 + \cdots + (2n + 1).$$

Sums such as $S_1, S_2, \ldots, S_n$ are called **partial sums** of the sequence. In the case of an arithmetic sequence, the nth partial sum can be expressed as

$$S_n = a + (a + d) + (a + 2d) + \cdots + [a + (n - 1)d].$$

Explicit representations for partial sums of arithmetic sequences are given in the next theorem.

Theorem 9.3   If a is the first term of an arithmetic sequence with common difference d, and $n \in N$, then the sum of the first n terms of the sequence is given by each of the formulas:

$$\text{I} \quad S_n = \frac{n}{2}(a + s_n); \qquad \text{II} \quad S_n = \frac{n}{2}[2a + (n - 1)d]. \qquad \textbf{(P)}$$

A proof of Theorem 9.3 can be constructed by mathematical induction (see Supplement M). Another approach to a proof of this theorem is illustrated next. (See also Exercise 9.2–I.)

EXAMPLE 1   For the finite arithmetic sequence 1, 3, 5, 7, 9:

$$S_5 = 1 + 3 + 5 + 7 + 9, \qquad\qquad \langle 1 \rangle$$

or

$$S_5 = 9 + 7 + 5 + 3 + 1. \qquad\qquad \langle 2 \rangle$$

In $<1>$, $a=1$ and $d=2$. In $<2>$, $a=9$ and $d=-2$. Hence, equations $<1>$ and $<2>$ can also be written, respectively, as

$$S_5 = 1 + (1 + 1 \cdot 2) + (1 + 2 \cdot 2) + (1 + 3 \cdot 2) + (1 + 4 \cdot 2), \qquad <3>$$

$$S_5 = 9 + [9 + 1(-2)] + [9 + 2(-2)] + [9 + 3(-2)] + [9 + 4(-2)]. \quad <4>$$

Adding corresponding members of $<3>$ and $<4>$, we have

$$2S_5 = [1 + 9] + [1 + 1 \cdot 2 + 9 + 1(-2)] + [1 + 2 \cdot 2 + 9 + 2(-2)]$$

$$+ [1 + 3 \cdot 2 + 9 + 3(-2)] + [1 + 4 \cdot 2 + 9 + 4(-2)]$$

$$2S_5 = (1 + 9) + (1 + 9) + (1 + 9) + (1 + 9) + (1 + 9) = 5(1 + 9)$$
$$S_5 = \tfrac{5}{2}(1 + 9). \qquad <5>$$

Observe that equation $<5>$ matches sum formula I of Theorem 9.3, with $n=5$, $a=1$ and $s_n=9$.

Sum formula I of Theorem 9.3 is probably used more frequently than the other. You may prefer to remember it in the form

$$S_n = n\left(\frac{a + s_n}{2}\right),$$

where the right member can be described as "the product of the number n of terms and the *average* of the first and last terms of the sequence." An advantage of sum formula II of Theorem 9.3 is that the last term of the sequence need not be directly computed.

EXAMPLE 2   The sum S_{20} of the first twenty terms of the arithmetic sequence 16, 12, 8, . . . , can be found by Theorem 9.3 in either of two ways. In each case we have that $a=16$, $d=-4$ and $n=20$. First, we use Theorem 9.1 to find s_{20}:

$$s_{20} = 16 + (20 - 1)(-4) = -60.$$

Then, by sum formula I of Theorem 9.3, we have

$$S_{20} = \tfrac{20}{2}[16 + (-60)] = -440.$$

Alternatively, by sum formula II we have

$$S_{20} = \tfrac{20}{2}[2(16) + (20 - 1)(-4)] = -440.$$

EXAMPLE 3   Find the sum of all the integers between 13 and 115 that are exactly divisible by 7.

SOLUTION   From Example 6 on page 291, we have the finite sequence

$$14, 21, 28, \ldots, 105, 112,$$

with $a=14$, $d=7$ and $n=15$. Hence, by Theorem 9.3,

$$S_{15} = \tfrac{15}{2}(14 + 112) = 945.$$

Sum of a Finite Geometric Sequence   An explicit representation of partial sums of geometric sequences is given in the following theorem.

Theorem 9.4   If a is the first term of a geometric sequence with common ratio r, $r \neq 1$ and $n \in N$, then the sum of the first n terms of the sequence is

$$S_n = \frac{a - ar^n}{1 - r}. \quad \text{(P)}$$

A proof of Theorem 9.4 can be constructed by mathematical induction (see Supplement M). Another approach to a proof of this theorem is illustrated next. (See also Exercise 9.4–II.)

EXAMPLE 4   For the finite geometric sequence 2, 6, 18, 54, 162:

$$S_5 = 2 + 6 + 18 + 54 + 162$$

or, with $a = 2$ and $r = 3$, we may write

$$S_5 = 2 + 2 \cdot 3^1 + 2 \cdot 3^2 + 2 \cdot 3^3 + 2 \cdot 3^4. \quad \langle 6 \rangle$$

Multiplying each member of $\langle 6 \rangle$ by the common ratio 3, we have

$$3S_5 = 2 \cdot 3^1 + 2 \cdot 3^2 + 2 \cdot 3^3 + 2 \cdot 3^4 + 2 \cdot 3^5. \quad \langle 7 \rangle$$

Subtracting corresponding members of $\langle 7 \rangle$ from $\langle 6 \rangle$, we obtain

$$(1 - 3)S_5 = 2 - 2 \cdot 3^1 + 2 \cdot 3^1 - 2 \cdot 3^2 + 2 \cdot 3^2 - 2 \cdot 3^3 + 2 \cdot 3^3 - 2 \cdot 3^4 + 2 \cdot 3^4 - 2 \cdot 3^5$$

$$S_5 = \frac{2 - 2 \cdot 3^5}{1 - 3}. \quad \langle 8 \rangle$$

Observe that equation $\langle 8 \rangle$ matches the conclusion of Theorem 9.4, with $a = 2$, $r = 3$ and $n = 5$.

EXAMPLE 5   For the geometric sequence 2, 6, 18, . . . , find S_6.

SOLUTION   We have $a = 2$, $r = 3$ and $n = 6$. Hence, by Theorem 9.4,

$$S_6 = \frac{2 - 2(3)^6}{1 - 3} = \frac{2(1 - 3^6)}{-2} = 728.$$

Sigma Notation   Sums of terms of sequences are encountered often enough in mathematics to motivate the introduction of a particular compact notation involving the capital Greek letter Σ (sigma). The partial sum

$$S_n = s_1 + s_2 + s_3 + \cdots + s_n \quad \langle 9 \rangle$$

can be denoted in sigma notation by

$$S_n = \sum_{k=1}^{n} s_k, \quad \langle 10 \rangle$$

read "the sum of all terms s_k from $k=1$ to $k=n$." The variable k is called the **index of summation***. The replacement set for k is called the **range of summation**. Sum $\langle 9 \rangle$ is said to be the *expanded form* of the indicated sum $\langle 10 \rangle$.

EXAMPLE 6   To find the expanded form of

$$\sum_{k=1}^{5}(-1)^k 2^k,$$

we successively substitute 1, 2, 3, 4 and 5 for k to obtain

$$S_5 = (-1)^1 2^1 + (-1)^2 2^2 + (-1)^3 2^3 + (-1)^4 2^4 + (-1)^5 2^5$$

or, more simply,

$$S_5 = -2 + 4 - 8 + 16 - 32.$$

The replacement set for the index of summation need not begin with 1.

EXAMPLE 7

$$\sum_{j=2}^{6}(2j-3) = (2 \cdot 2 - 3) + (2 \cdot 3 - 3) + (2 \cdot 4 - 3) + (2 \cdot 5 - 3) + (2 \cdot 6 - 3)$$
$$= 1 + 3 + 5 + 7 + 9.$$

Exercise Set 9.2

(A)   *Specify whether each sequence is arithmetic or geometric (assume it is one or the other); find the indicated partial sum. (See Examples 2 and 5.)*

1. 6, 12, 18, . . . ; S_{12}
2. 5, 13, 21, . . . ; S_9
3. 1, 3, 9, . . . ; S_5
4. 3, 6, 12, . . . ; S_8
5. 27, 18, 9, . . . ; S_{11}
6. 9, 3, -3, . . . ; S_{20}
7. 2, $\frac{3}{2}$, 1, . . . ; S_{19}
8. 3, $\frac{8}{3}$, $\frac{7}{3}$, . . . ; S_{22}
9. $\frac{1}{2}$, $\frac{1}{4}$, $\frac{1}{8}$, . . . ; S_7
10. $\frac{1}{3}$, $\frac{1}{9}$, $\frac{1}{27}$, . . . ; S_6

Each sigma expression represents a partial sum of an arithmetic or geometric sequence. Write the expanded form and use Theorem 9.3 or 9.4 to find the partial sum. (See Examples 6 and 7.)

11. $\displaystyle\sum_{k=1}^{8}(2k+5)$

12. $\displaystyle\sum_{j=2}^{8}(-3j+8)$

13. $\displaystyle\sum_{j=2}^{6}\left(\tfrac{1}{3}\right)^j$

14. $\displaystyle\sum_{k=3}^{9}\left(\tfrac{1}{2}\right)^k$

15. $\displaystyle\sum_{k=2}^{7}(-1)^k\frac{1}{2^k}$

16. $\displaystyle\sum_{k=1}^{4}(-1)^{k+1}\frac{1}{3^k}$

17. $\displaystyle\sum_{j=1}^{8}\left(\tfrac{1}{2}j-3\right)$

18. $\displaystyle\sum_{k=3}^{10}\left(\tfrac{2}{3}k-\tfrac{1}{9}\right)$

*Other letters, most often i and j, are also used for indices of summation.

19. Find the sum of all the even integers between 11 and 93.

20. Find the sum of all the integral multiples of 3 between 14 and 118.

21. Find the sum of all positive integers ending in 8 and less than 200.

22. Find the sum of all negative integers ending in 7 and greater than -100.

23. A gambler plays by the rule "double your bet each time you lose." If his first bet is $1 and he loses 9 consecutive bets, how much must the 10th bet be? How much will he lose in ten consecutive losing bets?

24. Answer the questions of Exercise 23 if the first bet is $5.

You may wish to use logarithms for Exercises 25 and 26.

(B) 25. Which method of payment for a 30-day job pays more?
1. Each day the pay is $100 more than the preceding day; $100 is paid for the first day.
2. Each day the pay is twice that of the preceding day; 1¢ is paid for the first day.

26. Employee A starts a job at $10,000 per year with a guarantee of a 10% raise each year. Employee B takes the same type of job but starts at $1 per year with the salary doubling each year. In 25 years, which employee earns more?

27. Show that the sum of the first k odd natural numbers is k^2.

28. Show that the sum of the first k even natural numbers is $k(k+1)$.

29. Find an expression for the sum of the first k natural number multiples of 3.

30. For a geometric sequence, show that $S_n = \dfrac{a}{1-r}(1-r^n)$.

(P) I. Prove Theorem 9.3 by considering the sum
$$S_n = a + (a+d) + (a+2d) + \cdots + [a+(n-1)d],$$
the same sum with the terms in reverse order
$$S_n = s_n + (s_n - d) + (s_n - 2d) + \cdots + [s_n - (n-1)d],$$
and following the method suggested by Example 1.

II. Prove Theorem 9.4 by considering the sums
$$S_n = a + ar + ar^2 + \cdots + ar^{n-1},$$
$$rS_n = ar + ar^2 + ar^3 + \cdots + ar^{n-1} + ar^n,$$
and following the method suggested by Example 4.

9.3   SUM OF AN INFINITE GEOMETRIC SEQUENCE

Sigma notation, together with the infinity symbol, ∞, can be used to indicate the sum of an infinite number of terms. The indicated sum of an

298   SEQUENCES AND SERIES

infinite number of terms is called an **infinite series**. In particular, the indicated sum of the terms of a geometric series

$$\sum_{k=1}^{\infty} ar^{k-1} = a + ar + ar^2 + \cdots + ar^{n-1} + \cdots$$

is called an **infinite geometric series**.

EXAMPLE 1   Two infinite geometric series are:

a. $\displaystyle\sum_{k=1}^{\infty} 3^k = 3 + 3^2 + 3^3 + \cdots + 3^n + \cdots$;

b. $\displaystyle\sum_{k=1}^{\infty} \frac{1}{3^k} = \frac{1}{3} + \frac{1}{3^2} + \frac{1}{3^3} + \cdots + \frac{1}{3^n} + \cdots$.

The problem of "finding the sum" of an infinite series, that is, of associating a real number with the sum of an infinite number of terms, is of great interest in mathematics. Essentially, the problem involves two parts. First, does a given infinite series have a sum? Second, if a given infinite series has a sum, how can it be obtained? In the case of an infinite geometric series, both parts of the problem can readily be solved with the aid of a concept from higher mathematics, as introduced next.

Consider the number r^n, where $r = \frac{1}{5}$ and $n \in N$. Table 9.1 displays values of $\left(\frac{1}{5}\right)^n$ as n is successively replaced by 1, 2, 3, 4, 5:

n	1	2	3	4	5
$\left(\frac{1}{5}\right)^n$	0.2	0.04	0.008	0.0016	0.00032

TABLE 9.1

The second row of the table suggests that $\left(\frac{1}{5}\right)^n$ takes on lesser values as n takes on greater values. We say that "$\left(\frac{1}{5}\right)^n$ approaches zero as n increases without bound" to describe this state of affairs. More generally, if r is a number between -1 and 1, that is, if $|r| < 1$, then we find that *r^n approaches zero as n increases without bound.*

Consider now the conclusion of Theorem 9.4,

$$S_n = \frac{a - ar^n}{1 - r} = \frac{a(1 - r^n)}{1 - r},$$

which can be written equivalently as

$$S_n = \frac{a}{1 - r}(1 - r^n),$$

and let us center our attention upon the factor $(1 - r^n)$. If $|r| < 1$, then r^n approaches zero as n increases without bound, from which we reason that $(1 - r^n)$ approaches $(1 - 0)$, or simply 1, as n increases without bound. In turn, this implies that

$$\frac{a}{1-r}(1-r^n) \qquad \text{approaches} \qquad \frac{a}{1-r} \cdot 1$$

as n increases without bound. This result suggests the next theorem, in which we use the symbol S_∞ as a label for the indicated sum.

Theorem 9.5   If $|r| < 1$, then

$$S_\infty = \sum_{k=1}^{\infty} ar^{k-1} = \frac{a}{1-r}.$$

In words, Theorem 9.5 asserts that the *sum of an infinite geometric series* is given by the number $\frac{a}{1-r}$, provided that r is between -1 and 1.

EXAMPLE 2   Find the sum (if it exists) of each infinite geometric series.

     a. $\displaystyle\sum_{k=1}^{\infty} \frac{1}{3^k}$ b. $\displaystyle\sum_{k=1}^{\infty} 3^k$

SOLUTION

     a. $\displaystyle\sum_{k=1}^{\infty} \frac{1}{3^k} = \frac{1}{3} + \frac{1}{9} + \frac{1}{27} + \cdots$. By inspection, $a = \frac{1}{3}$ and $r = \frac{1}{3}$. Because $-1 < \frac{1}{3} < 1$, we may apply Theorem 9.5:

$$S_\infty = \frac{a}{1-r} = \frac{\frac{1}{3}}{1-\frac{1}{3}} = \frac{1}{2}.$$

     b. $\displaystyle\sum_{k=1}^{\infty} 3^k = 3 + 9 + 27 + \cdots$. By inspection, $a = 3$ and $r = 3$. Because 3 is not between -1 and 1, the series has no sum.

In the next example, the "overbar" notation is used to mean a repeating group of one or more digits. Thus, $0.4\overline{4}$ means $0.4444\ldots$, $0.12\overline{12}$ means $0.121212\ldots$, etc.

EXAMPLE 3   Find the rational number whose nonterminating decimal expansion is $0.12\overline{12}$.

SOLUTION   The nonterminating decimal $0.12\overline{12}$ can be written as

$$0.12 + 0.0012 + 0.000012 + \cdots. \qquad\qquad \langle 1 \rangle$$

Sum $\langle 1 \rangle$ is seen to be an infinite geometric series with $a = 0.12$ and $r = 0.01$. But $|0.01| < 1$. Hence, by Theorem 9.5:

$$S_\infty = \frac{0.12}{1-0.01} = \frac{12}{99}, \qquad \text{or} \qquad \frac{4}{33}.$$

You may verify this result by dividing 4 by 33.

Exercise Set 9.3

(A) *Write in expanded form, showing the first four terms and the nth term. Use the ellipsis symbol as needed. (See Example 1.)*

1. $\displaystyle\sum_{j=1}^{\infty} \frac{1}{2^j}$

2. $\displaystyle\sum_{k=1}^{\infty} \frac{k+1}{3^{k-1}}$

3. $\displaystyle\sum_{j=1}^{\infty} (-1)^{j+1}\, 3^j$

4. $\displaystyle\sum_{k=1}^{\infty} (-1)^{k-1}\, 2^{k+1}$

Find the sum of each of the following infinite geometric series. If the series has no sum, tell why. (See Example 2.)

5. $25 + 10 + 4 + \cdots$

6. $12 + 6 + 3 + \cdots$

7. $4 + 10 + 25 + \cdots$

8. $3 + 6 + 12 + \cdots$

9. $1 + \frac{2}{3} + \frac{4}{9} + \cdots$

10. $1 + \frac{2}{5} + \frac{4}{25} + \cdots$

11. $1 + \frac{3}{2} + \frac{9}{4} + \cdots$

12. $1 + \frac{5}{2} + \frac{25}{4} + \cdots$

13. $3 - 1 + \frac{1}{3} - \cdots$

14. $5 - 1 + \frac{1}{5} - \cdots$

15. $\displaystyle\sum_{j=1}^{\infty} \left(\frac{3}{4}\right)^j$

16. $\displaystyle\sum_{k=1}^{\infty} (-1)^k \left(\frac{1}{3}\right)^k$

Find the rational number associated with each of the following nonterminating decimal expansions. (See Example 3.) Verify your answer.

17. $0.4\overline{4}$

18. $0.2\overline{2}$

19. $0.18\overline{18}$

20. $0.13\overline{13}$

21. $0.499\overline{9}$

22. $0.744\overline{4}$

23. $3.9\overline{18}$

24. $2.4\overline{28571}$

(B) 25. A ball is dropped from a height of 3 feet. Each time it strikes the floor, the ball bounces upwards $\frac{2}{3}$ of the distance it fell. Find the "distance" traveled by the ball up to the instant that it strikes the floor the fifth time. Find the "distance" traveled by the ball until coming to rest.

26. Find the error in this "proof" that -1 is equal to the sum of an infinite number of *positive* terms. By Theorem 9.5:

$$1 + x + x^2 + x^3 + \cdots = \frac{1}{1-x}.$$

Substituting 2 for x, we obtain: $1 + 2 + 4 + 8 + \cdots = -1$.

For what replacement set for x will each of the following infinite geometric series have a finite sum?

EXAMPLE

$$\frac{1}{x} + \frac{1}{x^2} + \frac{1}{x^3} + \cdots$$

SOLUTION   We determine that the common ratio r is equal to $\frac{1}{x}$. By Theorem 9.5, we require that $\left|\frac{1}{x}\right| < 1$, from which we have $\frac{1}{|x|} < 1 \leftrightarrow 1 < |x|$.

Solving the latter inequality, we obtain the required replacement set $\{x : x < -1 \text{ or } x > 1\}$.

27. $\dfrac{1}{x+1}+\dfrac{1}{(x+1)^2}+\dfrac{1}{(x+1)^3}+\cdots$

28. $\dfrac{3}{x+1}+\dfrac{9}{(x+1)^2}+\dfrac{27}{(x+1)^3}+\cdots$

29. $\dfrac{1}{3x-2}+\dfrac{2}{(3x-2)^2}+\dfrac{4}{(3x-2)^3}+\cdots$

30. $\dfrac{2}{3-4x}+\dfrac{4}{(3-4x)^2}+\dfrac{8}{(3-4x)^3}+\cdots$

(C)   I   The recursive definition of an arithmetic sequence is used in the following BASIC program to compute the first N terms of an arithmetic sequence with first term A and common difference D, as well as the sum of those terms. Run the program with the test data provided, then check the results by using a program you have written based upon the formulas of Theorem 9.3.

```
10 READ A,R,N
20 LET S=A/R
30 LET SUM=0
40 FOR K=1 TO N
50 LET S=S*R
60 LET SUM=SUM+S
70 PRINT S;
80 NEXT K
90 PRINT " "
100 PRINT "THE SUM IS " SUM
110 GO TO 10
120 DATA 1,2,10,1,-2,10,100,0.25,10
130 END
```

II   The recursive definition of a geometric sequence is used in the following BASIC program to compute the first N terms of a geometric sequence with first term A and common ratio R, as well as the sum of those terms. Run the program with the test data provided, then check the results by using a program you have written based upon the formula of Theorem 9.4.

```
10 READ A,N,D
20 LET S=A-D
30 LET SUM=0
40 FOR K=1 TO N
50 LET S=S+D
60 LET SUM=SUM+S
70 PRINT S;
80 NEXT K
90 PRINT " "
100 PRINT "THE SUM IS " SUM
110 GO TO 10
120 DATA 1,10,1,-10,11,2,0.23,20,0.04
130 END
```

9.4   THE BINOMIAL EXPANSION

In the sequence defined by $s_n=(a+b)^n$:

$$(a+b)^1,\ (a+b)^2,\ (a+b)^3,\ \ldots,\ (a+b)^n,\ \ldots$$

each term after the first consists of a product of binomial factors. By direct multiplication, each such term can be written in an *expanded form*. For example,

$$(a+b)^3 = a^3 + 3a^2b + 3ab^2 + b^3. \qquad \langle 1 \rangle$$

The right member of $\langle 1 \rangle$ is called the **expansion** of $(a+b)^3$. Such expansions are called **binomial expansions**. In this section our goal is to be able to obtain the expansion of $(a+b)^n$ without having to resort to repeated multiplications. First, however, we introduce a special notation that helps us to simplify the terms of binomial expansions.

Factorials   Products in which the factors are the consecutive natural numbers from 1 to n, inclusive, are often represented by the symbol $n!$, read "n factorial" or "factorial n." For example,

$$5! = 1 \cdot 2 \cdot 3 \cdot 4 \cdot 5 \quad \text{and} \quad 23! = 1 \cdot 2 \cdot 3 \cdots 21 \cdot 22 \cdot 23.$$

A formal definition of $n!$ follows.

Definition 9.3

I $0! = 1$; $1! = 1$.

II   For all $n \in N$, $n > 1$: $n! = 1 \cdot 2 \cdot 3 \cdots (n-2)(n-1)n$.

Definition 9.3 can be used to transform factorial expressions. For example:

$$7! = 1 \cdot 2 \cdot 3 \cdot 4 \cdot 5 \cdot 6 \cdot 7 = (1 \cdot 2 \cdot 3 \cdot 4 \cdot 5 \cdot 6)7 \quad \leftrightarrow \quad 7! = (6!)7.$$

In the same way: $7! = (5!) \cdot 6 \cdot 7$; $7! = (4!) \cdot 5 \cdot 6 \cdot 7$; etc. The preceding results suggest the next theorem.

Theorem 9.6   If n is a natural number, then $n! = [(n-1)!] \cdot n$.

In Theorem 9.6, note that if $n = 1$, then $(n-1)!$ becomes $0!$. This is one reason for defining $0!$ as in Definition 9.3–I.

EXAMPLE 1   Write $n!$ as a product in which $(n-3)!$ is one of the factors.

SOLUTION   By repeated applications of Theorem 9.6:

$$n! = [(n-1)!]n$$
$$n! = [(n-2)!](n-1)n$$
$$n! = [(n-3)!](n-2)(n-1)n$$

EXAMPLE 2   Simplify each expression.

a. $\dfrac{8!}{(3!)(5!)}$ b. $\dfrac{(n+2)!}{n!}$

SOLUTION

a. $\dfrac{8!}{(3!)(5!)} = \dfrac{(5!) \cdot 6 \cdot 7 \cdot 8}{1 \cdot 2 \cdot 3 \cdot (5!)} = \dfrac{6 \cdot 7 \cdot 8}{1 \cdot 2 \cdot 3} = 56.$

b. $\dfrac{(n+2)!}{n!} = \dfrac{(n!)(n+1)(n+2)}{n!} = (n+1)(n+2) = n^2 + 3n + 2.$

The Basic Binomial Expansion   Examine the following binomial expansions:

$$(a+b)^1 = a+b$$
$$(a+b)^2 = a^2 + 2ab + b^2$$
$$(a+b)^3 = a^3 + 3a^2b + 3ab^2 + b^3$$
$$(a+b)^4 = a^4 + 4a^3b + 6a^2b^2 + 4ab^3 + b^4$$
$$(a+b)^5 = a^5 + 5a^4b + 10a^3b^2 + 10a^2b^3 + 5ab^4 + b^5.$$

First, observe that for each exponent n:

1. *There are $n+1$ terms in the expansion.*
2. *The first term is a^n; the last term is b^n.*

Second, if we view a^n as a^nb^0 and b^n as a^0b^n, then each term in each expansion is of the form

$$Ca^sb^t,$$

where C is a numerical coefficient and s and t are nonnegative integers. Third, with respect to s and t note that

3. *In each term the sum of s and t is equal to n.*
4. *The exponent on the variable factor a decreases by 1 with each successive term.*
5. *The exponent on the variable factor b increases by 1 with each successive term.*

Except for the coefficient C of each term (to be considered next) we can now write a "basic form" of the expansion of $(a+b)^n$. For example, by applying the preceding five statements (you should verify each one) we obtain the basic form expansion when $n=6$:

$$(a+b)^6 = C_0a^6 + C_1a^5b + C_2a^4b^2 + C_3a^3b^3 + C_4a^2b^4 + C_5ab^5 + C_6b^6. \quad <2>$$

Note that we need not write the factors b^0 and a^0 in the first and last terms.

The Binomial Theorem   For $k=0, 1, 2, 3, \ldots, n$, it can be shown that the numerical coefficient C_k of the $(k+1)$st term in the expansion of $(a+b)^n$ is

$$C_k = \frac{n!}{k!\,(n-k)!}. \qquad <3>$$

EXAMPLE 3   The coefficient C_4 in the expansion of $(a+b)^6$ can be computed by using equation $<3>$ with $n=6$ and $k=4$:

$$C_4 = \frac{6!}{4!\,(6-4)!} = \frac{4! \cdot 5 \cdot 6}{4! \cdot 2} = 15.$$

Note that C_4 is the coefficient of the *fifth* term in the expansion of $(a+b)^6$. That is, the fifth term in expansion $<2>$ above is $15a^2b^4$.

Now, consider equation <3>. Note that if k is replaced by 0, we obtain

$$C_0 = \frac{n!}{0!\,(n-0)!} = \frac{n!}{n! \cdot 1} = 1,$$

which is the coefficient of the $(0+1)$st, or first term, a^n. If k is replaced by n, we obtain

$$C_n = \frac{n!}{n!\,(n-n)!} = \frac{n!}{n! \cdot 1} = 1,$$

which is the coefficient of the $(n+1)$st, or last term, b^n.

The results of the preceding discussion lead to the next theorem.

Theorem 9.7 (The Binomial Theorem)   If $n \in N$, then

$$(a+b)^n = a^n + C_1 a^{n-1}b + \cdots + C_k a^{n-k}b^k + \cdots + C_{n-1} ab^{n-1} + b^n,$$

where

$$C_k = \frac{n!}{k!\,(n-k)!}.$$

EXAMPLE 4   Write and simplify the expansions of

    a. $(x+2)^5$ b. $(2x-3y)^4$.

SOLUTION   Theorem 9.7 is applied in each case.

a. $(x+2)^5 = x^5 + \dfrac{5!}{1! \cdot 4!}x^4 2^1 + \dfrac{5!}{2! \cdot 3!}x^3 2^2 + \dfrac{5!}{3! \cdot 2!}x^2 2^3 + \dfrac{5!}{4! \cdot 1!}x^1 2^4 + 2^5$

     $(x+2)^5 = x^5 + 10x^4 + 40x^3 + 80x^2 + 80x + 32.$

b.   We view $(2x-3y)^4$ as $[2x+(-3y)]^4$. Then

$$(2x-3y)^4 = (2x)^4 + \frac{4!}{1! \cdot 3!}(2x)^3(-3y) + \frac{4!}{2! \cdot 2!}(2x)^2(-3y)^2 +$$

$$\frac{4!}{3! \cdot 1!}(2x)(-3y)^3 + (-3y)^4$$

     $(2x-3y)^4 = 16x^4 - 96x^3y + 216x^2y^2 - 216xy^3 + 81y^4.$

Exercise Set 9.4

(A)   *Write each product using factorial notation.*

     1. $1 \cdot 2 \cdot 3 \cdot 4 \cdot 5 \cdot 6 \cdot 7$ 2. $1 \cdot 2 \cdot 3 \cdot \;\cdots\; \cdot 31 \cdot 32 \cdot 33$

Write as a product without factorial notation. Show the first three factors, the last three factors and an ellipsis symbol.

     3. $19!$ 4. $112!$ 5. $(n-1)!$

     6. $(n+1)!$ 7. $(2n)!$ 8. $(2n+1)!$

For Exercises 9–12, see Example 1.

     9. Write $12!$ as a product in which $9!$ is one of the factors.

     10. Write $23!$ as a product in which $20!$ is one of the factors.

     11. Write $(n+1)!$ as a product in which $(n-1)!$ is one of the factors.

     12. Write $(n-1)!$ as a product in which $(n-3)!$ is one of the factors.

Simplify each expression. (See Example 2.)

13. $\dfrac{(8!)(0!)}{6!}$

14. $\dfrac{12!}{(9!)(0!)}$

15. $\dfrac{9 \cdot 8 \cdot 7 \cdot 6 \cdot 5}{5!}$

16. $\dfrac{16 \cdot 15 \cdot 14 \cdot 13 \cdot 12}{5!}$

17. $\dfrac{7!}{3!(7-3)!}$

18. $\dfrac{11!}{5!(11-5)!}$

19. $\dfrac{4!}{2!\,2!} \cdot \dfrac{2!\,11!}{13!}$

20. $\dfrac{13!}{5!\,8!} \cdot \dfrac{13!}{4!\,9!}$

21. $\dfrac{(n+1)!}{(n-1)!}$

22. $\dfrac{(2n-1)!}{(2n)!}$

23. $\dfrac{n!\,(2n)!}{(n-2)!(2n-1)!}$

24. $\dfrac{(n+2)!(n-3)!}{(n+3)!(n-2)!}$

Write, and simplify, the expansion of each binomial expression. (See Example 4.)

25. $(x+2)^6$

26. $(y-3)^5$

27. $(x+2y)^4$

28. $(2y-1)^5$

29. $(3x-2y)^4$

30. $\left(x-\dfrac{y}{2}\right)^4$

31. $\left(2a-\dfrac{b}{2}\right)^5$

32. $\left(\dfrac{x}{3}+2y\right)^3$

Write, and simplify, the first four terms of the expansion.

33. $(x+2y)^{28}$

34. $(x-\sqrt{3})^{12}$

35. $(x^2-\sqrt{2})^{18}$

36. $\left(a-\dfrac{b}{3}\right)^{20}$

(B) 37. Approximate $(1.02)^9$ to the nearest hundredth by considering the expansion of $(1+0.02)^9$.

38. Approximate $(0.98)^7$ to the nearest hundredth.

39. If \$1000 is invested at 6% compounded annually for n years, then $A = 1000(1+0.06)^n$. Find A to the nearest dollar if $n=5$ years.

40. In Exercise 39, find A if $n=10$ years.

Chapter 9 Self-Test
All sequences in this Self-Test are either arithmetic or geometric.

[9.1] *For each of the sequences defined below, specify the following information: the first four terms; whether arithmetic or geometric; the common difference d or the common ratio r.*

1. $s_n=2-3n$

2. $s_n=2(-3)^n$

3. $s_1=\frac{3}{2},\ s_{n+1}=\frac{1}{3}s_n$

4. $s_1=\frac{3}{2},\ s_{n+1}=\frac{1}{3}+s_n$

For each sequence, specify: whether arithmetic or geometric; the next three terms; the nth term; the tenth term.

5. $32x^5,\ 16x^4,\ 8x^3,\ \cdots$

6. $3x-6,\ 3x-12,\ 3x-18,\ \cdots$

7. Find the first term of: the *arithmetic* sequence with $s_4=81$ and $d=3$; the *geometric* sequence with $s_4=81$ and $r=3$.

8. Find x such that the sequence 10, x, 40 will be: arithmetic; geometric.

[9.2] *Find the indicated partial sum of each sequence.*

9. $\frac{2}{5}, \frac{2}{25}, \frac{2}{125}, \cdots; \, S_5$

10. $-\frac{7}{x}, -\frac{2}{x}, \frac{3}{x}, \cdots; \, S_{10}$

Write the expanded form and find the sum.

11. $\displaystyle\sum_{k=1}^{1000} kx^2$

12. $\displaystyle\sum_{j=2}^{7} \left(\tfrac{1}{2}\right)^j$

[9.3] *Find the sum (if it exists) of each infinite geometric series.*

13. $3 + 4 + \frac{16}{3} + \cdots$

14. $3 + \frac{9}{4} + \frac{27}{16} + \cdots$

15. $\displaystyle\sum_{k=1}^{\infty} \left(-\tfrac{2}{7}\right)^k$

16. $\displaystyle\sum_{k=1}^{\infty} \left(-\tfrac{7}{2}\right)^k$

17. Find the rational number associated with the nonterminating decimal numeral $0.18\overline{18}$.

[9.4] *Simplify each expression.*

18. $\dfrac{11!}{6! \,(11-6)!}$

19. $\dfrac{(n+2)!}{(n-2)!}$

Write and simplify the expansion of each binomial expression.

20. $(x+3)^5$

21. $\left(2y - \tfrac{1}{2}\right)^4$

22. Use a binomial expansion to compute $(1.06)^{10}$ to the nearest thousandth.

SUPPLEMENT M

MATHEMATICAL INDUCTION

The process of reasoning from a set of specific examples to a general conclusion is called *inductive reasoning*—a process that does not always lead to a valid conclusion. For example, consider the polynomial

$$P(n) = n^2 - n + 41^*.$$

If we compute $P(1)$, $P(2)$, ..., $P(40)$, we obtain *prime* numbers each time (the first six such primes are 41, 43, 47, 53, 61 and 71), a result that leads to the conjecture that $P(n)$ names a prime number for all natural number replacements of n. In fact, the conjecture is false because $P(41)$ is not prime:

$$P(41) = (41)^2 - 41 + 41 = (41)^2.$$

Now, suppose that we have an open sentence in the variable n and we want to prove that the sentence is a true statement for *all* $n \in N$. As the preceding discussion indicates, a *finite* number of cases does not constitute a proof. On the other hand, it is not possible to verify an *infinite* number of cases. A way out of this dilemma is provided by a method of proof, called **mathematical induction**, based upon the following axiom.

Axiom M If S is a set of natural numbers such that

I $1 \in S$;
II If $k \in S$ then $(k + 1) \in S$ for all $k \in N$;
    then S is equal to the set N of natural numbers.

Given an open sentence in n, if we want to prove by mathematical induction that the solution set S of the sentence is equal to N, we may proceed as follows:

I *Show that 1 is a solution of the open sentence.*
II *Show that whenever k is a solution then $(k + 1)$ is also a solution of the open sentence.*

If both parts are established then, by Axiom M, it follows that the solution set S includes all of the natural numbers.

EXAMPLE 1 Prove by mathematical induction that the sum of the first n consecutive odd natural numbers is equal to n^2.

SOLUTION In symbols, we want to prove that, for all $n \in N$,

$$1 + 3 + 5 + \cdots + (2n - 1) = n^2. \qquad \langle 1 \rangle$$

Part I Substituting 1 for n, we have $1 = 1^2$, a true statement. Hence, $1 \in S$, where S is the solution set of $\langle 1 \rangle$.

---

*In this section it is to be understood that the set N of natural numbers is the replacement set for the variables n or k.

Part II Assume that $k \in S$. That is, assume that

$$1 + 3 + 5 + \cdots + (2k - 1) = k^2 \qquad \langle 2 \rangle$$

is a true statement. We must use $\langle 2 \rangle$ to establish that $(k + 1) \in S$. Toward this end we may reason as follows. Equation $\langle 2 \rangle$ asserts that the sum of the first k odd natural numbers is k^2. To investigate the sum of the first $(k + 1)$ consecutive odd natural numbers, let us add $2k + 1$ (the next odd number after $2k - 1$) to each member of $\langle 2 \rangle$:

$$1 + 3 + 5 + \cdots + (2k - 1) + (2k + 1) = k^2 + (2k + 1)$$
$$1 + 3 + 5 + \cdots + (2k - 1) + (2k + 1) = (k + 1)^2. \qquad \langle 3 \rangle$$

Equation $\langle 3 \rangle$ is a true statement because it is equivalent to equation $\langle 2 \rangle$. But, as you can verify, $\langle 3 \rangle$ is precisely what is obtained if $k + 1$ is substituted for n in equation $\langle 1 \rangle$. It follows that $k + 1$ satisfies equation $\langle 1 \rangle$ and so $(k + 1) \in S$. By Axiom M, it follows that $S = N$. That is, the solution set of equation $\langle 1 \rangle$ includes all of the natural numbers. □

EXAMPLE 2 Prove Theorem 9.1 by mathematical induction.

SOLUTION We want to prove that the nth term of an arithmetic sequence is

$$s_n = a + (n - 1)d. \qquad \langle 4 \rangle$$

Part I For $n = 1$, we have $s_1 = a + 0d = a$, which is a true statement. Hence, $1 \in S$.

Part II Assume that $k \in S$. That is, assume that

$$s_k = a + (k - 1)d. \qquad \langle 5 \rangle$$

We must use equation $\langle 5 \rangle$ to establish that $k + 1$ is a solution of equation $\langle 4 \rangle$. By Definition 9.1:

$$s_{k+1} = s_k + d. \qquad \langle 6 \rangle$$

Substituting from equation $\langle 5 \rangle$ into equation $\langle 6 \rangle$, we have

$$s_{k+1} = a + (k - 1)d + d$$
$$s_{k+1} = a + kd. \qquad \langle 7 \rangle$$

But, as you can verify, equation $\langle 7 \rangle$ is what is obtained if $(k + 1)$ is substituted for n in equation $\langle 4 \rangle$. Because equation $\langle 7 \rangle$ is a true statement, it follows that $(k + 1) \in S$. □

In the next example we shall refer to the following *recursive* definition of a^n.

Definition A For $a \in R$ and $n \in N$: I $a^1 = a$; II $a^{n+1} = a^n \cdot a$.

EXAMPLE 3 Prove: For all $n \in N$ and $a, b \in R$, $(ab)^n = a^n b^n$.

(Continued on next page)

SOLUTION

Part I   If $n=1$ then, by Definition A–I, we have

$$(ab)^1 = ab = a^1 b^1.$$

Hence, $1 \in S$.

Part II   Assume that $k \in S$. That is, assume that

$$(ab)^k = a^k b^k. \qquad \langle 8 \rangle$$

Now, consider $(ab)^{k+1}$ and the following sequence of steps and reasons.

$(ab)^{k+1} = (ab)^k \cdot ab$	(Definition A–I)
$(ab)^{k+1} = (a^k b^k) \cdot ab$	(Equation $\langle 8 \rangle$)
$(ab)^{k+1} = (a^k \cdot a)(b^k \cdot b)$	(Axioms R–5, R–6)
$(ab)^{k+1} = a^{k+1} b^{k+1}$	(Definition A–II).

The last equation establishes that $(k+1) \in S$, and we conclude that $(ab)^n = a^n b^n$ for all $n \in N$.   □

Exercise Set M

All proofs in this exercise set are to be done by mathematical induction.

1. Complete part II of the following proof that, for all $n \in N$,

$$1 + 2 + 3 + \cdots + n = \frac{n}{2}(n+1).$$

Part I   For $n=1$, we have: $1 = \frac{1}{2}(1+1)$ $\leftrightarrow$ $1 = 1$, a true statement. Hence, $1 \in S$.

Part II   Assume that $k \in S$. That is, assume that

$$1 + 2 + 3 + \cdots + k = \frac{k}{2}(k+1). \qquad \langle 9 \rangle$$

Now, add $k+1$ to each member of $\langle 9 \rangle$ and complete the proof.

Prove that the equation in each of Exercises 2–10 is a true statement for all $n \in N$.

2. $3 + 5 + 7 + \cdots + (2n+1) = n(n+2)$.

3. $1 + 2 + 4 + \cdots + 2^{n-1} = 2^n - 1$.

4. $1^2 + 2^2 + 3^2 + \cdots + n^2 = \dfrac{n}{6}(n+1)(2n+1)$.

5. $1^3 + 2^3 + 3^3 + \cdots + n^3 = \dfrac{n^2}{4}(n+1)^2$.

6. $4 + 4^2 + 4^3 + \cdots + 4^n = \dfrac{4}{3}(4^n - 1)$.

7. $\dfrac{1}{1 \cdot 2} + \dfrac{1}{2 \cdot 3} + \dfrac{1}{3 \cdot 4} + \cdots + \dfrac{1}{n(n+1)} = \dfrac{n}{n+1}$.

8. $\dfrac{1}{1 \cdot 3} + \dfrac{1}{3 \cdot 5} + \dfrac{1}{5 \cdot 7} + \cdots + \dfrac{1}{(2n-1)(2n+1)} = \dfrac{n}{2n+1}$.

9. $(-1)^{2n}=1$.   (The product of an *even* number of factors -1 is 1.)

10. $(-1)^{2n-1}=-1$.   (The product of an *odd* number of factors -1 is -1.)

Use Definition A, page 309, for Exercises 11–13. (In Exercises 11 and 12, we say that "the induction is on n," which means that m is to be viewed as a constant.)

11. $a^m \cdot a^n = a^{m+n}$

12. $(a^m)^n = a^{mn}$

13. $\left(\dfrac{a}{b}\right)^n = \dfrac{a^n}{b^n}, \quad b \neq 0$

Prove each of the following theorems.

14. Theorem 9.2.   (Hint: Use Definition 9.2.)

15. Theorem 9.3–II.   (Use the result to establish Theorem 9.3–I.)

16. Theorem 9.4.

PROBABILITY

10.1   BASIC COUNTING PRINCIPLES

The theory of probability constitutes an important branch of mathematics with its roots in the study of gambling games. Today, probability has many applications to the physical and behavioral sciences, business administration and management, military stategies, etc., and more applications are being discovered all the time. A study of the elementary theory of probability requires the ability to do certain kinds of counting. In this section we consider two basic counting principles.

Union of Sets   Suppose that we know the number of members in each of two finite sets A and B, and we need to know the number of members in $A \cup B$, the union of the sets. There are two cases to consider: A and B have no common members ($A \cap B = \emptyset$); A and B have one or more common members ($A \cap B \neq \emptyset$). In the first case, we need only add the number of members in A to the number of members in B. That is, if $n(A)$ denotes the number of members in a finite set A, then

$$n(A \cup B) = n(A) + n(B) \qquad \langle 1 \rangle$$

holds for any finite sets A and B whenever $A \cap B = \emptyset$. In the second case, we must take into account $n(A \cap B)$, the number of common members. If we actually count all of the members in both sets, then the common members are counted twice and the sum is too great by the amount $n(A \cap B)$. Hence, in this case we can compute $n(A \cup B)$ by subtracting $n(A \cap B)$ from the sum computed in equation $\langle 1 \rangle$. These results suggest our first basic counting principle.

Theorem 10.1   For any two finite sets A and B,

$$n(A \cup B) = n(A) + n(B) - n(A \cap B).$$

Note that Theorem 10.1 includes both of the above cases because, if

$A \cap B = \emptyset$, then $n(A \cap B) = 0$ and the conclusion of Theorem 10.1 becomes the same as equation $\langle 1 \rangle$.

EXAMPLE 1   In a group of students, 22 are taking mathematics, 24 are taking science, 16 are taking both subjects, 28 are taking neither subject. How many students are taking mathematics or science (or both)? How many students are in the group?

SOLUTION   Let M be the set of mathematics students and S the set of science students. Then $M \cap S$ is the set of students taking both subjects and $M \cup S$ is the set taking either mathematics or science (or both); we want $n(M \cup S)$. We have $n(M) = 22$, $n(S) = 24$, $n(M \cap S) = 16$ and, by Theorem 10.1,

$$n(M \cup S) = n(M) + n(S) - n(M \cap S)$$
$$n(M \cup S) = 22 + 24 - 16 = 30.$$

The number of students in the group is the sum of $n(M \cup S)$ and the number who are taking *neither* of the two subjects. Hence, the group consists of $30 + 28 = 58$ students.

Cartesian Products   Another type of counting problem arises when we need to know the number of ways in which the members of one set A can be paired with the members of a second set B. But the very wording of the problem brings to mind the concept of a Cartesian product set $A \times B$ (See Section 2.1). For example, suppose that $n(A) = 4$ and $n(B) = 8$. Then each of the 4 members of A is to be paired with each of the 8 members of B to form $4 \cdot 8$ or 32 members of $A \times B$. This result suggests our second basic counting principle.

Theorem 10.2   For any two finite sets A and B,

$$n(A \times B) = n(A) \cdot n(B).$$

EXAMPLE 2   If a die (plural: dice) is rolled, the set D of possible outcomes can be listed as $\{1, 2, 3, 4, 5, 6\}$. If a coin is tossed, the set C of possible outcomes can be listed as $\{H, T\}$, where H and T denote heads and tails, respectively. The set $D \times C$ can be interpreted as the set of all possible pairings of the outcomes of both rolling a die and tossing a coin. To compute the number of such pairings, we have $n(D) = 6$, $n(C) = 2$ and, by Theorem 10.2,

$$n(D \times C) = n(D) \cdot n(C) = 6 \cdot 2 = 12.$$

You will find it instructive to list the members of $D \times C$ and verify the above result.

Theorem 10.2 can be generalized to apply to more than two sets. If we have k sets and n_j is the number of elements in the jth set, then the number of members in the Cartesian product of the k sets is given by

$$n(A_1 \times A_2 \times \cdots \times A_k) = n_1 \cdot n_2 \cdot \cdots \cdot n_k. \qquad \langle 2 \rangle$$

In practice, equation $\langle 2 \rangle$ is often interpreted from the point of view of performing a series of k tasks. That is, if the jth task can be performed in n_j ways, then the k tasks can be performed in $n_1 \cdot n_2 \cdot \cdots \cdot n_k$ ways.

EXAMPLE 3   In a certain state, automobile license plates are imprinted with two letters followed by three digits. If all possible arrangements are permitted, how many different license plates can be made?

SOLUTION   We view the problem as that of performing a series of 5 tasks, each consisting of filling one of the five spaces

____ ____ ____ ____ ____ .

Now, each of the first two spaces may be filled with any one of the 26 letters of the alphabet $(n_1 = n_2 = 26)$; each of the last three spaces can be filled by any one of the 10 digits from 0 through 9 $(n_3 = n_4 = n_5 = 10)$. Hence, by equation $\langle 2 \rangle$, the total number of different license plates is

$$26 \cdot 26 \cdot 10 \cdot 10 \cdot 10 = 676{,}000.$$

Exercise Set 10.1

(A)
1. In a group of 23 athletes, 10 are on the football team (set F), 7 are on the track team (set T), 2 are on both teams.
   a. Describe each set: $F \cap T$; $F \cup T$.
   b. Find: $n(F)$; $n(T)$; $n(F \cap T)$; $n(F \cup T)$.
   c. How many athletes are not on either team?

2. In a class of 50 students, 18 are sophomores (set S), 21 have jobs after school (set J), 10 sophomores work after school.
   a. Describe each set: $S \cap J$; $S \cup J$.
   b. Find: $n(S)$; $n(J)$; $n(S \cap J)$; $n(S \cup J)$.
   c. How many students are neither sophomores nor work after school?

3. A poll of residents in a certain neighborhood disclosed the following data: 62 residents owned a car (set C), 28 owned a boat (set B), 10 owned a boat and a car, 14 owned neither a boat nor a car. How many residents were polled?

4. A salesman contacts customers in their homes either by a telephone call or by sending a free sample or both. During one week, the salesman sent a certain number of free samples; made 80 telephone calls; sent a sample and made a telephone call to 10 homes; sent a sample or made a telephone call or both to 120 homes. How many free samples did he send?

5. On a 26 member baseball team, there are 3 men who can catch, 6 men who can pitch and 18 men who can do neither. How many men can both pitch and catch?

6. In a club of 24 members, 7 are on committee A, 8 on committee B and 9 are not on either committee. How many club members are on both committees?

7. In a certain group of students, 25 are enrolled in a speech class, 32 are enrolled in a business class and 7 are enrolled in both a speech and a business class. If all the students in this group are enrolled in at least one of the classes, how many students are there in the group?

8. On a basketball team of 14 men, 6 can play guard, 8 can play forward and 2 can play both positions. How many can play neither position?

9. In a class of 150 students, 83 participate in athletics and 47 participate in student clubs. If 32 students do not participate in either, how many participate in both?

10. In a survey of registered voters in a community, it was found that 48% had registered as Democrats at least once, 46% had registered as Republicans at least once, and 4% had registered in each party at least once. What percent of those surveyed had never registered as a Republican or as a Democrat?

11. How many different routes can you choose to drive from town A to town B to town C if there are 5 roads from A to B and 3 roads from B to C?

12. Jack has 2 hats, 4 pairs of trousers and 2 jackets. How many different outfits can Jack select to wear?

13. In how many ways can one consonant and one vowel be selected from the letters of the word PLASTIC?

14. In how many ways can 8 books be arranged side-by-side on a bookshelf?

15. How many 4-digit, 2-letter license plates can be made if the digits must precede the letters?

16. How many 10-digit telephone numbers are there if neither the first digit nor the fourth digit can be a 0 or a 1?

(B)   17. Pasquale's Pizza Parlor offers 5 different toppings for pizzas, each of which has a plain cheese base. How many different pizzas may be ordered if a customer may select 0, 1, 2, . . . , 5 different toppings?

18. Given 4 flags of different colors, how many signals can be made by arranging the flags on a vertical mast if at least 2 flags must be used for each signal?

19. How many 5-digit numerals beginning with 7 and naming an even number can be formed from the digits 0, 1, 2, 3, 4, 5, 6, 7, 8, 9:
a. If no digit may be repeated?
b. If each digit may be repeated up to 5 times?

20. In how many ways can 5 gifts, A, B, C, D, E be distributed to 5 persons if a specified person is to receive A and if no person is to receive more than one gift?

10.2   PERMUTATIONS

Given an automobile race between three cars, say a, b and c, consider the problem of counting all possible ways in which the cars can finish the race, assuming no ties. One way to accomplish this is to list all possible arrangements of the three letters,

$$abc \quad acb \quad bac \quad bca \quad cab \quad cba \qquad\qquad <1>$$

and then to count the results. In this case there are six arrangements. Now, two problems arise. One: are we sure that *all* possible arrange- have been listed? Two: if many objects are to be arranged, then the listing and counting method becomes lengthy and tedious. Thus, it would be helpful to have a formula for computing the total number of arrangements more efficiently, so we turn our attention to obtaining such a formula.

***n*-Membered Sets**   To begin, let us restate the problem. Given a set with n members, in how many different orders can the n members be arranged if each member is to be listed exactly once in each arrangement? Such *ordered arrangements* of n objects are called **permutations**. For example, listing $<1>$ shows the six permutations of the members of $\{a, b, c\}$. As an indication of how the problem can be solved, consider the task of filling each of the three blanks

$$\underline{\qquad} \quad \underline{\qquad} \quad \underline{\qquad}$$

with one of the members of $\{a, b, c\}$, with no repetitions.* Now, the first blank can be filled in three ways, the second blank in two ways and the third blank in one way. Hence, the number of permutations of the members of the given 3-membered set is $3 \cdot 2 \cdot 1 = 3!$. Note that $3! = 6$ and so this result agrees with that obtained from listing $<1>$. The preceding argument can be generalized to lead to the conclusion of the following theorem.

Theorem 10.3   The number of permutations, denoted by $_nP_n$, of the members of an n-membered set is: $_nP_n = n!$.

EXAMPLE 1   The number of six-digit numerals that can be formed from the members of $\{1, 2, 3, 4, 5, 6\}$, if no digit may be repeated, is the same as the number of permutations of the members of a 6-membered set: $_6P_6 = 6! = 720$. Thus, 720 such numerals can be formed.

***r*-Membered Subsets of *n*-Membered Sets**   Suppose that each member of an 8-membered basketball squad is qualified to play each position (right or left forward, center, right or left guard) and consider the problem of counting how many different 5-membered teams can be formed from the 8 players. Rather than thinking of 5 blanks, we think of 5

---

*"No repetitions" means that the same member does not appear more than once in each permutation.

positions, and reason as follows. The first position can be filled in 8 ways, the second in 7 ways, the third in 6 ways, the fourth in 5 ways and the fifth in 4 ways. Hence, the total number of 5-membered teams that can be formed is $8 \cdot 7 \cdot 6 \cdot 5 \cdot 4 = 6720$. Now, observe that

$$8 \cdot 7 \cdot 6 \cdot 5 \cdot 4 = \frac{8 \cdot 7 \cdot 6 \cdot 5 \cdot 4 \cdot (3 \cdot 2 \cdot 1)}{(3 \cdot 2 \cdot 1)} = \frac{8!}{3!}.$$

More generally, observe that for natural numbers n and r, $r \leq n$,

$$n(n-1)(n-2) \cdots (n-r+1)$$

$$= \frac{n(n-1)(n-2) \cdots (n-r+1)[(n-r)(n-r-1) \cdots 2 \cdot 1]}{[(n-r)(n-r-1) \cdots 2 \cdot 1]},$$

$$n(n-1)(n-2) \cdots (n-r+1) = \frac{n!}{(n-r)!}. \qquad \langle 2 \rangle$$

The preceding discussion, together with equation $\langle 2 \rangle$, suggests a theorem that enables us to count the possible permutations of the n members of a set if the n members are to be arranged r at a time, $r \leq n$, with no repetitions.

Theorem 10.4   The number of permutations, denoted by $_nP_r$, of r-membered subsets of an n-membered set is

$$_nP_r = \frac{n!}{(n-r)!}.$$

EXAMPLE 2   The number of five-digit numerals (no repetitions) that can be formed from the nine members of $\{1, 2, 3, 4, 5, 6, 7, 8, 9\}$ is

$$_9P_5 = \frac{9!}{(9-5)!} = \frac{9!}{4!} = 15{,}120.$$

Repetitions of Members   Up to this point we have been counting permutations in which repetitions of members do not occur. We next turn our attention to the case that repetitions *do* occur.

EXAMPLE 3   If repetitions are permitted, then the number of three-digit numerals that can be formed from the members of $\{6, 7, 8\}$ is $3 \cdot 3 \cdot 3 = 27$, because each of the three digits can be selected for each of the three positions.

Now consider the number of distinguishable permutations of the letters of the word ELEMENT. Note that the word contains three E's. If we assign subscripts to each E, then we have 7 distinct letters:

$$E_1, L, E_2, M, E_3, N, T,$$

that can be permuted in 7! ways. For each of these permutations, if we leave the positions of the other letters unchanged, we can permute the 3 letters E_1, E_2 and E_3 in 3! ways. Hence, if we let P denote the total number of distinguishable permutations of the letters E, L, E, M, E, N, T, then

$$7! = P \cdot 3! \quad \text{or} \quad P = \frac{7!}{3!}. \qquad \langle 3 \rangle$$

In the word ELEMENT only one letter is repeated, whereas in the word MISSISSIPPI more than one letter is repeated. The reasoning that leads to equations $\langle 3 \rangle$ can be generalized to apply to the case in which more than one member is repeated, as stated in the following theorem.

Theorem 10.5   The number of distinguishable permutations P of a set of n objects in which k objects occur more than once, and such that there are n_1 of the first object, n_2 of the second object, . . . , and n_k of the kth object, is given by

$$P = \frac{n!}{n_1! \cdot n_2! \cdots n_k!}.$$

EXAMPLE 4   In the word MISSISSIPPI, the letters S and I are repeated 4 times each and the letter P is repeated twice. Hence, the number of distinguishable permutations of the 11 letters of the word is

$$P = \frac{11!}{4!4!2!} = 34{,}650.$$

Exercise Set 10.2

(A)   1. List all the permutations of the members of $\{1, 2, 3\}$.

2. List all the permutations of the members of $\{1, 2, 3, 4\}$.

Compute each of the following. (See Examples 1 and 2.)

3. $_7P_7$ 4. $_5P_5$ 5. $_8P_3$ 6. $_8P_5$

7. $_9P_2$ 8. $_9P_7$ 9. $_{10}P_4$ 10. $_{10}P_6$

11. How many 5-digit numerals can be written using the digits 1, 2, 3, 4, 5:   a. without repetitions?   b. with repetitions?

12. How many 6-digit numerals can be written using the digits 1, 2, 3, 4, 5, 6:   a. without repetitions?   b. with repetitions?

13. How many 5-digit numerals can be written using the digits 0, 1, 2, 3, 4, 5, 6, 7, 8, 9:   a. without repetitions?   b. with repetitions?

14. How many 6-digit numerals can be written using the digits 0, 1, 2, 3, 4, 5, 6, 7, 8, 9:   a. without repetitions?   b. with repetitions?

15. In Exercise 13a (no repetitions), how many of the numerals are less than 50,000? How many of these are even?

16. In Exercise 14a (no repetitions), how many of the numerals are less than 500,000? How many of these are odd?

17. A certain bicycle padlock is opened by turning each of three rings until the correct permutation of 3 digits from {0, 1, 2, 3, 4, 5, 6, 7, 8, 9} is set. How many different padlock permutations are available?

18. An office safe is opened by turning a dial to each of three different numbers in a given sequence. If there are 40 numbers on the dial, how many different safe lock permutations are available?

19. How many different "words" can be made by using three different letters of the word SQUARE?

20. How many "words" can be made by using five different letters of the word TRIANGLE?

21. Find the number of distinguishable permutations of the letters of OCCURRENCE. (See Example 4.)

22. Find the number of distinguishable permutations of the letters of MATHEMATICS?

(B)   23. How many "words" that begin and end with a vowel can be formed from all of the letters of COUNTABLE?

24. How many permutations are there of the letters in PRODUCTS that begin with P and end with S?

25. How many arrangements of the letters of FEINT can be made if the vowels are not to be separated?

26. Find the number of distinguishable permutations of the letters of DISTINGUISHABLE. In how many of the permutations are the three I's together?

27. In how many ways can 10 books be arranged on a shelf if two specified books must be side by side?

28. In how many ways can 5 different keys be arranged on a key ring?

(C)   I   The following BASIC program uses a recursion definition for computing $_NP_K = P(N, K)$, the number of permutations of N objects taken K at a time. Run the program with the test data given and check the results by using Theorem 10.4. Write another BASIC program based upon Theorem 10.4.

```
10 READ N,K
20 LET P=1
30 FOR I=N-K+1 TO N
40 LET P=P*I
50 NEXT I
60 PRINT "P("N","R")="P
70 GO TO 10
80 DATA 6,6,10,7,9,5
90 END
```

10.3   COMBINATIONS

In Section 10.2 we counted the number of ordered arrangements, or permutations, of the members of a given set. In this section we are concerned with counting groupings of objects where the order in which they are arranged is *not* to be considered.

Counting Combinations   A small market has 4 check-out employees, $\{a, b, c, d\}$ and 3 cash-registers at which customers can be checked out. How many different groupings of 3 out of the 4 employees can be assigned to check-out duty at a given time if the order in which the employees are assigned to registers is not to be considered? For convenience, we shall refer to each different grouping in which the order is not considered as a *combination*. To answer the question of how many combinations of 3 out of 4 can be formed, consider the following list of the 24 permutations of 4 employees $\{a, b, c, d\}$ taken 3 at a time:

1. *abc   acb   bac   bca   cab   cba*
2. *abd   adb   bad   bda   dab   dba*
3. *adc   acd   dac   dca   cad   cda*
4. *dbc   dcb   bdc   bcd   cdb   cbd.*

Table 1

At first glance, it may appear that there are 24 combinations. A closer look, however, shows that each row consists of one combination permuted 6 times. Thus, there are only 4 combinations of employees available, and the question is answered. Before proceeding further, note that 4, the number of combinations counted, is also the number of distinct 3-membered subsets of the 4-membered set of employees.

As in the case of counting permutations, the process of listing and counting is lengthy and tedious, a formula would be a great advantage. Before developing such a formula, we specify more formally what is meant by a combination. A **combination** is an *r*-membered subset of an *n*-membered set. We use the symbol $\binom{n}{r}$ to name the number of combinations of *n* objects taken *r* at a time. (The symbol $_nC_r$ is also commonly used.)

EXAMPLE 1   Each of the following subsets of $\{a, b, c\}$ is a combination:

$\emptyset$, $\{a\}$, $\{b\}$, $\{c\}$, $\{a, b\}$, $\{a, c\}$, $\{b, c\}$, $\{a, b, c\}$.

By inspection, we see that

$$\binom{3}{0} = 1, \qquad \binom{3}{1} = 3, \qquad \binom{3}{2} = 3, \qquad \binom{3}{3} = 1.$$

Now, let us generalize the procedure followed above with respect to the market problem. (You may find it helpful to refer to Table 1 as you read.) Given a set with *n* members, we want to compute $\binom{n}{r}$, the number of *r*-membered subsets. First, we know that there is a total of $_nP_r$ permutations possible. Second, we know that these permutations include $_rP_r$ permutations of each distinct r-membered subset. That is, to each distinct r-membered subset there are $r!$ permutations. Hence, the total number of permutations is equal to the product of the number of combinations $\binom{n}{r}$ and $r!$, as stated in the next theorem.

Theorem 10.6   The number of combinations of n objects taken r at a time is

$$\binom{n}{r} = \frac{_nP_r}{r!}.$$

EXAMPLE 2   The number of 13-card hands that can be dealt from an ordinary bridge deck of 52 cards is

$$\binom{52}{13} = \frac{_{52}P_{13}}{13!}.$$

By Theorem 10.4: $_{52}P_{13} = \frac{52!}{(52-13)!} = \frac{52!}{39!}.$

Hence,

$$\binom{52}{13} = \frac{\frac{52!}{39!}}{13!} = \frac{52!}{39!\,13!} = 615{,}013{,}559{,}600. \qquad \langle 1 \rangle$$

The computations in Example 2, particularly equations $\langle 1 \rangle$, suggest the next theorem, an alternate form of Theorem 10.6.

Theorem 10.7   The number of combinations of n objects taken r at a time is

$$\binom{n}{r} = \frac{n!}{(n-r)!\,r!}.$$

EXAMPLE 3   The number of ways of selecting a 3-person committee from a club of 12 members is

$$\binom{12}{3} = \frac{12!}{9!\,3!} = 220.$$

Binomial Coefficients   Consider the right member of the conclusion of Theorem 10.7 and note that, if r is replaced by k, we have an expression identical to the right member of equation $\langle 3 \rangle$ on page 304. Hence, C_k, the numerical coefficient of the $(k+1)^{\text{st}}$ term in the expansion of $(a+b)^n$, is also the number of combinations of n objects taken k at a time. You may prefer to remember the Binomial Theorem (Theorem 9.7) in the form

$$(a+b)^n = \binom{n}{0}a^n + \binom{n}{1}a^{n-1}b^1 + \binom{n}{2}a^{n-2}b^2 + \cdots + \binom{n}{k}a^{n-k}b^k + \cdots + \binom{n}{n}b^n.$$

EXAMPLE 4   The fifth term in the expansion of $(x+1)^7$ can be found as follows. We have: $n=7$; $k+1=5$ from which $k=4$. Hence, the fifth term is

$$\binom{7}{4}x^{7-4}(1)^4=\frac{7!}{3!\,4!}x^3=35x^3.$$

Applications   We close this section with a consideration of some counting problems that involve the use of more than one of our counting formulas.

EXAMPLE 5   How many 5-card poker hands consisting of 3 kings and 2 queens can be formed from an ordinary bridge deck?

SOLUTION   We view the problem as the task of filling two blanks—one with three kings and the other with 2 queens. Now, there are 4 kings and 4 queens in a bridge deck. Hence, there are $\binom{4}{3}$ combinations of 3 kings and $\binom{4}{2}$ combinations of 2 queens. Thus, the total number of such hands is

$$\binom{4}{3}\cdot\binom{4}{2}=\frac{4!}{3!\,(4-3)!}\cdot\frac{4!}{2!\,(4-2)!}=24.$$

EXAMPLE 6   A drawing of 2 marbles is made from an urn that contains 7 red and 3 white marbles. How many drawings of two marbles will include at least one red?

SOLUTION   The number of drawings that include at least one red is the sum of the number of drawings that include exactly one red and the number of drawings that include exactly two reds. If exactly one red is drawn, then there must be exactly one white to complete the pair. The number of ways to draw one red and one white is

$$\binom{7}{1}\cdot\binom{3}{1}=7\cdot3=21.$$

If exactly two reds are drawn, then no whites can appear. The number of ways to draw 2 reds and 0 whites is

$$\binom{7}{2}\cdot\binom{3}{0}=21\cdot1=21.$$

Hence, the total number of ways is $21+21=42$. An alternate method consists of finding how many possible pairs of marbles can be drawn from the given 10, and subtracting from that number the number of pairs that include 0 reds and 2 whites. Thus:

$$\binom{10}{2}-\binom{7}{0}\cdot\binom{3}{2}=45-3=42.$$

Exercise Set 10.3

(A)   *Compute each of the following.*

1. $\binom{9}{2}$ 2. $\binom{9}{7}$ 3. $\binom{6}{1}$ 4. $\binom{6}{5}$

5. $\binom{12}{4}$ 6. $\binom{12}{8}$ 7. $\binom{20}{18}$ 8. $\binom{20}{2}$

9. In how many ways can a committee of 4 be chosen from a group of 10 people?

10. In how many ways can 3 books be selected from a set of 9 different books?

11. How many different 5-card poker hands can be dealt from a 52-card bridge deck?

12. How many different 2-card blackjack hands can be dealt from a 52-card bridge deck?

13. Eight points lie on a circle. How many chords are determined by these points? How many triangles are determined by these points as vertices? How many quadrilaterals?

14. In how many ways can a set of one or more points be selected from a set of 6 points?

15. In how many ways can a committee of 2 men and 3 women be chosen from a group of 5 men and 7 women?

16. In how many ways can 2 chemistry texts and 4 laboratory manuals be chosen from 6 chemistry texts and 8 laboratory manuals?

17. A set of 3 marbles is drawn at random from an urn containing 6 red, 8 white and 7 blue marbles. In how many ways can the set be chosen if the set is to contain: a. At least one white marble? b. No more than one white marble?

18. In how many ways can a selection of 5 books be made from a set of 10 books if: a. 2 certain books must be chosen? b. 2 certain books must not be chosen?

19. Write and simplify the seventh term of $(x-2)^{10}$. (See Example 4.)

20. Write and simplify the fourth term of $(2x+3)^9$.

(B)   21. In how many ways can a committee of 4 persons be selected from 5 married couples if: a. A couple may not serve together on the same committee? b. No two couples may serve on the same committee?

22. Show that: $\binom{n}{n-r}=\binom{n}{r}$.

23. Show that: $\binom{n+1}{r+1}=\binom{n}{r+1}+\binom{n}{r}$.

24. If $\binom{n}{2}=\frac{3}{28}\cdot\binom{2n}{3}$, find n.

25. If $\binom{n}{5}=\binom{n}{8}$, find $\binom{n}{10}$ and $\binom{15}{n}$.

26. Show that $\binom{n}{0} + \binom{n}{1} + \binom{n}{2} + \cdots + \binom{n}{n} = 2^n$. [Hint: Consider the expansion of $(1 + 1)^n$.]

27. Show that the number of ways of selecting at least one of n objects is $2^n - 1$. (Hint: See Exercise 26.)

28. How many nonempty subsets of a 20-membered set can be found?

(C) I   The following BASIC program uses a recursion definition to compute $\binom{N}{K} = C(N, K)$, the number of combinations of N objects taken K at a time. Run the program with the test data given and check the results. Write a program to compute $\binom{N}{K}$ based upon the formula of Theorem 10.7.

```
10 READ N,K
20 LET M=N
30 LET C=1
40 FOR I=1 TO K
50 LET C=C*M/I
60 LET M=M-1
70 NEXT I
80 PRINT "C("N","K")="C
90 GO TO 10
100 DATA 5,3,10,4,6,6
110 END
```

10.4   PROBABILITY

Now that we have developed some techniques for counting, we are ready to turn our attention to elementary probability theory. As stated earlier, the study of probability had its roots in gambling problems. Even today, we find that illustrations from the field of gambling are helpful in developing some of the basic concepts.

Sample Spaces and Events   We use the word "experiment" to refer to some process or procedure that can be repeated in such a manner that its results, or *outcomes*, can be studied. Thus, rolling a pair of dice, tossing a coin, selecting a lottery number, may be referred to as experiments. A **sample space** for an experiment is a set of all possible outcomes of the experiment. An **event** is a subset of a sample space.

EXAMPLE 1   An experiment consists of tossing a coin and a die. An outcome (head or tail) for the coin is any member of the set $\{H, T\}$; an outcome for the die is any member of $\{1, 2, 3, 4, 5, 6\}$. The sample space for the experiment is

$$\{H, T\} \times \{1, 2, 3, 4, 5, 6\} = \{(H, 1), (H, 2), \ldots, (T, 5), (T, 6)\}.$$

Three of the many possible events are:

1. Obtaining a tail: $\{(T, 1), (T, 2), (T, 3), (T, 4), (T, 5), (T, 6)\}$.
2. Obtaining a head and an even number: $\{(H, 2), (H, 4), (H, 6)\}$.
3. Obtaining a head or an even number:

$\{(H, 1), (H, 2), (H, 3), (H, 4), (H, 5), (H, 6), (T, 2), (T, 4), (T, 6)\}$.

Equally Likely Outcomes   In some experiments we can expect each of the outcomes to be "equally likely." That is, if the experiment is performed a great number of times, we can expect to obtain each outcome the same number of times. For such experiments we give a special name to the ratio of $n(E)$, the number of outcomes in the event E, to $n(S)$, the number of outcomes in the sample space S.

Definition 10.1   The **probability** $P(E)$ of an event E in a nonempty sample space S of equally likely outcomes is

$$P(E) = \frac{n(E)}{n(S)}.$$

EXAMPLE 2   A die is tossed. The sample space is $\{1, 2, 3, 4, 5, 6\}$. The event E that "not more than three dots appear on the uppermost face" is $E = \{3, 4, 5, 6\}$. Assuming that the die is "fair" so that each outcome is equally likely, the probability of the event E is

$$P(E) = \frac{n(E)}{n(S)} = \frac{4}{6} = \frac{2}{3}.$$

EXAMPLE 3   Two cards are drawn from an ordinary 52-card bridge deck. Let us compute the probability of the event E of "drawing 2 aces." The sample space S consists of all possible pairs (2-membered subsets) of a 52-membered set. Hence, by Theorem 10.7, $n(S) = \binom{52}{2}$. Because there are 4 aces in the deck, we have $n(E) = \binom{4}{2}$. Thus, assuming that the cards are drawn with equal likelihood,

$$P(E) = \frac{n(E)}{n(S)} = \frac{\binom{4}{2}}{\binom{52}{2}} = \frac{\frac{4!}{2!\,2!}}{\frac{52!}{50!\,2!}} = \frac{1}{221}.$$

Union of Two Events   The definition of probability can be combined with Theorem 10.1, our first basic counting principle, to obtain an expression for the probability of the union of two events, $P(E_1 \cup E_2)$. That is, the probability that event E_1 occurs, or event E_2 occurs, or both events E_1 and E_2 occur.

Theorem 10.8   For each two events E_1 and E_2 of a nonempty sample space S,

$$P(E_1 \cup E_2) = P(E_1) + P(E_2) - P(E_1 \cap E_2).$$

The notation $P(E_1 \cap E_2)$ in the conclusion of Theorem 10.8 means "the probability that both events E_1 and E_2 occur."

EXAMPLE 4   Two dice are tossed. Let E_1 be the event "a 2 shows" and E_2 the event "the sum of the two numbers is 4 or 5." Let us compute $P(E_1 \cup E_2)$, the probability that either "a 2 shows" or "the sum of the two numbers is 4 or 5." The sample space S is equal to $\{1, 2, 3, 4, 5, 6\} \times \{1, 2, 3, 4, 5, 6\}$ and, by Theorem 10.2, $n(S) = 36$. An inspection of the graph of S in Figure 10.4-1 shows that $n(E_1) = 11$, $n(E_2) = 7$ and $n(E_1 \cap E_2) = 3$. Hence,

$$\begin{aligned}
P(E_1 \cup E_2) &= P(E_1) + P(E_2) - P(E_1 \cap E_2) \\
&= \frac{n(E_1)}{n(S)} + \frac{n(E_2)}{n(S)} - \frac{n(E_1 \cap E_2)}{n(S)} \\
&= \tfrac{11}{36} + \tfrac{7}{36} - \tfrac{3}{36} \\
&= \tfrac{5}{12}.
\end{aligned}$$

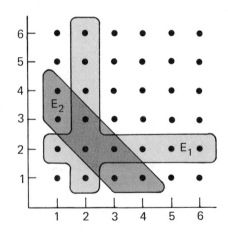

Figure 10.4-1

Special Events   Consider again the experiment of Example 1. In particular, let E_1 be the event "a head and an even number" and let E_2 be the event "a tail and an even number," and note that

$$E_1 \cap E_2 = \{(H, 2), (H, 4), (H, 6)\} \cap \{(T, 2), (T, 4), (T, 6)\} = \emptyset.$$

Pairs of events with no common members, such as E_1 and E_2 above, are characterized in the next definition.

Definition 10.2   Two events E_1 and E_2 are **mutually exclusive** if and only if $E_1 \cap E_2 = \emptyset$.

Of special interest is a pair of mutually exclusive events whose union is the entire sample space.

Definition 10.3   Two events E_1 and E_2 are **complementary events** in a sample space S if and only if

$$E_1 \cup E_2 = S \qquad \text{and} \qquad E_1 \cap E_2 = \emptyset.$$

The symbol E' is used to name the *complement* of E. Thus, if E_1 and E_2 are complementary events, then $E_2 = E_1'$ and $E_1 = E_2'$.

EXAMPLE 5   If 3 coins are tossed, then a sample space S is

$$\{\text{HHH, HHT, HTH, THH, HTT, THT, TTH, TTT}\}.$$

Now, if E is the event "at most two coins land tails," then the complementary event E' is the event $\{\text{TTT}\}$, "All 3 coins land tails."

In Example 5, note that $E \cup E' = S$. That is, events E and E' form the entire sample space S. Hence,

$$P(E \cup E') = P(S) = \frac{n(S)}{n(S)} = 1. \qquad \langle 1 \rangle$$

Furthermore, because $E \cap E' = \emptyset$, it follows that

$$P(E \cap E') = \frac{n(\emptyset)}{n(S)} = \frac{0}{n(S)} = 0. \qquad \langle 2 \rangle$$

From $\langle 1 \rangle$ and $\langle 2 \rangle$ and Theorem 10.8, we have

$$P(E \cup E') = P(E) + P(E') - P(E \cap E')$$
$$1 = P(E) + P(E') - 0,$$

as stated more formally in the next theorem.

Theorem 10.9   For any two complementary events E and E' of a nonempty sample space S: $P(E) + P(E') = 1$.

EXAMPLE 6   For the two complementary events of Example 5, we have that $P(E') = \frac{1}{8}$ and, by Theorem 10.9, $P(E) = 1 - \frac{1}{8} = \frac{7}{8}$. You may wish to verify this result directly from S in Example 5.

Exercise Set 10.4

(A)   *If two dice are tossed, then a sample space S consists of the 36 ordered pairs of numbers in $\{1, 2, 3, 4, 5, 6\} \times \{1, 2, 3, 4, 5, 6\}$. The components of each ordered pair are the two numbers that appear on the uppermost faces of the dice. We use "sum" to mean "the sum of the two numbers that appear." Exercises 1–20 refer to "dice tossing."*

1. List the event R "the two numbers on the dice are the same." Specify $n(R)$. Describe the event R'. Specify $n(R')$.

2. List the event T "a 3 shows." Specify $n(T)$. Describe the event T'. Specify $n(T')$.

3. List the event E "the sum is 7." Specify $n(E)$.

4. List the event F "the sum is 11." Specify $n(F)$.

5. In the following table, the first row lists all possible sums, the second row specifies the number n of ways that each sum can occur. Fill in the missing details.

sum	2	3	4	5	6	7	8	9	10	11	12
n	?	?	?	?	?	6	?	?	?	2	?

6. If C is the event "sum 2 or 3 or 12," specify $n(C)$.

7. Find $P(R)$ and $P(R')$ for the events of Exercise 1.

8. Find $P(T)$ and $P(T')$ for the events of Exercise 2.

Compute the probability of each sum (or sums).
9. $P(7)$ 10. $P(11)$ 11. $P(2, 3$ or $12)$
12. $P(7$ or $11)$ 13. $P(2$ or $12)$ 14. $P(2, 3, 4, 9, 10, 11$ or $12)$

Let E be the event "3 appears on one die only" and F the event "the sum is less than 5." Compute each probability.
15. $P(E)$ 16. $P(F)$ 17. $P(E')$
18. $P(F')$ 19. $P(E \cup F)$ 20. $P(E' \cup F')$

A sack contains 5 red, 6 white and 3 blue marbles. Two marbles are drawn. Let RR be the event that both marbles are red, WW that both marbles are white, and BB that both marbles are blue. Let RW, RB, and WB be the events that a red and a white, a red and a blue, and a white and a blue, respectively, are drawn. Find each of the following.
21. $P(RR)$ 22. $P(WW)$ 23. $P(BB)$
24. $P(RW)$ 25. $P(RB)$ 26. $P(WB)$

27. Find the probability that neither marble is red.

28. Find the probability that at least one marble is blue.

29. Find the probability that either both marbles are white or at least one is blue.

30. Find the probability that both marbles are different colors.

Two cards are selected from a standard bridge deck of 52 cards. Find the probability of each of the following.
31. Each card is the same color.

32. One card is a face card and the other is not.

33. One card is a heart and the other is a king.

34. Each card has the same face value.

(B)   *A sack contains 3 blue and 7 green marbles. A second sack contains 5 blue and 5 green marbles. Two marbles are drawn from each sack. Find the probability of drawing the following:*

35. 4 blue marbles.

36. 4 green marbles.

37. 2 marbles of each color.

38. 1 marble of one color and 3 of the other color.

39. Show that if E_1 and E_2 are events and $E_1 \subseteq E_2$, then $P(E_1) \leq P(E_2)$.

40. Show that if E_1 and E_2 are events, then $P(E_1 \cap E_2) \leq P(E_1 \cup E_2)$.

10.5   CONDITIONAL PROBABILITY

There are some experiments in which one event may be related to another. That is, the occurrence of one event may affect the probability of a subsequent event.

Independent and Dependent Events   Intuitively, we refer to two events as being *independent* if the likelihood of either event has no bearing on the likelihood of the other, and as being *dependent* if there is an effect. However, we define this concept mathematically in terms of probabilities, using the symbol $P(E_1 \cap E_2)$ to denote the probability that both events E_1 *and* E_2 occur, as stated in Section 10.4.

Definition 10.4   Two events E_1 and E_2 are **independent** if and only if

$$P(E_1 \cap E_2) = P(E_1) \cdot P(E_2).$$

The events are **dependent** if and only if they are not independent.

EXAMPLE 1   A red die and a green die are tossed. Let E_1 be the event that the red die shows a 3, and E_2 be the event that the green die shows a 6. From Figure 10.5−1 on page 330, we observe that

$$n(E_1) = 6, \qquad n(E_2) = 6, \qquad n(E_1 \cap E_2) = 1 \qquad \text{and} \qquad n(S) = 36.$$

Thus,

$$P(E_1) = P(E_2) = \tfrac{6}{36} = \tfrac{1}{6}, \qquad P(E_1 \cap E_2) = \tfrac{1}{36},$$

and the product

$$P(E_1) \cdot P(E_2) = \tfrac{1}{6} \cdot \tfrac{1}{6} = \tfrac{1}{36}.$$

Hence, the events E_1 and E_2 are *independent* because

$$P(E_1 \cap E_2) = P(E_1) \cdot P(E_2).$$

This result agrees with our intuitive feeling that the outcome obtained from one die should have no effect upon the outcome obtained from the other die.

EXAMPLE 2   In the experiment of Example 1, let E_3 be the event that the sum of the red and green dice outcomes is greater than 8. We can picture the

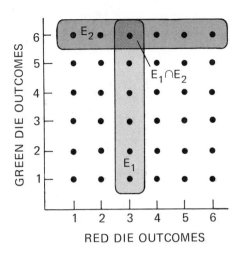

Figure 10.5-1

Figure 10.5-2

situation as in Figure 10.5-2. Observe that $n(E_2) = 6$, $n(E_3) = 10$, $n(E_2 \cap E_3) = 4$ and $n(S) = 36$. Hence,

$$P(E_2) = \tfrac{1}{6}, \qquad P(E_3) = \tfrac{5}{18}, \qquad \text{and} \qquad P(E_2) \cdot P(E_3) = \tfrac{1}{6} \cdot \tfrac{5}{18} = \tfrac{5}{108}.$$

Furthermore,

$$P(E_2 \cap E_3) = \frac{n(E_2 \cap E_3)}{n(S)} = \tfrac{4}{36} = \tfrac{1}{9}.$$

But $\tfrac{5}{108} \neq \tfrac{1}{9}$. Hence, $P(E_2) \cdot P(E_3) \neq P(E_2 \cap E_3)$ and so E_2 and E_3 are *dependent* events. This result also agrees with our intuition that suggests that the sum of the numbers is affected by the outcome on the green die.

Conditional Probability   Consider again the experiment of Example 2 with the following changes. Suppose that event E_2 has occurred. What then is the probability of event E_3 occurring? In other words, given that the green die shows a 6, what is the probability that the sum of the numbers showing is greater than 8? Looking again at Figure 10.5-2, we observe that of the 36 possible outcomes, only six outcomes, the elements of E_2, show a 6 on the green die. Hence E_2 constitutes a "reduced" sample space for determining the probability that the sum of the numbers showing is greater than 8. Thus, we wish to determine the number of elements in E_3 that are *also* in E_2, and compute

$$\frac{n(E_2 \cap E_3)}{n(E_2)}. \qquad\qquad \langle 1 \rangle$$

Since only 4 of the elements of E_3 are also in E_2, the probability we seek is $\tfrac{4}{6}$, or $\tfrac{2}{3}$.

We can place this problem in a more general context as follows. Suppose that in a sample space S we are interested in the probability that an event A occurs, given that an event B has already occurred. We need only consider those outcomes of A that are *also* outcomes of B, namely those outcomes in $A \cap B$, since we know that event B has already occurred. (See Figure 10.5-3.) Thus we can limit our sample space to B because, by hypothesis, the outcomes of B are the only possible outcomes that the experiment can produce.

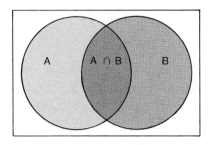

Figure 10.5-3

Definition 10.5   The **conditional probability** $P(A \mid B)$ of an event A, given that B has occurred, is

$$P(A \mid B) = \frac{n(A \cap B)}{n(B)}, \qquad n(B) \neq 0,$$

provided that the sample space consists of equally likely outcomes.

The symbol $P(A \mid B)$ in Definition 10.5 is read "the probability of A, given that B has occurred."

EXAMPLE 3   From the discussion preceding expression $\langle 1 \rangle$, the probability of obtaining a sum greater than 8, given that a green 6 already shows, is

$$P(E_3 \mid E_2) = \frac{n(E_3 \cap E_2)}{n(E_2)} = \frac{4}{6} = \frac{2}{3}.$$

An immediate consequence of Definition 10.5 is the next theorem.

Theorem 10.10   The conditional probability $P(A \mid B)$ of the event A, given that B has occurred, is

$$P(A \mid B) = \frac{P(A \cap B)}{P(B)}, \qquad P(B) \neq 0.$$

EXAMPLE 4   Referring once more to the events of Example 2, where

$$P(E_2) = \tfrac{1}{6} \quad \text{and} \quad P(E_2 \cap E_3) = \tfrac{1}{9},$$

we have, by Theorem 10.10,

$$P(E_3 \mid E_2) = \frac{P(E_3 \cap E_2)}{P(E_2)} = \frac{\tfrac{1}{9}}{\tfrac{1}{6}} = \frac{2}{3}.$$

In many situations we may be able to determine the conditional probability of an event and use the result to compute the probability that *both* of two events occur.

EXAMPLE 5   An urn contains 6 red and 4 blue marbles. Two marbles are drawn in succession, the first marble not being replaced before the second drawing. To find the probability that *both marbles are red*, we may proceed as follows. If B is the event that a red marble is drawn, then $P(B) = \tfrac{6}{10}$. After B has occurred, that is, *after* one red marble has been drawn, 5 red marbles remain among a total of 9 marbles. Hence, if A is the event of obtaining a red marble on the second draw after one red marble has been drawn, then $P(A \mid B) = \tfrac{5}{9}$. Since we seek $P(A \cap B)$, it follows from Theorem 10.10 that

$$P(A \cap B) = P(B) \cdot P(A \mid B) = \tfrac{6}{10} \cdot \tfrac{5}{9} = \tfrac{1}{3}.$$

Figure 10.5-4 shows a diagrammatic way to help think about problems such as that in Example 5. For instance, we can observe that the probability that the two drawings result in marbles of different colors is $\tfrac{4}{15} + \tfrac{4}{15} = \tfrac{8}{15}$, and the probability that both marbles drawn are the same color is $\tfrac{1}{3} + \tfrac{2}{15} = \tfrac{7}{15}$.

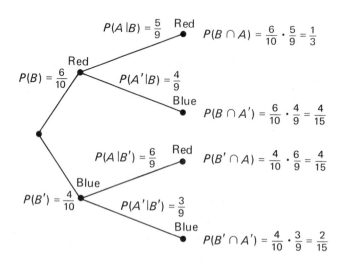

Figure 10.5-4

Exercise Set 10.5

(A)
1. A coin is tossed twice. Let A be the event "the first toss is a tail," let B be the event "the second toss is a tail," let C be the event "exactly two tosses are tails." Are A and B dependent or independent? Answer the same question for: A and C; B and C.

2. A coin is tossed three times. Let A be the event "the first toss is a tail," let B be the event "the second toss is a tail," let C be the event "exactly two tails in a row appear." Are A and B dependent or independent? Answer the same question for: A and C; B and C.

One sack contains 3 red and 5 green marbles. A second sack contains 6 red and 4 green marbles. A sack is selected at random and a marble is drawn from that sack. Find the probability of drawing each of the following:

3. A red marble, given that the first sack is chosen.

4. A green marble, given that the second sack is chosen.

5. A red marble.

6. A green marble.

A sack contains 8 red and 10 green marbles. Two marbles are drawn in succession, the first not being replaced before drawing the second. Find the probability of each of the following:

7. Both marbles are green.

8. The first marble is green and the second marble is red.

9. The marbles are of different colors.

10. The marbles are the same color.

11. Both marbles are red, given that the first marble is red.

12. The marbles are of different colors, given that the first marble is green.

A coin is tossed four consecutive times. Find the probability of each of the following:

13. The second toss is a head.

14. The fourth toss is a tail.

15. The second and fourth tosses are heads.

16. The second and fourth tosses are heads, but the first and third tosses are tails.

17. Exactly two heads turn up.

18. At least three tails turn up.

(B)
19. Four cards are drawn from a bridge deck of 52 cards. Find the probability that they consist of an ace, a king, a queen, and a jack.

20. The probability that A will be alive a year from now is 95% and the probability that B will be alive a year from now is 80%. What is the probability that either A or B will be alive a year from now?

21. In a room containing five people, what is the probability that at least two of them have the same birthday? [Hint: Consider the complementary event.]

22. In the experiment of Exercises 3-6 of this set, what is the probability that the second sack was chosen, given that a green marble was chosen?

Chapter 10   Self-Test

[**10.1**] 1. In a survey of 400 college students, 310 drank coffee, 205 drank tea, 180 drank both coffee and tea. How many students drank either coffee or tea or both? How many students drank neither coffee nor tea?

2. At one time, telephone numbers consisted of two out of 24 letters of the alphabet followed by four of the digits from 0 to 9. At most, how many such telephone numbers were available? How many more telephone numbers became available when a fifth digit was inserted between the two letters and the four digits?

[**10.2**] 3. Compute: $_6P_6$; $_6P_3$.

4. How many numerals can be written with the digits 6, 7, 8 and 9:
a. without repetitions? b. with repetitions?

5. A soap company contest required that a "word" formed from all of the letters of the brand name KLEENKLEAR be written on a Kleen-klear label and mailed in; a certain one of the words is to be the grand winner. How many words on wrappers would have to be mailed in to guarantee winning the contest?

[**10.3**] 6. Compute: $\binom{4}{1} \cdot \binom{13}{5} \div \binom{52}{3}$.

7. A committee of three students is to be selected at random from a student government containing 6 sophomores, 8 juniors and 7 seniors. In how many ways can the committee be selected if it must contain: a. At least one senior? b. No more than one senior? c. Exactly one senior?

8. Write and simplify the eighth term: $(y - 1)^{14}$.

[**10.4**] 9. Four cards are drawn from a 52-card bridge deck. Find the probability that all four cards are of the same: a. suit; b. denomination (face value).

10. Three marbles are drawn from a sack containing 5 red, 3 white and 4 blue marbles. What is the probability that all three marbles are of different colors?

[**10.5**] 11. Two dice are tossed. What is the probability that one die shows 3, given that the sum is 8?

12. In a certain city on a certain day, the probability that it will shower is 0.35, while the probability that it will shower and a rainbow will be visible is 0.005. What is the probability that a rainbow will be visible, given that it showers?

SUPPLEMENT N

BINOMIAL PROBABILITIES

In this section we consider experiments characterized by the following properties:

1. Each time the experiment is repeated, exactly one of two possible outcomes must result. We refer to one of the outcomes as "success" and the other as "failure."
2. If the probability of success is p, then the probability of failure is $q = 1 - p$.
3. The probability p of success remains unchanged for all trials of the experiment.

EXAMPLE 1   Over a period of many years, records show that a certain breed of white mice produces offspring in such a way that the probability of a female is 0.4; the probability of a male is 0.6; a typical litter consists of 8 offspring. We may view the "birth of an offspring" as an experiment; a "female" as a success (the choice is arbitrary); a "male" as a failure. Referring to the above three properties, note that:

1. The outcome of each experiment must be a female or a male.
2. The probability p of a success is 0.4; the probability q of a failure is $1 - 0.4 = 0.6$.
3. The sex of a given offspring is independent of the sex of the offspring born before or after it. Hence, in any 8 trials of the experiment the possibility of a success is unchanged from trial to trial.

For the litters referred to in Example 1, let us consider the problem of computing the probability of each of the possible combinations of females and males, e.g.: 0 females, 8 males; 1 female, 7 males; etc. In particular, let us consider a litter of 3 females and 5 males. Because the sex of any one mouse is independent of the sex of any other mice in the litter, by Definition 10.4 it follows that the probability of any *one* of the possible litters of 3 females and 5 males is

$$p^3 q^5 = (0.4)^3 (0.6)^5 = 0.00497664.$$

However, a litter of 8 offspring that includes 3 females can occur in $\binom{8}{3} =$ 56 ways. (Alternatively, the 5 males can appear in $\binom{8}{5}$ ways, but $\binom{8}{5} = \binom{8}{3}$, as you can verify.) Hence, the total probability of obtaining 3 females and 5 males in a litter of 8 is

$$\binom{8}{3} p^3 q^5 = 0.27869184.$$

The preceding discussion suggests the following generalization. For a litter of n offsprings in which the probability of a female is p and the probability of a male is $q = 1 - p$, the probability of obtaining exactly k

females and $n-k$ males is given by

$$\binom{n}{k}p^k q^{n-k}. \tag{<1>}$$

Note that this expression is equivalent to the general term in the expansion of $(q+p)^n$.

Returning to the problem of computing the probability of each possible combination of females and males in a litter of 8, by applying expression <1> we have*:

Females	Males	Probabilities
0	8	$\binom{8}{0}p^0 q^8 = 1 \cdot 1 \cdot (0.6)^8 = 0.0168$
1	7	$\binom{8}{1}p^1 q^7 = 8(0.4)^1(0.6)^7 = 0.0896$
2	6	$\binom{8}{2}p^2 q^6 = 28(0.4)^2(0.6)^6 = 0.2090$
3	5	$\binom{8}{3}p^3 q^5 = 56(0.4)^3(0.6)^5 = 0.2787$
4	4	$\binom{8}{4}p^4 q^4 = 70(0.4)^4(0.6)^4 = 0.2322$
5	3	$\binom{8}{3}p^5 q^3 = 56(0.4)^5(0.6)^3 = 0.1239$
6	2	$\binom{8}{2}p^6 q^2 = 28(0.4)^6(0.6)^2 = 0.0413$
7	1	$\binom{8}{1}p^7 q^1 = 8(0.4)^7(0.6)^1 = 0.0079$
8	0	$\binom{8}{0}p^8 q^0 = 1(0.4)^8 \cdot 1 = 0.0007$

Table 1

The probabilities listed in Table 1 can be used to determine probabilities of other events with respect to litters of mice.

EXAMPLE 2

a. The probability that a litter contains less than 3 females, $P(k < 3)$, is
$$P(k < 3) = P(0 \text{ females}) + P(1 \text{ female}) + P(2 \text{ females})$$
$$P(k < 3) = 0.0168 + 0.0896 + 0.2090$$
$$P(k < 3) = 0.3154.$$

b. The probability of 3 or more females, $P(k \geq 3)$, can be computed in at least two ways. One: we can add all of the probabilities listed in Table 1, starting with $P(3 \text{ females})$. Alternatively, using the result of Part (a):
$$P(k \geq 3) = 1 - P(k < 3) = 1 - 0.3154 = 0.6846.$$

Expression <1> is applicable to any set of repeated experiments that satisfy the three properties listed at the start of this section.

*Probabilities have been rounded off to 4 places.

EXAMPLE 3 A die is tossed 8 times (equivalently, 8 dice are tossed). What is the probability that a 5 or a 6 appears exactly 3 times?

SOLUTION Let A be the event that a 5 or a 6 appears (success) and B be the event that a 1, 2, 3 or 4 appears (failure). Then either A or B must occur and the number that appears at each toss is independent of the number that appears at any other toss. Also,

$$P(A) = \frac{2}{6} = \frac{1}{3} \quad \text{and} \quad P(B) = \frac{4}{6} = \frac{2}{3},$$

from which we have

$$p = \frac{1}{3} \quad \text{and} \quad q = 1 - p = \frac{2}{3}.$$

Hence, applying expression $\langle 1 \rangle$ with $n = 8$ and $k = 3$:

$$P(k = 3) = \binom{8}{3}\left(\frac{1}{3}\right)^3\left(\frac{2}{3}\right)^5 = \frac{1792}{6561}.$$

Exercise Set N

For Exercises 1 and 2, see Examples 1 and 2 and use Table 1.

1. What is the probability that a litter of 8 mice will include less than 2 females or more than 6 females?

2. What is the probability that the number of females is between 2 and 6, inclusive ($2 \le k \le 6$)? Can you do this exercise in two ways?

3. If the die of Example 3 is tossed 8 times, what is the probability that a 5 or a 6: a. Appears every time? b. Never appears? c. Appears at least once? d. Appears at least 6 times?

4. Ten coins are tossed. What is the probability that: a. Exactly 4 heads appear? b. No more than 4 heads appear? c. All heads appear? d. No heads appear?

5. Boys and girls are born in a certain family with equal likelihood. If the family has 5 children, what is the probability that: a. There are more girls than boys? b. There are all girls? c. There are no more than 2 girls?

6. Answer the questions of Exercise 5 for a family with 5 children if $P(\text{boy}) = \frac{2}{3}$ and $P(\text{girl}) = \frac{1}{3}$.

7. If the probability of a marriage ending in divorce within twenty years is $\frac{1}{3}$, what is the probability that in a group of 10 couples at least 7 couples will be divorced within twenty years?

8. A history quiz consists of 20 multiple choice questions. For each question there are 5 possible choices, only one of which is correct. If a student guesses on all 20 questions, what is the probability that the student will get at least 90% correct?

9. A meteorologist estimates the probability of a long term drought in each of 7 regions to be 0.4. What is the probability that more than one-half of the regions will suffer long term droughts?

10. If the probability that a certain student will answer each of 10 questions correctly is 0.9, what is the probability that the student will miss at least 4 questions?

APPENDIX

Appendix

Table 1   Common Logarithms

N	0	1	2	3	4	5	6	7	8	9
1.0	.0000	.0043	.0086	.0128	.0170	.0212	.0253	.0294	.0334	.0374
1.1	.0414	.0453	.0492	.0531	.0569	.0607	.0645	.0682	.0719	.0755
1.2	.0792	.0828	.0864	.0899	.0934	.0969	.1004	.1038	.1072	.1106
1.3	.1139	.1173	.1206	.1239	.1271	.1303	.1335	.1367	.1399	.1430
1.4	.1461	.1492	.1523	.1553	.1584	.1614	.1644	.1673	.1703	.1732
1.5	.1761	.1790	.1818	.1847	.1875	.1903	.1931	.1959	.1987	.2014
1.6	.2041	.2068	.2095	.2122	.2148	.2175	.2201	.2227	.2253	.2279
1.7	.2304	.2330	.2355	.2380	.2405	.2430	.2455	.2480	.2504	.2529
1.8	.2553	.2577	.2601	.2625	.2648	.2672	.2695	.2718	.2742	.2765
1.9	.2788	.2810	.2833	.2856	.2878	.2900	.2923	.2945	.2967	.2989
2.0	.3010	.3032	.3054	.3075	.3096	.3118	.3139	.3160	.3181	.3201
2.1	.3222	.3243	.3263	.3284	.3304	.3324	.3345	.3365	.3385	.3404
2.2	.3424	.3444	.3464	.3483	.3502	.3522	.3541	.3560	.3579	.3598
2.3	.3617	.3636	.3655	.3674	.3692	.3711	.3729	.3747	.3766	.3784
2.4	.3802	.3820	.3838	.3856	.3874	.3892	.3909	.3927	.3945	.3962
2.5	.3979	.3997	.4014	.4031	.4048	.4065	.4082	.4099	.4116	.4133
2.6	.4150	.4166	.4183	.4200	.4216	.4232	.4249	.4265	.4281	.4298
2.7	.4314	.4330	.4346	.4362	.4378	.4393	.4409	.4425	.4440	.4456
2.8	.4472	.4487	.4502	.4518	.4533	.4548	.4564	.4579	.4594	.4609
2.9	.4624	.4639	.4654	.4669	.4683	.4698	.4713	.4728	.4742	.4757
3.0	.4771	.4786	.4800	.4814	.4829	.4843	.4857	.4871	.4886	.4900
3.1	.4914	.4928	.4942	.4955	.4969	.4983	.4997	.5011	.5024	.5038
3.2	.5051	.5065	.5079	.5092	.5105	.5119	.5132	.5145	.5159	.5172
3.3	.5185	.5198	.5211	.5224	.5237	.5250	.5263	.5276	.5289	.5302
3.4	.5315	.5328	.5340	.5353	.5366	.5378	.5391	.5403	.5416	.5428
3.5	.5441	.5453	.5465	.5478	.5490	.5502	.5514	.5527	.5539	.5551
3.6	.5563	.5575	.5587	.5599	.5611	.5623	.5635	.5647	.5658	.5670
3.7	.5682	.5694	.5705	.5717	.5729	.5740	.5752	.5763	.5775	.5786
3.8	.5798	.5809	.5821	.5832	.5843	.5855	.5866	.5877	.5888	.5899
3.9	.5911	.5922	.5933	.5944	.5955	.5966	.5977	.5988	.5999	.6010
4.0	.6021	.6031	.6042	.6053	.6064	.6075	.6085	.6096	.6107	.6117
4.1	.6128	.6138	.6149	.6160	.6170	.6180	.6191	.6201	.6212	.6222
4.2	.6232	.6243	.6253	.6263	.6274	.6284	.6294	.6304	.6314	.6325
4.3	.6335	.6345	.6355	.6365	.6375	.6385	.6395	.6405	.6415	.6425
4.4	.6435	.6444	.6454	.6464	.6474	.6484	.6493	.6503	.6513	.6522
4.5	.6532	.6542	.6551	.6561	.6571	.6580	.6590	.6599	.6609	.6618
4.6	.6628	.6637	.6646	.6656	.6665	.6675	.6684	.6693	.6702	.6712
4.7	.6721	.6730	.6739	.6749	.6758	.6767	.6776	.6785	.6794	.6803
4.8	.6812	.6821	.6830	.6839	.6848	.6857	.6866	.6875	.6884	.6893
4.9	.6902	.6911	.6920	.6928	.6937	.6946	.6955	.6964	.6972	.6981
5.0	.6990	.6998	.7007	.7016	.7024	.7033	.7042	.7050	.7059	.7067
5.1	.7076	.7084	.7093	.7101	.7110	.7118	.7126	.7135	.7143	.7152
5.2	.7160	.7168	.7177	.7185	.7193	.7202	.7210	.7218	.7226	.7235
5.3	.7243	.7251	.7259	.7267	.7275	.7284	.7292	.7300	.7308	.7316
5.4	.7324	.7332	.7340	.7348	.7356	.7364	.7372	.7380	.7388	.7396
N	0	1	2	3	4	5	6	7	8	9

Table 1 *(continued)*

N	0	1	2	3	4	5	6	7	8	9
5.5	.7404	.7412	.7419	.7427	.7435	.7443	.7451	.7459	.7466	.7474
5.6	.7482	.7490	.7497	.7505	.7513	.7520	.7528	.7536	.7543	.7551
5.7	.7559	.7566	.7574	.7582	.7589	.7597	.7604	.7612	.7619	.7627
5.8	.7634	.7642	.7649	.7657	.7664	.7672	.7679	.7686	.7694	.7701
5.9	.7709	.7716	.7723	.7731	.7738	.7745	.7752	.7760	.7767	.7774
6.0	.7782	.7789	.7796	.7803	.7810	.7818	.7825	.7832	.7839	.7846
6.1	.7853	.7860	.7868	.7875	.7882	.7889	.7896	.7903	.7910	.7917
6.2	.7924	.7931	.7938	.7945	.7952	.7959	.7966	.7973	.7980	.7987
6.3	.7993	.8000	.8007	.8014	.8021	.8028	.8035	.8041	.8048	.8055
6.4	.8062	.8069	.8075	.8082	.8089	.8096	.8102	.8109	.8116	.8122
6.5	.8129	.8136	.8142	.8149	.8156	.8162	.8169	.8176	.8182	.8189
6.6	.8195	.8202	.8209	.8215	.8222	.8228	.8235	.8241	.8248	.8254
6.7	.8261	.8267	.8274	.8280	.8287	.8293	.8299	.8306	.8312	.8319
6.8	.8325	.8331	.8338	.8344	.8351	.8357	.8363	.8370	.8376	.8382
6.9	.8388	.8395	.8401	.8407	.8414	.8420	.8426	.8432	.8439	.8445
7.0	.8451	.8457	.8463	.8470	.8476	.8482	.8488	.8494	.8500	.8506
7.1	.8513	.8519	.8525	.8531	.8537	.8543	.8549	.8555	.8561	.8567
7.2	.8573	.8579	.8585	.8591	.8597	.8603	.8609	.8615	.8621	.8627
7.3	.8633	.8639	.8645	.8651	.8657	.8663	.8669	.8675	.8681	.8686
7.4	.8692	.8698	.8704	.8710	.8716	.8722	.8727	.8733	.8739	.8745
7.5	.8751	.8756	.8762	.8768	.8774	.8779	.8785	.8791	.8797	.8802
7.6	.8808	.8814	.8820	.8825	.8831	.8837	.8842	.8848	.8854	.8859
7.7	.8865	.8871	.8876	.8882	.8887	.8893	.8899	.8904	.8910	.8915
7.8	.8921	.8927	.8932	.8938	.8943	.8949	.8954	.8960	.8965	.8971
7.9	.8976	.8982	.8987	.8993	.8998	.9004	.9009	.9015	.9020	.9025
8.0	.9031	.9036	.9042	.9047	.9053	.9058	.9063	.9069	.9074	.9079
8.1	.9085	.9090	.9096	.9101	.9106	.9112	.9117	.9122	.9128	.9133
8.2	.9138	.9143	.9149	.9154	.9159	.9165	.9170	.9175	.9180	.9186
8.3	.9191	.9196	.9201	.9206	.9212	.9217	.9222	.9227	.9232	.9238
8.4	.9243	.9248	.9253	.9258	.9263	.9269	.9274	.9279	.9284	.9289
8.5	.9294	.9299	.9304	.9309	.9315	.9320	.9325	.9330	.9335	.9340
8.6	.9345	.9350	.9355	.9360	.9365	.9370	.9375	.9380	.9385	.9390
8.7	.9395	.9400	.9405	.9410	.9415	.9420	.9425	.9430	.9435	.9440
8.8	.9445	.9450	.9455	.9460	.9465	.9469	.9474	.9479	.9484	.9489
8.9	.9494	.9499	.9504	.9509	.9513	.9518	.9523	.9528	.9533	.9538
9.0	.9542	.9547	.9552	.9557	.9562	.9566	.9571	.9576	.9581	.9586
9.1	.9590	.9595	.9600	.9605	.9609	.9614	.9619	.9624	.9628	.9633
9.2	.9638	.9643	.9647	.9652	.9657	.9661	.9666	.9671	.9675	.9680
9.3	.9685	.9689	.9694	.9699	.9703	.9708	.9713	.9717	.9722	.9727
9.4	.9731	.9736	.9741	.9745	.9750	.9754	.9759	.9763	.9768	.9773
9.5	.9777	.9782	.9786	.9791	.9795	.9800	.9805	.9809	.9814	.9818
9.6	.9823	.9827	.9832	.9836	.9841	.9845	.9850	.9854	.9859	.9863
9.7	.9868	.9872	.9877	.9881	.9886	.9890	.9894	.9899	.9903	.9908
9.8	.9912	.9917	.9921	.9926	.9930	.9934	.9939	.9943	.9948	.9952
9.9	.9956	.9961	.9965	.9969	.9974	.9978	.9983	.9987	.9991	.9996
N	0	1	2	3	4	5	6	7	8	9

Table 2 *Natural Logarithms*

n	ln n	n	ln n	n	ln n
		5.0	1.6094	10	2.3026
0.1	7.6974 *	5.1	1.6292	11	2.3979
0.2	8.3906	5.2	1.6487	12	2.4849
0.3	8.7960	5.3	1.6677	13	2.5649
0.4	9.0837	5.4	1.6864	14	2.6391
0.5	9.3069	5.5	1.7047	15	2.7081
0.6	9.4892	5.6	1.7228	16	2.7726
0.7	9.6433	5.7	1.7405	17	2.8332
0.8	9.7769	5.8	1.7579	18	2.8904
0.9	9.8946	5.9	1.7750	19	2.9444
1.0	0.0000	6.0	1.7918	20	2.9957
1.1	0.0953	6.1	1.8083	25	3.2189
1.2	0.1823	6.2	1.8245	30	3.4012
1.3	0.2624	6.3	1.8405	35	3.5553
1.4	0.3365	6.4	1.8563	40	3.6889
1.5	0.4055	6.5	1.8718	45	3.8067
1.6	0.4700	6.6	1.8871	50	3.9120
1.7	0.5306	6.7	1.9021	55	4.0073
1.8	0.5878	6.8	1.9169	60	4.0943
1.9	0.6419	6.9	1.9315	65	4.1744
2.0	0.6931	7.0	1.9459	70	4.2485
2.1	0.7419	7.1	1.9601	75	4.3175
2.2	0.7885	7.2	1.9741	80	4.3820
2.3	0.8329	7.3	1.9879	85	4.4427
2.4	0.8755	7.4	2.0015	90	4.4998
2.5	0.9163	7.5	2.0149	95	4.5539
2.6	0.9555	7.6	2.0281	100	4.6052
2.7	0.9933	7.7	2.0412	105	4.6540
2.8	1.0296	7.8	2.0541	110	4.7005
2.9	1.0647	7.9	2.0669	115	4.7449
3.0	1.0986	8.0	2.0794	120	4.7875
3.1	1.1314	8.1	2.0919	125	4.8283
3.2	1.1632	8.2	2.1041	130	4.8675
3.3	1.1939	8.3	2.1163	135	4.9053
3.4	1.2238	8.4	2.1282	140	4.9416
3.5	1.2528	8.5	2.1401	145	4.9767
3.6	1.2809	8.6	2.1518	150	5.0106
3.7	1.3083	8.7	2.1633	155	5.0434
3.8	1.3350	8.8	2.1748	160	5.0752
3.9	1.3610	8.9	2.1861	165	5.1060
4.0	1.3863	9.0	2.1972	170	5.1358
4.1	1.4110	9.1	2.2083	175	5.1648
4.2	1.4351	9.2	2.2192	180	5.1930
4.3	1.4586	9.3	2.2300	185	5.2204
4.4	1.4816	9.4	2.2407	190	5.2470
4.5	1.5041	9.5	2.2513	195	5.2730
4.6	1.5261	9.6	2.2618	200	5.2983
4.7	1.5476	9.7	2.2721	205	5.3230
4.8	1.5686	9.8	2.2824	210	5.3471
4.9	1.5892	9.9	2.2925	215	5.3706

*Append - 10 (see page 267, Example 1).

Table 3 *Exponential Functions*

x	e^x	e^{-x}	x	e^x	e^{-x}
0.00	1.0000	1.0000	2.0	7.3891	0.1353
0.01	1.0101	0.9901	2.1	8.1662	0.1225
0.02	1.0202	0.9802	2.2	9.0250	0.1108
0.03	1.0305	0.9705	2.3	9.9742	0.1003
0.04	1.0408	0.9608	2.4	11.023	0.0907
0.05	1.0513	0.9512	2.5	12.182	0.0821
0.06	1.0618	0.9418	2.6	13.464	0.0743
0.07	1.0725	0.9324	2.7	14.880	0.0672
0.08	1.0833	0.9331	2.8	16.445	0.0608
0.09	1.0942	0.9139	2.9	18.174	0.0550
0.10	1.1052	0.9048	3.0	20.086	0.0498
0.11	1.1163	0.8958	3.1	22.198	0.0450
0.12	1.1275	0.8869	3.2	24.533	0.0408
0.13	1.1388	0.8781	3.3	27.113	0.0369
0.14	1.1503	0.8694	3.4	29.964	0.0334
0.15	1.1618	0.8607	3.5	33.115	0.0302
0.16	1.1735	0.8521	3.6	36.598	0.0273
0.17	1.1853	0.8437	3.7	40.447	0.0247
0.18	1.1972	0.8353	3.8	44.701	0.0224
0.19	1.2092	0.8270	3.9	49.402	0.0202
0.20	1.2214	0.8187	4.0	54.598	0.0183
0.21	1.2337	0.8106	4.1	60.340	0.0166
0.22	1.2461	0.8025	4.2	66.686	0.0150
0.23	1.2586	0.7945	4.3	73.700	0.0136
0.24	1.2712	0.7866	4.4	81.451	0.0123
0.25	1.2840	0.7788	4.5	90.017	0.0111
0.30	1.3499	0.7408	4.6	99.484	0.0101
0.35	1.4191	0.7047	4.7	109.95	0.0091
0.40	1.4918	0.6703	4.8	121.51	0.0082
0.45	1.5683	0.6376	4.9	134.29	0.0074
0.50	1.6487	0.6065	5.0	148.41	0.0067
0.55	1.7333	0.5769	5.1	164.02	0.0061
0.60	1.8221	0.5488	5.2	181.27	0.0055
0.65	1.9155	0.5220	5.3	200.34	0.0050
0.70	2.0138	0.4966	5.4	221.41	0.0045
0.75	2.1170	0.4724	5.5	244.69	0.0041
0.80	2.2255	0.4493	5.6	270.43	0.0037
0.85	2.3396	0.4274	5.7	298.87	0.0034
0.90	2.4596	0.4066	5.8	330.30	0.0030
0.95	2.5857	0.3867	5.9	365.04	0.0027
1.0	2.7183	0.3679	6.0	403.43	0.0025
1.1	3.0042	0.3329	6.5	665.14	0.0015
1.2	3.3201	0.3012	7.0	1096.6	0.0009
1.3	3.6693	0.2725	7.5	1808.0	0 0006
1.4	4.0552	0.2466	8.0	2981.0	0.0003
1.5	4.4817	0.2231	8.5	4914.8	0.0002
1.6	4.9530	0.2019	9.0	8103.1	0.0001
1.7	5.4739	0.1827	9.5	13360	0.00007
1.8	6.0496	0.1653	10.0	22026	0.00005
1.9	6.6859	0.1496			

Table 4 Squares, Square Roots, and Prime Factors

No.	Sq.	Sq. Rt.	Factors	No.	Sq.	Sq. Rt.	Factors
1	1	1.000		51	2,601	7.141	$3 \cdot 17$
2	4	1.414	2	52	2,704	7.211	$2^2 \cdot 13$
3	9	1.732	3	53	2,809	7.280	53
4	16	2.000	2^2	54	2,916	7.348	$2 \cdot 3^3$
5	25	2.236	5	55	3,025	7.416	$5 \cdot 11$
6	36	2.449	$2 \cdot 3$	56	3,136	7.483	$2^3 \cdot 7$
7	49	2.646	7	57	3,249	7.550	$3 \cdot 19$
8	64	2.828	2^3	58	3,364	7.616	$2 \cdot 29$
9	81	3.000	3^2	59	3,481	7.681	59
10	100	3.162	$2 \cdot 5$	60	3,600	7.746	$2^2 \cdot 3 \cdot 5$
11	121	3.317	11	61	3,721	7.810	61
12	144	3.464	$2^2 \cdot 3$	62	3,844	7.874	$2 \cdot 31$
13	169	3.606	13	63	3,969	7.937	$3^2 \cdot 7$
14	196	3.742	$2 \cdot 7$	64	4,096	8.000	2^6
15	225	3.873	$3 \cdot 5$	65	4,225	8.062	$5 \cdot 13$
16	256	4.000	2^4	66	4,356	8.124	$2 \cdot 3 \cdot 11$
17	289	4.123	17	67	4,489	8.185	67
18	324	4.243	$2 \cdot 3^2$	68	4,624	8.246	$2^2 \cdot 17$
19	361	4.359	19	69	4,761	8.307	$3 \cdot 23$
20	400	4.472	$2^2 \cdot 5$	70	4,900	8.367	$2 \cdot 5 \cdot 7$
21	441	4.583	$3 \cdot 7$	71	5,041	8.426	71
22	484	4.690	$2 \cdot 11$	72	5,184	8.485	$2^3 \cdot 3^2$
23	529	4.796	23	73	5,329	8.544	73
24	576	4.899	$2^3 \cdot 3$	74	5,476	8.602	$2 \cdot 37$
25	625	5.000	5^2	75	5,625	8.660	$3 \cdot 5^2$
26	676	5.099	$2 \cdot 13$	76	5,776	8.718	$2^2 \cdot 19$
27	729	5.196	3^3	77	5,929	8.775	$7 \cdot 11$
28	784	5.292	$2^2 \cdot 7$	78	6,084	8.832	$2 \cdot 3 \cdot 13$
29	841	5.385	29	79	6,241	8.888	79
30	900	5.477	$2 \cdot 3 \cdot 5$	80	6,400	8.944	$2^4 \cdot 5$
31	961	5.568	31	81	6,561	9.000	3^4
32	1,024	5.657	2^5	82	6,724	9.055	$2 \cdot 41$
33	1,089	5.745	$3 \cdot 11$	83	6,889	9.110	83
34	1,156	5.831	$2 \cdot 17$	84	7,056	9.165	$2^2 \cdot 3 \cdot 7$
35	1,225	5.916	$5 \cdot 7$	85	7,225	9.220	$5 \cdot 17$
36	1,296	6.000	$2^2 \cdot 3^2$	86	7,396	9.274	$2 \cdot 43$
37	1,369	6.083	37	87	7,569	9.327	$3 \cdot 29$
38	1,444	6.164	$2 \cdot 19$	88	7,744	9.381	$2^3 \cdot 11$
39	1,521	6.245	$3 \cdot 13$	89	7,921	9.434	89
40	1,600	6.325	$2^3 \cdot 5$	90	8,100	9.487	$2 \cdot 3^2 \cdot 5$
41	1,681	6.403	41	91	8,281	9.539	$7 \cdot 13$
42	1,764	6.481	$2 \cdot 3 \cdot 7$	92	8,464	9.592	$2^2 \cdot 23$
43	1,849	6.557	43	93	8,649	9.644	$3 \cdot 31$
44	1,936	6.633	$2^2 \cdot 11$	94	8,836	9.695	$2 \cdot 47$
45	2,025	6.708	$3^2 \cdot 5$	95	9,025	9.747	$5 \cdot 19$
46	2,116	6.782	$2 \cdot 23$	96	9,216	9.798	$2^5 \cdot 3$
47	2,209	6.856	47	97	9,409	9.849	97
48	2,304	6.928	$2^4 \cdot 3$	98	9,604	9.899	$2 \cdot 7^2$
49	2,401	7.000	7^2	99	9,801	9.950	$3^2 \cdot 11$
50	2,500	7.071	$2 \cdot 5^2$	100	10,000	10.000	$2^2 \cdot 5^2$

ANSWERS

ANSWERS FOR SELECTED EXERCISES

Diagnostic Test   (page 1)

1. 12 is a member of $\{10, 12, 14\}$ **2.** x and y are real numbers   **3.** $\{5, 7, 9, 11\}$
4. a. $\{2, 3, 4, 6, 8, 9, 10, 12\}$ **b.** $\{6, 12\}$ **5.** commutativity of multiplication   **6.** commutativity of addition   **7.** associativity of multiplication   **8.** additive inverse   **9.** mutiplicative inverse   **10.** distributivity   **11.** HYP: x is a real number; CONC: x^2 is nonnegative
12. a. 0 is the additive identity **b.** $a + (-a) = 0$ **c.** commutativity and associativity of addition
d. $a + (-a) = 0$ **e.** 0 is the additive identity   **13.** (Figure 13 follows)

13. a.

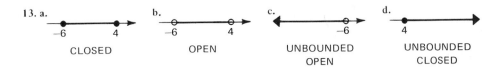

CLOSED	OPEN	UNBOUNDED OPEN	UNBOUNDED CLOSED

14. (Figure 14 follows)

14. a.

b.

a. $\{x: -6 \le x \le 4\}$ **b.** $\{x: -6 < x < 4\}$ **15.** b **16.** $-3 + 9i$ **17.** $11 + 2i$ **18.** 256
19. 9   **20.** $-2x$ **21.** 7 or -7 **22.** -5 **23.** $11\sqrt{3x}$ **24.** $4 + \sqrt{3}$ **25.** $\dfrac{x^2 - 2x\sqrt{3} + 3}{x^2 - 3}$
26. $-9 + 0i$ **27.** $\frac{11}{17} + \frac{7}{17}i$ **28.** $-1; -15; -5x^3 + 3x^2 + 7x - 1; -4 - 12i$
29. $3x^3 + 9x^2 - 43x + 11$ **30.** $3x^2(5x - 1)(x + 4)$ **31.** $(5x - 12y)(5x + 12y)$
32. $(2y - 5)(4y^2 + 10y + 25)$ **33.** $(x + y)(x^2 - xy + y^2)$ **34.** $(x - 2y)(3x^2 + y)$
35. not factorable   **36.** $\left(x - \frac{4}{9}\right)^2$ **37.** $(x - 3i)(x + 3i)$ **38.** $\dfrac{x - 5}{x + 3}$ **39.** $\dfrac{2x^2 - 4x + 20}{(x + 2)(x - 6)}$
40. $4x - 14 + \dfrac{33x + 62}{x^2 + 4x + 5}$ **41.** $x^2 + 5x + 25$ **42.** (Figure 42 follows)   **43.** positive x-coordinates and negative y-coordinates   **44.** (Figure 44 follows)

42.   **44.**

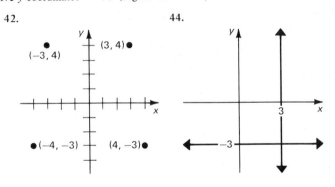

Exercise Set 0.1   (page 6)

1. 3 is a member of $\{1, 2, 3\}$ **3.** 16 is a natural number   **5.** x is a real number   **7.** a and b are natural numbers   **9.** $\sqrt{5}$ is not a rational number   **11.** the set of integers is a proper subset of the set of reals   **13.** the irrationals and rationals have no common members

15. $\{1, 2, 3, 4, 5, 6, 8, 10\}$ **17.** $\emptyset$ **19.** $\{1, 2, 3, 4, 5, 6, 7, 8, 9, 10\}$ **21.** $\{1, 2, 3, 4, 5\}$

27. $\{\pm 6, \pm 3, \pm 2, \pm 1\}$ **29.** $\{\pm 8, \pm 4, \pm 2, \pm 1\}$ **31.** $\{\pm 6, \pm 3, \pm 2, \pm 1, \pm\frac{3}{2}, \pm\frac{3}{4}, \pm\frac{1}{2}, \pm\frac{3}{8}, \pm\frac{1}{4}, \pm\frac{1}{8}\}$

33.

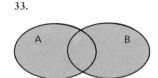

35.

37.

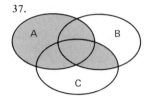

Exercise Set 0.2   (page 9)

1. E-1   **3.** E-4   **5.** R-3   **7.** R-1   **9.** R-7   **11.** R-4   **13.** R-9   **15.** O-2
17. O-3   **19.** O-1

Exercise Set 0.3   (page 14)

I *Part 1*

1. $a, b, c \in R$; $a=b$; $c \neq 0$	1. Hyp.
2. $ac = ac$	2. Ax. E-1
3. $ac = bc$	3. Ax. E-4

Part 2

1. $a, b, c \in R$; $ac=bc$; $c \neq 0$	1. Hyp.
2. $(ac)\dfrac{1}{c} = (bc)\dfrac{1}{c}$	2. Thm. 0.2, part 1
3. $a\left(c \cdot \dfrac{1}{c}\right) = b\left(c \cdot \dfrac{1}{c}\right)$	3. Ax. R-6
4. $c \cdot \dfrac{1}{c} = 1$	4. Ax. R-8
5. $a \cdot 1 = b \cdot 1$	5. Ax. E-4
6. $a \cdot 1 = a$, $b \cdot 1 = b$	6. Ax. R-7
7. $a = b$	7. Ax. E-4

III a. R-3 **b.** 0.2 **c.** R-9 **d.** R-3 **e.** 0.1   **V** $(-a)(b)=(-1a)b$ by Thm. 0.7-I; $(-1a)b=-1(ab)$ by Ax. R-6; $-1(ab)=-ab$ by Thm. 0.7-I

VII *Part 1*

1. $a, b, c, d \in R$; $b \neq 0$, $d \neq 0$; $ad=bc$	1. Hyp.
2. $ad \cdot \dfrac{1}{b} \cdot \dfrac{1}{d} = bc \cdot \dfrac{1}{b} \cdot \dfrac{1}{d}$	2. Thm. 0.2
3. $\left(a \cdot \dfrac{1}{b}\right)\left(d \cdot \dfrac{1}{d}\right) = \left(b \cdot \dfrac{1}{b}\right)\left(c \cdot \dfrac{1}{d}\right)$	3. Ax. R-5, R-6
4. $a \cdot \dfrac{1}{b} = c \cdot \dfrac{1}{d}$	4. Ax. R-8, R-7
5. $\dfrac{a}{b} = \dfrac{c}{d}$	5. Def. 0.2

Part 2

1. $a, b, c, d \in R$; $b \neq 0$, $d \neq 0$; $\dfrac{a}{b} = \dfrac{c}{d}$	1. Hyp.
2. $a \cdot \dfrac{1}{b} \cdot bd = c \cdot \dfrac{1}{d} \cdot bd$	2. Def. 0.2, Thm. 0.2
3. $(ad)\left(b \cdot \dfrac{1}{b}\right) = (bc)\left(d \cdot \dfrac{1}{d}\right)$	3. Ax. R-5, R-6
4. $ad \cdot 1 = bc \cdot 1$	4. Ax. R-8
5. $ad = bc$	5. Ax. R-7

IX a. 0.2 **b.** $R-9$ **c.** 0.2   **XI** By Def. 0.2 and Ax. $R-5$ and $R-6$: $\dfrac{a}{b}\cdot\dfrac{c}{d}=a\cdot\dfrac{1}{b}\cdot c\cdot\dfrac{1}{d}=ac\cdot\dfrac{1}{b}\cdot\dfrac{1}{d}.$

By Theorem A and Def. 0.2: $ac\cdot\dfrac{1}{b}\cdot\dfrac{1}{d}=ac\cdot\dfrac{1}{bd}=\dfrac{ac}{bd}.$ Hence, by Ax. $E-4$: $\dfrac{a}{b}\cdot\dfrac{c}{d}=\dfrac{ac}{bd}.$

XIII By Def. 0.4, if $a+c>b+c$, then $a+c=(b+c)+p$, where $p>0$. By Ax. $R-1$ and $R-2$: $a+c=(b+p)+c$. By Thm. 0.1, $a=b+p$, from which, by Def. 0.4, we have $a>b$.

XV By Def. 0.4, if $a>b$, then $b+p=a$, $p>0$. By Thm. 0.2, $(b+p)c=ac$ and, by Ax. $R-9$, $bc+pc=ac$ from which $bc=ac+(-pc)$. Now, if $p>0$ and $c<0$, then $pc<0$ and so $-pc>0$. Hence, by Def. 0.4, $bc>ac$.

Exercise Set 0.4   (page 18)

1. half-open, (Figure 1 follows)   **3.** closed   **5.** open, unbounded   **7.** closed, unbounded, (Figure 7 follows)   **9.** $\{x: x\le-1\}\cup\{x: x>4\}$; $\{x: x\le-1$ or $x>4\}$

11. $\{x: x<-3\}\cup\{x: 2\le x<6\}$; $\{x: x<-3$ or $2\le x<6\}$

13. $\{x: 1<x<5\}\cup\{x: 8\le x\le11\}$; $\{x: 1<x<5$ or $8\le x\le11\}$

15. $\{x: x\le-8\}\cup\{x: -5\le x\le-1\}\cup\{x: x>3\}$; $\{x: x\le-8$ or $-5\le x\le-1$ or $x>3\}$

17. $\{x: -5\le x<5\}$, (Figure 17 follows)   **19.** $\{x: x<-6\}$, (Figure 19 follows)   **21.** $\emptyset$

23. $\{x: 2\le x\le6\}$ **25.** $|-6(-x)|=|6x|$ (Thm. 0.7–III)   **27.** $\{x: x>0\}=R^+$
$\qquad\qquad\qquad\qquad\qquad\qquad =|6||x|$ (Thm. 0.20)
$\qquad\qquad\qquad\qquad\qquad\qquad =6(-x)$ (Def. 0.4)
$\qquad\qquad\qquad\qquad\qquad\qquad =-6x$ (Thm. 0.7–II)

1. **7.** **17.** **19.**

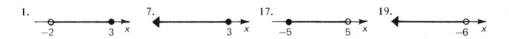

I *Case 1:* If $r>0$, then $-r<0$. Hence: $|r|=r$; $|-r|=-(-r)=r$; and $|r|=|-r|$.   *Case 2:* If $r=0$, then $-r=0$ and $0=|r|=|-r|$.   *Case 3:* If $r<0$, then $-r>0$. Hence: $|r|=-r$; and $|r|=|-r|$.

III *Case i:* If $a\ge0$ and $b\ge0$, then $ab\ge0$ by Ax. 0.3. Then, $|a|=a$, $|b|=b$ and $|ab|=ab$. Hence, $ab=|ab|=|a||b|$.   *Case ii:* If $a\ge0$ and $b<0$, then $ab\le0$ by Thm. 0.7–II. Then $|a|=a$, $|b|=-b$, and $|ab|=-ab$. Hence, $-ab=|ab|=|a||b|$.   *Case iii:* If $a<0$ and $b\ge0$, then $ab<0$. Then $|a|=-a$, $|b|=b$ and $|ab|=-ab$. Hence, $-ab=|ab|=|a||b|$.
Case iv: If $a<0$ and $b<0$, then $ab>0$. Then $|a|=-a$, $|b|=-b$ and $|a||b|=(-a)(-b)=ab$. Also, $|ab|=ab$. Thus $ab=|ab|=|a||b|$.

Exercise Set 0.5   (page 22)

1. 2, 3   **3.** -3, 5   **5.** $y=6$, $x=-2$ **7.** $x=-x=0$, $y=-3$ **9.** $-2+3i$ **11.** $9+2i$
13. 13   **15.** $-10+12i$ **17.** 1   **19.** -9 **21.** 16

23. $(a+bi)(c+di)=(ac-bd)+(ad+bc)i$
$\qquad\qquad\qquad\quad =(ca-db)+(cb+da)i$
$\qquad\qquad\qquad\quad =(c+di)(a+bi)$

25. $(a+bi)+(-a-bi)=[a+(-a)]+[b+(-b)]i=0+0i$

27. $(a+bi)(a-bi)=(a^2+b^2)+(-ab+ab)i$
$\qquad\qquad\qquad\quad =(a^2+b^2)+0i$
$\qquad\qquad\qquad\quad =(a^2+b^2)\in R$ (Closure of add. and mult. in R)

Exercise Set 0.6   (page 26)

1. 2   **3.** $\frac{1}{9}$ **5.** $\frac{4}{81}$ **7.** $\frac{24}{25}$ **9.** $\frac{9}{8}$ **11.** $\frac{8}{125}$ **13.** ±5 **15.** 3   **17.** ±3 **19.** -2
21. $(3i)(3i)=9i^2=-9$; $(-3i)(-3i)=9i^2=-9$
23. 2 imag. sq. roots;   1 real, 2 imag. cube roots;   4 imag. 4th roots
25. 2 real sq. roots;   1 real, 2 imag. cube roots;   2 real, 2 imag. 4th roots   **27.** 2   **29.** 3

31. -4 **33.** -8 **35.** 8   **37.** 27   **39.** $\frac{1}{125}$ **41.** $\frac{1}{256}$ **43.** $\dfrac{x^6}{y^6}$ **45.** $3x^2y$ **47.** $\dfrac{y}{9x^2}$

49. $3x$ **51.** $|x|$ **53.** $9|x|$ **55.** $3|x+5|$

Exercise Set 0.7 (page 29)

1. 6 **3.** 3 **5.** undef. **7.** 2 **9.** -4 **11.** $\frac{5}{11}$ **13.** $4xy^2\sqrt{2}$ **15.** $2y\sqrt[3]{4x^2y}$

17. $2xy\sqrt[4]{x^2y}$ **19.** $\frac{\sqrt{2x}}{3}$ **21.** $32\sqrt[3]{2}$ **23.** $\frac{x\sqrt{xy}}{y}$ **25.** $12|x|y\sqrt{y}$ **27.** $3|x+4|\sqrt{2}$

29. $2|x|$ **31.** $2\sqrt{2x}$ **33.** $\sqrt{6}-2\sqrt{3}$ **35.** $-4+2\sqrt{15}$ **37.** x^2-y **39.** $3x+8$

41. $\frac{2x^2-3x\sqrt{5}+5}{x^2-5}$ **43.** $\frac{9-x}{x+8\sqrt{x}+15}$ **45.** $(-1+\sqrt{3}\,i)(-1+\sqrt{3}\,i)(-1+\sqrt{3}\,i)=$

$(-2-2\sqrt{3}\,i)(-1+\sqrt{3}\,i)=8;\quad (-1-\sqrt{3}\,i)(-1-\sqrt{3}\,i)(-1-\sqrt{3}\,i)=(-2+2\sqrt{3}\,i)(-1-\sqrt{3}\,i)=8$

Exercise Set 0.8 (page 32)

1. $6i$ **3.** $-7i$ **5.** $\sqrt{-64}$ **7.** $-\sqrt{-256}$ **9.** $7i$ **11.** $9i\sqrt{2}$ **13.** 4 **15.** $-7-2\sqrt{10}$

17. $-4-3\sqrt{15}$ **19.** $3-4i$ **21.** $4i$ **23.** $-i$ **25.** $\frac{1}{5}+\frac{2}{5}i$ **27.** $\frac{3}{5}+\frac{1}{5}i$ **29.** $\frac{4}{5}-\frac{3}{5}i$

31. $-\frac{9}{61}+\frac{38}{61}i$ **33.** i **35.** i **37.** -1 **39.** $\frac{a-bi}{a^2+b^2}$

Exercise Set 0.9 (page 35)

1. 2nd **3.** 0 **5.** 6 **7.** 30 **9.** x^2+5x+6 **11.** 29 **13.** $2x^2-xy-11x+7y+29$

15. $4x^2-6x-3$ **17.** $-2x^2+10x-7$ **19.** $3x+2$ **21.** $2x+1$ **23.** $x^2-2x-15$

25. $3x^2+13x-10$ **27.** $x-29$ **29.** x^3+8 **31.** $x^{5/6}+x^{3/4}$ **33.** $x-x^{1/2}$ **35.** zero

Exercise Set 0.10 (page 38)

1. $3x^2(1+2x^2)$ **3.** $(x-6)(y-3)$ **5.** $(x-7)(x+4)$ **7.** $(2x+y)(x-4y)$ **9.** $(2x-3)(2x+3)$

11. $(x+2)(x^2-2x+4)$ **13.** $x(x-1)^2$ **15.** $y^2(xy-8)(xy+8)$ **17.** $(x^2-2)(x+4)$

19. $(x+b)(bx+1)$ **21.** $(2x-1)^2(2x+1)^2$ **23.** $(x-4)(x+4)(2x^2+3)$ **25.** $\left(x-\frac{3}{4}\right)^2$

27. $\left(y+\frac{2}{3}\right)\left(y^2-\frac{2}{3}y+\frac{4}{9}\right)$ **29.** $(3x-\sqrt{5})(3x+\sqrt{5})$ **31.** $(x+8i)(x-8i)$

Exercise Set 0.11 (page 42)

1. $\frac{y}{2x^3}$ **3.** $\frac{x-y}{x+y}$ **5.** $\frac{x+2}{x-3}$ **7.** x^2+3x **9.** $\frac{9x^2-2x+10}{75x^2}$ **11.** $\frac{2}{y-6}$ **13.** $\frac{3x-2}{x}$

15. $\frac{3x^2-10x-8}{(2x-1)(x-5)(x+3)}$ **17.** $\frac{x^2+6x+9}{x^2-x-20}$ **19.** $\frac{2x+3}{x-6}$ **21.** $\frac{xy}{y-x}$ **23.** $\frac{xy}{x^2y^2+1}$

25. $(2x-2)+\frac{5}{3x+1}$ **27.** $(2x-5)+\frac{15x+5}{x^2+x-1}$ **29.** $x^5+x^4+x^3+x^2+x+1$

Exercise Set 0.12 (page 45)

1.

3.

5.

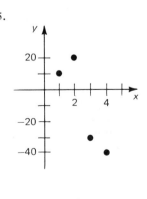

7. II **9.** neg. x-axis **11.** IV **13.** pos. x-axis **15. b.** zero y-coordinate
17. Both coordinates pos. **19.** Both coordinates neg. **21.** pos. x-coordinate
23. neg. y-coordinate **25.** $A_2(2, -1)$; $B_2(6, -8)$; $C_2(3, 2)$; $D_2(-3, 4)$
27. y-coordinates opposite in sign

Exercise Set 1.1   (page 49)

1. $\{-3\}$ **3.** $\{-\frac{1}{2}\}$ **5.** $\{3\}$ **7.** $\emptyset$ **9.** $\{5\}$ **11.** $-\frac{1}{2}$ **13.** $t = \dfrac{d}{r}$ **15.** $x = \dfrac{-(r+t)}{2}$

17. $y = \dfrac{d}{a-c}$ **19.** $x = \dfrac{ab}{a+b}$ **21.** $y = -4x - 17$ **23.** $S = \dfrac{n}{2}[2a + (n-1)d]$ **25.** 25, 69

27. 35 rockers, 45 folders   **29.** 30, and 88 m.p.h.

Exercise Set 1.2   (page 54)

1. $\{-4, 3\}$ **3.** $\{\frac{1}{2}, 6\}$ **5.** $\{-\frac{1}{2}, 2\}$ **7.** $\{\pm 4\}$ **9.** $\{8 \pm 3i\}$ **11.** $\left\{\dfrac{3 \pm \sqrt{11}}{4}\right\}$

13. $\{7, -1\}$ **15.** $\{-1, \frac{5}{3}\}$ **17.** $\left\{\dfrac{5 \pm \sqrt{13}}{2}\right\}$ **19.** $\left\{\dfrac{3 \pm i\sqrt{23}}{4}\right\}$ **21.** $y = \pm\frac{3}{4}\sqrt{x^2 - 16}$

23. $y = \pm\frac{2}{5}\sqrt{25 - x^2}$ **25.** $y = \pm\dfrac{b}{a}\sqrt{a^2 - x^2}$ **27.** $y = \dfrac{-3 \pm \sqrt{5}}{2}x$ **29.** $x = y$, or $x = \dfrac{y}{4}$

31. $\{2, -\frac{1}{2}, -4\}$ **33.** $\{0, \frac{2}{3}, -1\}$ **35.** $\{\pm 2, \pm 3\}$ **37.** $x = -r \pm \sqrt{r^2 - s}$

Exercise Set 1.3   (page 58)

1. conjugate imaginaries   **3.** real, unequal, irrational   **5.** real, unequal, rational
7. real, rational, multiplicity two   **9.** $3x^2 - 16x + 5 = 0$ **11.** $16x^2 + 40x + 25 = 0$
13. $x^2 + 4 = 0$ **15.** $(x - 1 + \sqrt{2})(x - 1 - \sqrt{2})$ **17.** $(x + 3 - \sqrt{5})(x + 3 + \sqrt{5})$
19. $(x + 3 - \sqrt{3}i)(x + 3 + \sqrt{3}i)$ **21.** $(x + 18)(x - 8)$ **23.** $(2x - 15)(x + 14)$ **25.** not factorable
27. $k = \pm 8$ **29.** 11″   **31. a.** 2 sec. **b.** $\sqrt{10}$ sec.

33. $r_1 + r_2 = \dfrac{-b + \sqrt{b^2 - 4ac}}{2a} + \dfrac{-b - \sqrt{b^2 - 4ac}}{2a} = -\dfrac{b}{a}$; $r_1 \cdot r_2 = \dfrac{-b + \sqrt{b^2 - 4ac}}{2a} \cdot \dfrac{-b - \sqrt{b^2 - 4ac}}{2a} = \dfrac{c}{a}$

Exercise Set 1.4   (page 62)

1. $\{9\}$ **3.** $\{2\}$ **5.** $\{12\}$ **7.** $\emptyset$ **9.** $\{23\}$ **11.** $\{81\}$ **13.** $\{-64, 343\}$ **15.** $\{1, 64\}$
17. $\{\pm 3, \pm 2i\}$ **19.** $\{-\sqrt[3]{2}, -\sqrt[3]{5}\}$ **21.** $\{-1\}$ **23.** $z = \pm\sqrt{-3x}$ or $z = \pm\sqrt{-2x}$; z real if
$\{x: x \in R^-\}$, z imaginary if $\{x: x \in R^+\}$

Exercise Set 1.5   (page 68)

1. $\{x: x > 3\}$, (Figure 1 follows)   **3.** $\{x: x \geq \frac{15}{2}\}$, (Figure 3 follows)   **5.** $\{x: -15 \leq x \leq 9\}$,
(Figure 5 follows)   **7.** $\{x: -\frac{7}{3} < x < \frac{13}{3}\}$, (Figure 7 follows)   **9.** $k < -\frac{4}{3}$ **11.** $\{x: -1 \leq x \leq 4\}$

1.
3

3.
$\frac{15}{2}$

5.
-15 9

7.
$-\frac{7}{3}$ $\frac{13}{3}$

13. $\{x: x < -\frac{1}{2} \text{ or } x > 3\}$ **15.** $\{x: x \leq -\frac{2}{3} \text{ or } x \geq \frac{1}{2}\}$, (Figure 15 follows)
17. $\{x: x < -1 \text{ or } x > 5\}$ **19.** $\{x: x \leq -5 \text{ or } x > -1\}$ **21.** $\{x: x < -10 \text{ or } -5 < x < 5\}$,
(Figure 21 follows)   **23.** $\{x: -2 < x \leq -1 \text{ or } 1 < x \leq 4\}$ **25.** $\{x: 0 < x < 3 \text{ or } x > 5\}$
27. $\{x: x \leq 2 - 2\sqrt{2} \text{ or } x \geq 2 + 2\sqrt{2}\}$, (Figure 27 follows)

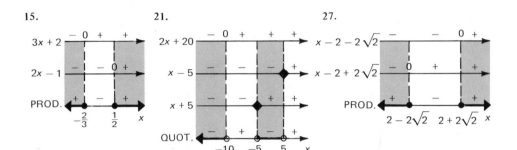

Exercise Set 1.6   (page 72)

1. $\{-2, 8\}$ **3.** $\{-\frac{22}{7}, 2\}$ **5.** $\{-4, 16\}$ **7.** $\{x: -5 < x < 5\}$ **9.** $\{x: x < -4 \text{ or } x > 2\}$,
(Figure 9 follows)   **11.** $\{x: -6 < x < 4\}$ **13.** R **15.** $\{x: -\frac{7}{2} \leq x \leq \frac{1}{2}\}$, (Figure 15 follows)
17. $\{x: x \leq -\frac{7}{2} \text{ or } x \geq \frac{1}{2}\}$ **19.** $\{x: x < -2 \text{ or } x > 5\}$ **21.** $\{x: x \leq -3 \text{ or } x \geq \frac{9}{2}\}$
23. $\{x: x \leq -13 \text{ or } x \geq 11\}$, (Figure 23 follows)   **25.** $\{x: x < -5 \text{ or } x > 11\}$
27. $\{x: \frac{5}{2} \leq x \leq \frac{7}{2}\}$, (Figure 27 follows)   **29.** $\{x: -13 < x < 11\}$ **31.** $\{\frac{2}{3}, 2\}$ **33.** $\{-\frac{1}{5}, 1\}$
35. $|x - a| < k \leftrightarrow -k < x - a < k \leftrightarrow a - k < x < a + k; \quad a$

Chapter 1   Self-Test   (page 73)

1. $\{-12\}$ **2.** $\{4\}$ **3.** $y = -\frac{a}{b}x - \frac{c}{b}$ **4.** 120 at \$3, 90 at \$4, 60 at \$5   **5.** $\{-2, 9\}$
6. $\left\{\frac{-4 \pm \sqrt{86}}{2}\right\}$ **7.** $y = \pm\frac{2}{5}\sqrt{x^2 - 25}$ **8.** real, unequal, rational   **9.** conjugate imaginary
10. 6:00 P.M.   **11.** $\{9\}$ **12.** $\{\frac{8}{27}, \frac{27}{8}\}$ **13.** $\{x: -5 < x \leq 10\}$, (Figure 13 follows)
14. $\{x: x < 0 \text{ or } \frac{1}{3} < x < 1\}$, (Figure 14 follows)   **15.** $\{x: -2 \leq x \leq 5\}$
16. $\{x: x < -\frac{5}{2} \text{ or } x > 3\}$, (Figure 16 follows)

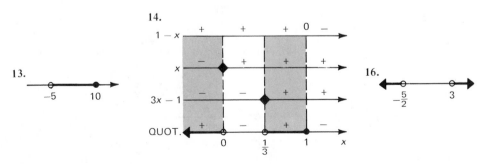

Exercise Set 2.1 (page 80)

1. $\{(1, 3), (1, 4), (1, 5), (2, 3), (2, 4), (2, 5), (3, 3), (3, 4), (3, 5)\}$
3. $\{(1, 1), (1, 2), (1, 3), (2, 1), (2, 2), (2, 3), (3, 1), (3, 2), (3, 3)\}$
5. $\{(1, 1), (2, 1), (2, 2), (3, 1), (3, 2), (3, 3)\}$; $\{1, 2, 3\}$; $\{1, 2, 3\}$
7. $\{x: x \neq -5, 3\}$ 9. $\{x: x \leq -5 \text{ or } x \geq 3\}$ 11. $\{x: x < -5 \text{ or } x > 3\}$ 13. R; $\{y: y \geq 0\}$
15. $\{x: x \geq 1\}$; $\{y: y \geq 0\}$ 17. $\{x: x \leq -4 \text{ or } x \geq 4\}$; $\{y: y \geq 0\}$ 19. 20   21. 2
23. $x^2 + 3x + 2$ 25. $x^2 - (2a + 3)x + a^2 + 3a + 2$ 27. x 29. $3 \neq 2$, (Figure 29 follows)
31. $2 \neq 3 - 4$, (Figure 31 follows)   33. $3 \neq |2|$, (Figure 33 follows)   35. 0   37. 2, 6   39. 1
41. $-1, \dfrac{-5}{2}, \dfrac{-11}{2}$ 43. **a.** is **b.** is not

29.

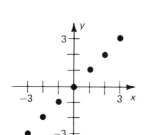

31.

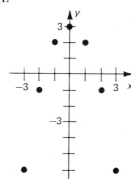

33.

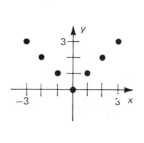

Exercise Set 2.2 (page 88)

1. x-intercept: -2, y-intercept: -10 3. x-intercept: $-\frac{8}{3}$, y-intercept: 4   5. y-intercept: -6
7. (Figure 7 follows)   9. (Figure 9 follows)

7.

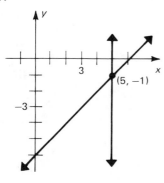

9.

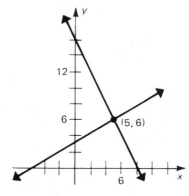

11. $\sqrt{106}$; $\dfrac{-9}{5}$ 13. 11; 0   15. 11; undef.   17. $10 + \sqrt{10}$

19. $y = -\dfrac{3}{5}x + \dfrac{6}{5}$; $m = \dfrac{-3}{5}$; $b = \dfrac{6}{5}$ 21. $y = x + 2$; $m = 1$; $b = 2$ 23. not possible

25. $\dfrac{x}{2} + \dfrac{y}{5} = 1$ 27. $\dfrac{x}{3b} + \dfrac{y}{b} = 1$ 29. $\dfrac{x}{2} + \dfrac{y}{3} = 1$; $a = 2$; $b = 1$ 31. $\dfrac{x}{2} + \dfrac{y}{-3} = 1$; $a = 2$; $b = -3$

33. parallel   35. perpendicular   37. neither   39. $F = 50 + 10x$; $1.60, 53\frac{1}{3}$¢/mi;
$2.00, 50$¢/mi; $2.40, 48$¢/mi   41. $m = 1.4$, $b = 20$, $y = \$363$
43. If $a - b \geq 0$, then $|a - b| = (a - b)$; $|a - b|^2 = (a - b)^2$.
If $a - b < 0$, then $|a - b| = -(a - b)$; $|a - b|^2 = [-(a - b)]^2 = (a - b)^2$.

Exercise Set 2.3   (page 93)

3.

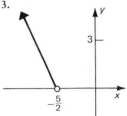

9.

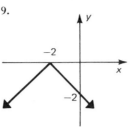

13.

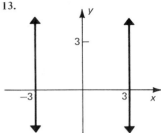

19.

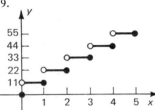

25.

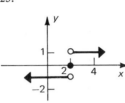

31.

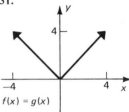

$f(x) = g(x)$

Exercise Set 2.4   (page 97)

1. vertex: $(0, 3)$, axis: $x = 0$ **3.** vertex: $(0, -3)$, axis: $x = 0$ **5.** vertex: $(-3, 0)$, axis: $x = -3$
7. vertex: $(3, 3)$, axis: $x = 3$ (Figure 7 follows)   **9.** vertex: $(1, -1)$, axis: $x = 1$ **11.** vertex:
$(-2, 4)$, axis: $x = -2$ **13.** vertex: $(2, -4)$, axis: $x = 2$ **15.** vertex: $(\frac{3}{2}, 3)$, axis: $x = \frac{3}{2}$
17. vertex: $(-\frac{3}{2}, \frac{13}{2})$, axis: $x = -\frac{3}{2}$ **19.** vertex: $(1, \frac{4}{3})$, axis: $x = 1$ **21.** $(-2, 4)$, $(2, 4)$ (Figure
21 follows)   **23.** $(2, 5)$, $(5, 8)$ **25.** Figure 25 follows   **31.** 1250 sq. ft.;   25 by 50   **33.** $\frac{1}{2}$
35. $(6 - .05x)$; $(6 - .05x)(40 + x)$;   40

7.

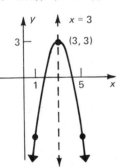

21.

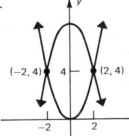

25.

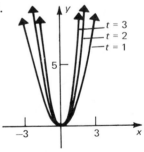

Exercise Set 2.5   (page 101)

1. $P = ks$ **3.** $t = kd^{3/2}$ **5.** $F = \dfrac{k}{d^2}$ **7.** $T = kPV$ **9.** $f = \dfrac{km_1 m_2}{r^2}$ **11.** 216   **13.** $\frac{4}{3}$

15. $\pm\dfrac{\sqrt{6}}{4}$ **17.** $2\frac{1}{2}$ sec.;   36 sec.   **19.** 40 cc

Chapter 2 Self-Test   (page 102)

1. 14; 3; x; x **2.** (Figure 2 follows)   **3.** $\{x: x \le -5 \text{ or } x \ge 5\}$; $\{y: y \ge 0\}$
4. $y = -\dfrac{3}{5}x + 3$, $m = -\dfrac{3}{5}$; $\dfrac{x}{5} + \dfrac{y}{3} = 1$, $a = 5$, $b = 3$;   **5.** 25; $\dfrac{-24}{7}$; $24x + 7y - 45 = 0$
6. (Figure 6 follows)   **7.** (Figure 7 follows)   **8.** vertex: $(-3, -5)$, axis: $x = -3$
9. $(-2, 0)$, $(1, 3)$ **10.** $\{t: 0 \le t \le 25\}$;   2500 ft.   **11.** 5   **12.** 30

2.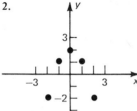

6.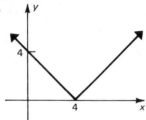

7.

Exercise Set A (page 106)

1. $y = mx - 4$ **3.** $\dfrac{x}{-4} + \dfrac{y}{b} = 1$ **5.** $y = b$ **7.** $y - 2 = m(x - 5)$ **9.** $\dfrac{x}{a} + \dfrac{y}{\frac{12}{a}} = 1$

11. (Figure 11 follows) **13.** $\dfrac{x}{2} + \dfrac{y}{b} = 1$, (Figure 13 follows) **15.** $y - 2 = m(x - 3)$, (Figure 15

follows) **17.** $y = -\frac{3}{2}x + b$; $y = -\frac{3}{2}x + \frac{19}{2}$ **19.** $(1 + 3t)x + (4 + 2t)y + (-6 - 9t) = 0$
21. $49x - 4y - 114 = 0$ **23.** $9x + 16y - 36 = 0$ **25.** $5x - 12 = 0$

11.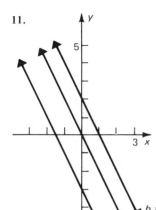

13.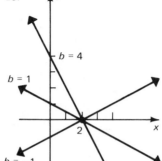

15.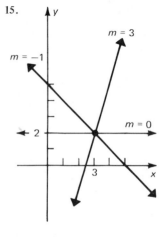

Exercise Set B (page 110)

1. $8x_1 + 4\Delta x - 7$ **3.** $3x_1^2 + 3x_1\Delta x + (\Delta x)^2 + 12x_1 + 6\Delta x$ **5.** $\dfrac{-3}{x_1(x_1 + \Delta x)}$

7. 149; 85; 70.6; 69.16; 69.016 **9.** 43; 7; 3.31; 3.0301; 3.003001 **11.** $2ax_1 + a\Delta x + b$;
$v = 2ax_1 + b$

Exercise Set C (page 114)

1. a. $3x + 1$; $x \in R$ **b.** $-x - 1$; $x \in R$ **c.** $2x^2 + x$; $x \in R$ **d.** $\dfrac{x}{2x + 1}$; $\left\{x: x \neq -\frac{1}{2}\right\}$ **e.** $\dfrac{2x + 1}{x}$;

$\{x: x \neq 0\}$ **f.** $2x + 1$; $x \in R$ **g.** $2x + 1$; $x \in R$ **3. a.** $x^2 + 3x - 1$; $x \in R$ **b.** $x^2 - 3x - 1$; $x \in R$

c. $3x^3 - 3x$; $x \in R$ **d.** $\dfrac{x^2 - 1}{3x}$; $\{x: x \neq 0\}$ **e.** $\dfrac{3x}{x^2 - 1}$; $\{x: x \neq \pm 1\}$ **f.** $9x^2 - 1$; $x \in R$ **g.** $3x^2 - 3$;

$x \in R$ **5. a.** $\dfrac{10}{3}x - \dfrac{4}{3}$; $x \in R$ **b.** $\dfrac{8}{3}x - \dfrac{8}{3}$; $x \in R$ **c.** $x^2 + \dfrac{4}{3}x - \dfrac{4}{3}$; $x \in R$ **d.** $\dfrac{9x - 6}{x + 2}$; $\{x: x \neq -2\}$

e. $\dfrac{x + 2}{9x - 6}$; $\left\{x: x \neq \frac{2}{3}\right\}$ **f.** x; $x \in R$ **g.** x; $x \in R$ **7. a.** $x^2 + \sqrt{x}$; $\{x: x \geq 0\}$ **b.** $x^2 - \sqrt{x}$; $\{x: x \geq 0\}$

c. $x^{5/2}$; $\{x: x \geq 0\}$ **d.** $x^{3/2}$; $\{x: x > 0\}$ **e.** $x^{3/2}$; $\{x: x \geq 0\}$ **f.** x; $\{x: x > 0\}$ **g.** $|x|$; $x \in R$

9. a. x^2+4; $x \in R$ **b.** $3x^2-2$; $x \in R$ **c.** $-2x^4+5x^2+3$; $x \in R$ **d.** $\dfrac{2x^2+1}{3-x^2}$; $\{x: x \neq \pm\sqrt{3}\}$

e. $\dfrac{3-x^2}{2x^2+1}$; $x \in R$ **f.** $2x^4-12x^2+19$; $x \in R$ **g.** $-4x^4-4x^2+2$; $x \in R$ **11. a.** $|x^2-4|+x+2$;

$x \in R$ **b.** $|x^2-4|-x-2$; $x \in R$ **c.** $|x^2-4|(x+2)$; $x \in R$ **d.** $\dfrac{|x^2-4|}{x+2}$; $\{x: x \neq -2\}$ **e.** $\dfrac{x+2}{|x^2-4|}$;

$\{x: x \neq \pm 2\}$ **f.** $|x^2+4x|$; $x \in R$ **g.** $|x^2-4|+2$; $x \in R$ **13.** The intersection of the domains is empty.

Exercise Set 3.1   (page 119)

1. $P_1(4, 3)$, $P_2(-4, -3)$, $P_3(-4, 3)$ **3.** $P_1(-6, 0)$, $P_2(6, 0)$, $P_3(6, 0)$ **5.** $P_1(3, -\sqrt{7})$,
$P_2(-3, \sqrt{7})$, $P_3(-3, -\sqrt{7})$ **7. a.** no **b.** no **c.** yes **9. a.** yes **b.** yes **c.** yes **11. a.** no
b. yes **c.** no **13.** $(0, \pm 2)$, $(4, 0)$; symmetric to x-axis; $\{x: x \leq 4\}$; $y \in R$ (Figure 13
follows) **15.** $(\pm 3, 0)$; symmetric to y-axis; $\{x: x \leq -3 \text{ or } x \geq 3\}$; $\{y: y \geq 0\}$
17. $(0, \pm\frac{9}{2})$, $(9, 0)$; symmetric to x-axis; $\{x: 0 \leq x \leq 9\}$; $\{y: -\frac{9}{2} \leq y \leq \frac{9}{2}\}$
19. no intercepts; symmetric to origin; $\{x: x \neq 0\}$; $\{y: y \neq 0\}$ (Figure 19 follows)
21. no **23.** no **25.** $P_1(-4, 3)$, $Q_1(5, 5)$, $R_1(3, 2)$ **27.** yes **29.** no **31.** no
33. $(0, 0)$, $(\pm 3, 0)$; symmetric to origin; $x, y \in R$ **35.** $(0, 0)$; no symmetry; $\{x: x \geq 0\}$;
$\{y: y \geq 0\}$ **37.** $(0, 4)$, $(\pm 2, 0)$; symmetric to y-axis; $x \in R$; $\{y: y \geq 0\}$ (Figure 37 follows)

13.   **19.**   **37.**

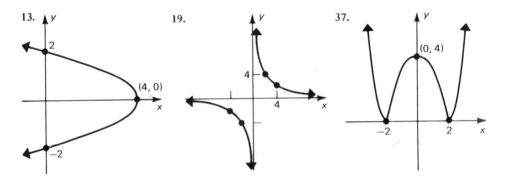

Exercise Set 3.2   (page 126)

1. vertex: $(0, -2)$, axis: $y=-2$ **3.** vertex: $(-4, 2)$, axis: $y=2$ **5.** vertex: $(-4, -2)$,
axis: $y=-2$ **7.** $r=4$ **9.** $r=5$ **11.** $(0, 3)$, $(0, -3)$, $(4, 0)$, $(-4, 0)$ **13.** $(0, 6)$, $(0, -6)$,
$(2, 0)$, $(-2, 0)$ **15.** $(-4, 0)$, $(4, 0)$, $y=-\frac{3}{4}x$, $y=\frac{3}{4}x$ **17.** $(0, 5)$, $(0, -5)$, $y=\frac{5}{4}x$, $y=-\frac{5}{4}x$
19. (Figure 19 follows) **23.** ellipse, $(0, 7)$, $(0, -7)$, $(2, 0)$, $(-2, 0)$, (Figure 23 follows)
25. hyperbola, $(7, 0)$, $(-7, 0)$, $y=-\frac{2}{7}x$, $y=\frac{2}{7}x$, (Figure 25 follows) **27.** circle, $r=2$, (Figure 27
follows) **29.** parabola, vertex: $(4, 0)$, axis: $y=0$ **31.** rectangular hyperbola, (Figure 31
follows) **33.** hyperbola $(0, \frac{4}{3})$, $(0, -\frac{4}{3})$, $y=-\frac{1}{3}x$, $y=\frac{1}{3}x$ **35.** ellipse, $(0, \frac{5}{3})$, $(0, -\frac{5}{3})$, $(\frac{5}{2}, 0)$,
$(-\frac{5}{2}, 0)$ **39.** (Figure 39 follows) **43.** $\{x: -r \leq x \leq r\}$, $\{y: -r \leq y \leq r\}$
45. $\{x: -a \leq x \leq a\}$, $\{y: -b \leq y \leq b\}$ **47.** $\{x: x \leq -a \text{ or } x \geq a\}$, $y \in R$

Exercise Set 3.3   (page 132)

1. ellipse; $(3, -1)$ **3.** circle; $(4, -1)$; $r=5$ **5.** hyperbola; $(-2, 1)$; $2x-3y+7=0$,
$2x+3y+1=0$, (Figure 5 follows) **7.** parabola; $(2, -1)$; $y=-1$, (Figure 7 follows)
9. circle; $(\frac{3}{4}, 2)$; $r=\sqrt{6}$ **11.** parabola; $(-\frac{1}{2}, 3)$; $x=-\frac{1}{2}$ **13.** ellipse; $(\frac{3}{2}, 0)$, (Figure 13
follows)
15. hyperbola; $(-2, -3)$; $4x-y+5=0$, $4x+y+11=0$ **17.** $(x-2)^2+(y+6)^2=16$
19. $(x-5)^2+(y-\frac{3}{2})^2=5$ **23.** (Figure 23 follows) **25.** (Figure 25 follows)
31. (Figure 31 follows)

Figures for Exercise Set 3.2

19.

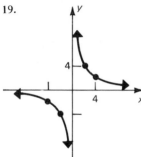

23.

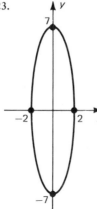

ellipse

25.

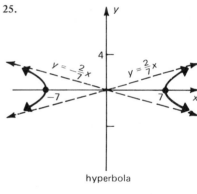

hyperbola

27.

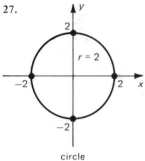

circle

31.

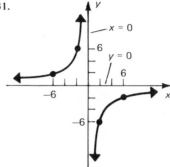

rectangular hyperbola

39.

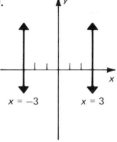

Figures for Exercise Set 3.3

5.

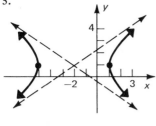

7.

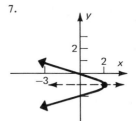

13.

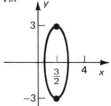

23.

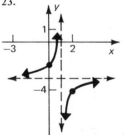

25.

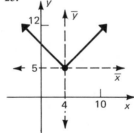

31.

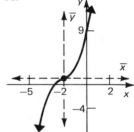

Exercise Set 3.4   (page 136)

11.

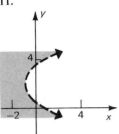

13.

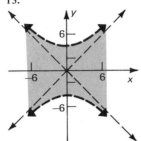

17.

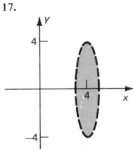

23.

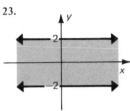

29.

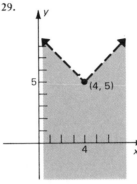

35.

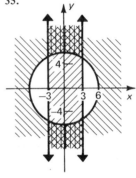

Exercise Set 3.5   (page 141)

1. a. $y = \frac{1}{3}x + \frac{4}{3}$ **b.** (Figure 1b follows) $F: x \in R,\ y \in R;$ $F^{-1}: x \in R,\ y \in R$ **c.** F^{-1} is a function
3. a. $y = 3x + 5$ **b.** $F: x \in R,\ y \in R;$ $F^{-1}: x \in R,\ y \in R$ **c.** F^{-1} is a function   **5. a.** $x = 2$
b. $F: x \in R,\ \{y: y = 2\};$ $F^{-1}: \{x: x = 2\},\ y \in R$ **c.** F^{-1} not a function   **7. a.** $y = -3$
b. $F: \{x: x = -3\},\ y \in R;$ $F^{-1}: x \in R,\ \{y: y = -3\}$ **c.** F^{-1} is a function   **9. a.** $y = \pm\sqrt{x + 4}$
b. (Figure 9b follows) $F: x \in R,\ \{y: y \geq -4\};$ $F^{-1}: \{x: x \geq -4\},\ y \in R$ **c.** F^{-1} not a function
11. a. $y = x^{1/3}$ **b.** (Figure 11b follows) $F: x \in R,\ y \in R;$ $F^{-1}: x \in R,\ y \in R$ **c.** F^{-1} is a
function   **13. a.** $y = x^2$ **b.** (Figure 13b follows) $F: \{x: x \geq 0\},\ \{y: y \geq 0\};$ $F^{-1}: \{x: x \geq 0\},$
$\{y: y \geq 0\}$ **c.** F^{-1} is a function   **15.** is not   **17.** is   **19.** is not   **21. a.** $\{x: x \geq 0\},$
$\{y: y \geq 0\}$ **b.** $y = \sqrt{x};\ \{x: x \geq 0\},\ \{y: y \geq 0\}$ **23. a.** (Figure 23a follows) $\{x: x \geq -2\},$
$\{y: y \geq 0\}$ **b.** no inverse function   **25. a.** (Figure 25a follows) $\{x: x \leq 0\},\ \{y: y \geq 2\}$
b. $y = -\sqrt{\dfrac{x-2}{3}};\ \{x: x \geq 2\},\ \{y: y \leq 0\}$ **27. a.** $\{x: x \geq 0\},\ \{y: y \geq -4\}$ **b.** $y = \frac{1}{2}(x + 4);$
$\{x: x \geq -4\},\ \{y: y \geq 0\}$ **29.** are inverses   **31.** not inverses   **33.** are inverses

1. b.

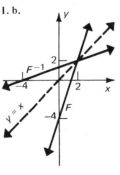

9. b.

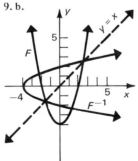

11. b.

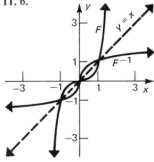

358

13. b.

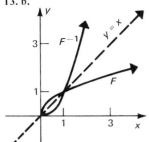

23. a.

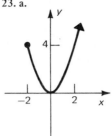

25. a.

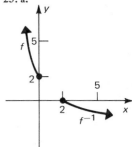

Chapter 3 Self-Test (page 142)

1. to origin **2.** to x-axis, y-axis, origin **3.** $(\pm 1, 0)$; to y-axis; $\{x: x \leq -1 \text{ or } x \geq 1\}$; $\{y: y \geq 0\}$, (Figure 3 follows) **4.** ellipse; $(0, 0)$, (Figure 4 follows) **5.** circle; $(1, -2)$; $r = 2$ **6.** hyperbola; $(0, 0)$; $y = \pm \frac{1}{6}x$ **7.** parabola; $\left(\frac{1}{2}, \frac{7}{2}\right)$; $x = \frac{1}{2}$ **8.** ellipse; $(1, -2)$ **9.** hyperbola; $(1, -2)$; $3x - 2y - 5 = 0$, $3x + 2y - 1 = 0$ **11.** (Figure 11 follows) **14.** $y = \frac{2}{3}x + 2$; $f: x \in R$, $y \in R$; $f^{-1}: x \in R$, $y \in R$; f^{-1} is a function **15.** $y = -\sqrt{x + 1}$; $f: \{x: x \leq 0\}$, $\{y: y \geq -1\}$; $f^{-1}: \{x: x \geq -1\}$, $\{y: y \leq 0\}$

3.

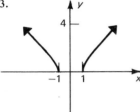

4.

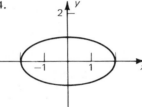

11.

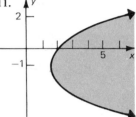

Exercise Set D (page 149)

3. a.

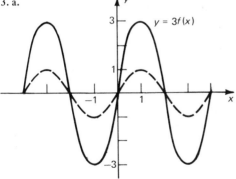

$y = 3f(x)$

3. b.

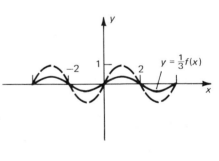

$y = \frac{1}{3}f(x)$

9.

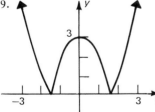

11.

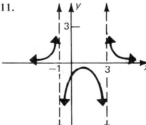

13.

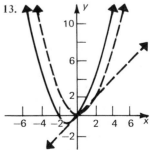

Exercise Set 4.1   (page 156)

1. a. $(x-2)(x^2-4x-6)-8$ b. $(x+2)(x^2-8x+18)-32$ 3. a. $(x-1)(2x^3-x^2-9x-14)-17$
b. $(x+1)(2x^3-5x^2-3x-2)-1$ 5. a. $(x-4)(3x^4+12x^2+2x^2+8x-1)-6$
b. $(x+4)(3x^4-12x^3+2x^2-8x-1)+2$ 7. a. $(x-2i)[3x^2+(-2+6i)x-4i]+3$
b. $(x+2i)[3x^2+(-2-6i)x+4i]+3$ 9. a. $(x-\frac{1}{2})(4x^2-4x+1)+\frac{19}{2}$ b. $(x-\frac{3}{2})(4x^2+3)+\frac{27}{2}$
11. a. $(x-\frac{1}{4})(16x^3+4x^2+64x+16)$ b. $(x+2i)(16x^3-32ix^2-x+2i)$ 13. -70; $-\frac{65}{4}$
15. 29; $5+2i$ 17. $(x+2)(x+3)(x-3)$; $-2,-3,3$ 19. $3(x-1)(x-\frac{2}{3})(x+2)$; $1, \frac{2}{3}, -2$
21. $4(x+7)(x+\frac{1}{2})(x-\frac{1}{2})$; $-7, -\frac{1}{2}, \frac{1}{2}$ 23. 2, -1; $(x-2)$, $(x+1)$ 25. 1, 0, -2; $(x-1)$, x,
$(x+2)$ 27. 18   29. $\begin{array}{c|ccccc} & 1 & 0 & 0 & \cdots & 0 & -1 \\ \hline 1 & 1 & 1 & 1 & \cdots & 1 & 0 \end{array}$ $r=0 \leftrightarrow (x-1)$ is a factor

I $A(x)=(x-c)Q(x)+r$; $A(c)=(c-c)Q(c)+r=r$

Exercise Set 4.2   (page 163)

1. $(x+3)^2(x-2)(x-1)$; -3(double), 2, 1   3. $(x+2)^2(x-i)(x+i)$; -2(double), $i, -i$
5. $(x-i)(x+i)(x-5)$; $i, -i, 5$ 7. $[x-(1+i)][x-(1-i)](x-3)$; $1+i, 1-i, 3$
9. $\begin{array}{ccc} + & - & \text{Imag.} \\ 2 & 1 & 0 \\ 0 & 1 & 2 \end{array}$ 11. $\begin{array}{ccc} + & - & \text{Imag.} \\ 0 & 4 & 0 \\ 0 & 2 & 2 \\ 0 & 0 & 4 \end{array}$ 13. $\begin{array}{ccc} + & - & \text{Imag.} \\ 0 & 1 & 4 \end{array}$ 15. 2; -2 17. 4; -3

19. 2; -3 21. $A(-1)=-3<0$, $A(-2)=2>0$ 23. $A(-2)=-9<0$, $A(-3)=34>0$
25. $A(3)=58>0$, $A(2)=-26<0$ 27. 2, 1; 1, 0; $-1, -2$ 29. By hypothesis, $a_i>0$ for
$i=0, 1, \ldots, n$. Consider any c such that $c>0$. Then $c^n>0$ and so $a_ic^i>0$ for $i=0, 1, \ldots, n$.
Hence, $A(c)=a_nc^n+\cdots+a_1c+a_0>0$. But if $A(c)>0$ then $A(c)\neq0$ and c is not a zero of $A(x)$.
Thus, $A(x)$ has no positive zeros.

Exercise Set 4.3   (page 170)

1. $\begin{array}{ccc} + & - & \text{Imag.} \\ 2 & 1 & 0 \\ 0 & 1 & 2 \end{array}$ $\{\pm4, \pm2, \pm1\}$; $\{2, 1, -2\}$ 3. $\begin{array}{ccc} + & - & \text{Imag.} \\ 0 & 3 & 0 \\ 0 & 1 & 2 \end{array}$ $\{\pm6, \pm3, \pm2, \pm1\}$; $\{-1, -2, -3\}$
5. $\begin{array}{ccc} + & - & \text{Imag.} \\ 1 & 2 & 0 \\ 1 & 0 & 2 \end{array}$ $\{\pm12, \pm6, \pm4, \pm3, \pm2, \pm1, \pm\frac{4}{3}, \pm\frac{2}{3}, \pm\frac{1}{3}\}$; $\{\frac{4}{3}, -1, -3\}$
7. $\begin{array}{ccc} + & - & \text{Imag.} \\ 0 & 3 & 0 \\ 0 & 1 & 2 \end{array}$ $\{\pm8, \pm4, \pm2, \pm1, \pm\frac{8}{3}, \pm\frac{4}{3}, \pm\frac{2}{3}, \pm\frac{1}{3}\}$; $\{-\frac{2}{3}, 2i, -2i\}$
9. $\begin{array}{ccc} + & - & \text{Imag.} \\ 2 & 1 & 0 \\ 0 & 1 & 2 \end{array}$ $\{\pm1, \pm\frac{1}{12}, \pm\frac{1}{6}, \pm\frac{1}{4}, \pm\frac{1}{3}, \pm\frac{1}{2}\}$; $\{\frac{1}{2}, \frac{1}{3}, -\frac{1}{2}\}$
11. $\begin{array}{ccc} + & - & \text{Imag.} \\ 2 & 2 & 0 \\ 2 & 0 & 2 \\ 0 & 2 & 2 \\ 0 & 0 & 4 \end{array}$ $\{\pm12, \pm6, \pm4, \pm3, \pm2, \pm1\}$; $\{2, 1, -2, -3\}$
13. $\begin{array}{ccc} + & - & \text{Imag.} \\ 3 & 1 & 0 \\ 1 & 1 & 2 \end{array}$ $\{\pm10, \pm5, \pm2, \pm1, \pm\frac{10}{3}, \pm\frac{5}{3}, \pm\frac{2}{3}, \pm\frac{1}{3}\}$; $\{\frac{2}{3}, -1, 2+i, 2-i\}$
15. $\{2, -2, -3\}$ 17. $\{1, -2, -3\}$ 19. None of the members of $\{\pm3, \pm1, \pm\frac{3}{2}, \pm\frac{1}{2}\}$ are
solutions   21. 3 or $\dfrac{7-\sqrt{33}}{2}$ in.   23. The only possible rational solutions of $x^2-3=0$ are ±3,
±1, none of which satisfy the equation. But $\sqrt{3}$ is a solution. Hence, $\sqrt{3}$ is not rational.
25. Let $x=\sqrt[3]{r}$, from which $x^3-r=0$. Because r is prime, the only possible rational solutions of
$x^3-r=0$ are $\pm r, \pm1$, none of which satisfy the equation. But $\sqrt[3]{r}$ is a solution. Hence, $\sqrt[3]{r}$ is
irrational.   27. $-1, \frac{1}{2}\pm\frac{1}{2}i\sqrt{3}$

Exercise Set 4.4   (page 173)

1. (Figure 1 follows)   3. $-2, 3, 5$ 5. $-2, -1$; 0, 1; 2, 3   7. 0, 1, (Figure 7 follows)
9. 1, 2   11. $-3, -2$; $-1, 0$; 0, 1; 2, 3, (Figure 11 follows)   13. $-5, -4$; 0, 1
15. (Figure 15 follows)   21. (Figure 21 follows)   23. $y=x(x-4)(x-5)$
25. $y=(x-2)(x+2)(x-3)(x+3)$ 27. $y=(x-1)(x-2)(x+1)(x+3)$, (Figure 27 follows)

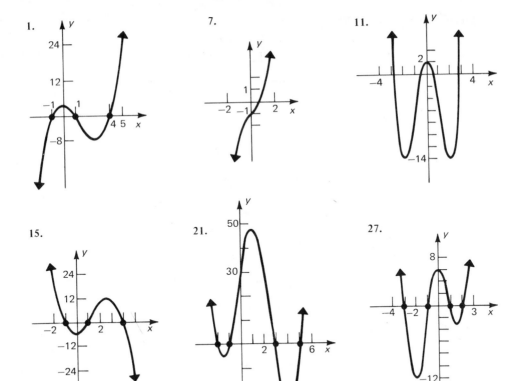

Exercise Set 4.5 (page 180)

1. $V: x=3, x=-3$; $H: y=0$ **3.** $V: x=-5$; $OB: y=4x-20$ **5.** $H: y=\frac{5}{6}$ **7.** $V: x=1$,
$x=-1$; $OB: y=x+6$ **9.** SYMM. to origin; $\{x: x \neq 0\}$; $\{y: y \neq 0\}$; $H: x$-axis
11. x-int.: -4; y-int.: $\frac{4}{3}$; $\{x: x \neq -3\}$; $\{y: y \neq 1\}$; $V: x=-3$; $H: y=1$ **13.** x-int.: $\frac{1}{2}$;
y-int.: 1; $\{x: x \neq 1, -1\}$; $y \in R$; $V: x=1, x=-1$; $H: y=0$, (Figure 13 follows)
15. x-int.: 2; y-int.: -3; $x \in R$; $H: y=0$ **17.** y-int.: $-\frac{9}{4}$; $\{x: x \neq 4\}$; $V: x=4$; $OB: y=x$,
(Figure 17 follows)   **19.** x-int.: $-2, 0, 2$; y-int.: 0; SYMM. to origin; $\{x: x \neq -1, 1\}$; $y \in R$;
$V: x=-1, x=1$; $OB: y=x$ **21.** x-int.: $-2, 2$; y-int.: $-\frac{8}{9}$; $\{x: x \neq -3, 3\}$; $V: x=3, x=-3$;
$H: y=-2$; SYMM. to y-axis   **23.** SYMM. to origin; $\{x: x < -1$ or $x > 1\}$; $\{y: y < -1$ or
$y > 1\}$; $V: x=-1, x=1$; $H: y=-1, y=1$, (Figure 23 follows)   **25.** x-int.: 0; y-int.: 0;
SYMM. to origin; $x \in R$; $\{y: -1 < y < 1\}$; $H: y=-1, y=1$

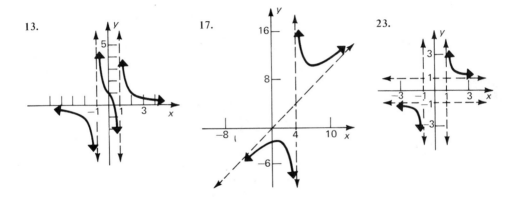

Chapter 4 Self-Test (page 180)

1. $(x-3)(x^3+x^2-10x-16)-24$ **2.** $0; 0$ **3.** $(x+4)(x-1)(x-3)(x+2); -4, 1, 3, -2$
4. $2; -3$ **5.** $+ \quad - \quad$ imag. **6.** $-3, -2; 0, 1; 1, 2$ **7.** $\{3, \frac{3}{2}, -1, -\frac{1}{2}\}$ **8.** $-5, -4; -1, 0;$

$$\begin{array}{ccc} 1 & 3 & 0 \\ \hline 1 & 1 & 2 \end{array}$$

1, 2; (Figure 8 follows) **9. a.** $V: x=-1, x=1; OB: y=2x+1$ **b.** $V: x=-5, x=2; H: y=1$
10. y-int.: $-\frac{9}{4}$; SYMM. to y-axis; $\{x \neq -2, 2\}$; $\{y: y \leq -\frac{9}{4} \text{ or } y \geq 0\}$; $V: x=-2, x=2$;
H: x-axis, (Figure 10 follows)

8.

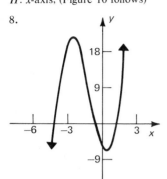

10.

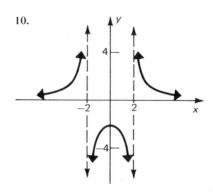

Exercise Set E (page 183)

1. 1.19 **3.** 0.3 **5.** -1.4 **7.** 2.8 **9.** 1.26

Exercise Set 5.1 (page 190)

1. $\{(1, 1)\}$ **3.** $\left\{\left(-27, \frac{40}{3}\right)\right\}$ **5.** $\{(x, y): x-2y=11\}$ **7.** $\{(9, 6)\}$ **9.** $\left\{\left(-\frac{23}{97}, \frac{25}{97}\right)\right\}$

11. $\emptyset$ **13.** $\{(1, -2, 2)\}$ **15.** $\{(2, -1, 0)\}$ **17.** $\left\{\left(\frac{2}{3}, -3, 3\right)\right\}$ **19.** $\left\{\left(\frac{1}{4}, \frac{1}{2}\right)\right\}$

21. $x^2+y^2-5x-y+4=0$ **23.** $y=x^2-7x+6$ **25.** $\frac{2}{13}, \frac{2}{3}$ **27.** \$9000, \$6000
29. $\{(-4, -3), (-2, 1)\}$ **31.** $\{(1, -1), (-1, 1)\}$

Exercise Set 5.2 (page 197)

1. $\{(3, -2)\}$ **3.** $\left\{\left(\frac{3}{20}, \frac{12}{5}\right)\right\}$ **5.** $\begin{bmatrix} 1 & 1 \\ 1 & -1 \end{bmatrix}; \begin{bmatrix} 1 & 1 & | & 8 \\ 1 & -1 & | & 4 \end{bmatrix}$

7. $\begin{bmatrix} 1 & 1 & -3 \\ 3 & 2 & 1 \\ -2 & 1 & -1 \end{bmatrix}; \begin{bmatrix} 1 & 1 & -3 & | & 3 \\ 3 & 2 & 1 & | & 7 \\ -2 & 1 & -1 & | & 0 \end{bmatrix}$ **9.** $\begin{bmatrix} 1 & 1 & 0 \\ 0 & 1 & 1 \\ 1 & 0 & -2 \end{bmatrix}; \begin{bmatrix} 1 & 1 & 0 & | & 2 \\ 0 & 1 & 1 & | & 1 \\ 1 & 0 & -2 & | & -2 \end{bmatrix}$

11. $x+2y=3$ **13.** $x+3y=4$ **15.** $x=2$
$\quad\quad 3y=4$ $\quad\quad 2x-6y+z=-3$ $\quad\quad y=\frac{1}{3}$
$\quad\quad\quad\quad -2y+z=0$ $\quad\quad z=-\frac{1}{2}$

17. $\left\{\left(\frac{1}{2}, \frac{2}{3}\right)\right\}$ **19.** $\{(1, 2, 0)\}$ **21.** $\{(4, -2, 3)\}$ **23.** $\{(1, -3, 2)\}$ **31.** $\{(0, 2, -2, 2)\}$

Exercise Set 5.3 (page 200)

1. $\{(x, y, z): x=-\frac{5}{4}z+\frac{7}{8}; y=\frac{3}{4}z+\frac{11}{8}; z \in R\}$ **3.** $\emptyset$
5. $\{(x_1, x_2, x_3, x_4): x_1=2x_4+1; x_2=2x_4+2; x_3=x_4; x_4 \in R\}$
7. $\{(x_1, x_2, x_3, x_4): x_1=7+x_2+14x_4; x_3=\frac{8}{3}x_4+\frac{4}{3}; x_2, x_4 \in R\}$
9. $\{(x_1, x_2, x_3, x_4): x_1=-\frac{3}{7}x_4; x_2=x_3=\frac{2}{7}x_4; x_4 \in R\}$ **11.** $\emptyset$
13. $\{(x_1, x_2, x_3, x_4): x_1=\frac{5}{7}x_3+\frac{4}{7}; x_2=\frac{1}{7}x_3-x_4-\frac{2}{7}; x_3, x_4 \in R\}$
15. $\{(x, y, z): x=3-2z; y=z-3; z \in R\}$ **17.** $\{(x, y, z): x=3+\frac{1}{2}z; y=\frac{1}{2}z; z \in R\}$
19. $(5, 34, 11), (10, 25, 15), (15, 16, 19), (20, 7, 23), (0, 43, 7)$

Exercise Set 5.4 (page 206

1. $\{(4, -3), (-3, 4)\}$ **3.** $\{(0, 4), (2, 0)\}$ **5.** $\{(4, 0), (0, -1)\}$, (Figure 5 follows)

7. $\{(0, 0), (1, 1)\}$ **9.** $\left\{\left(\dfrac{2\sqrt{6}}{3}, \dfrac{8}{3}\right), \left(-\dfrac{2\sqrt{6}}{3}, \dfrac{8}{3}\right)\right\}$ **11.** $\emptyset$ **13.** $\{(5, 0), (-5, 0)\}$

15. $\{(4, 3), (-4, 3), (4, -3), (-4, -3)\}$ **17.** $\emptyset$, (Figure 17 follows)   **19.** $\{(2, -2), (-2, 2)\}$

21. $\left\{\left(\dfrac{2}{3}, \dfrac{1}{3}\right), \left(-\dfrac{2}{3}, -\dfrac{1}{3}\right), \left(-\dfrac{4}{\sqrt{3}}, \dfrac{1}{\sqrt{3}}\right), \left(\dfrac{4}{\sqrt{3}}, -\dfrac{1}{\sqrt{3}}\right)\right\}$

23. $\{(2\sqrt{7}, \sqrt{7}), (-2\sqrt{7}, -\sqrt{7}), (-4, 1), (4, -1)\}$

25. $\left\{(3, 4), (3, 2), \left(\dfrac{2\sqrt{30}}{3}, \dfrac{\sqrt{30}}{6}\right), \left(\dfrac{-2\sqrt{30}}{3}, \dfrac{-\sqrt{30}}{6}\right)\right\}$ **27.** $\left\{(0, 0), \left(0, \dfrac{-1}{3}\right), \left(\dfrac{3}{5}, 0\right), (1, -1)\right\}$

29. $\left\{(0, 0), \left(0, \dfrac{-13}{3}\right), \left(\dfrac{4}{5}, 0\right), (2, -2)\right\}$ **31.** $\{(0, 0), (2, 0), (-2, -8)\}$

33. $\{(0, 0), (1, 1)\}$, (Figure 33 follows)   **35.** $\{(4, 4)\}$

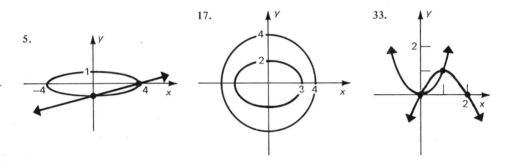

5.

17.

33.

Chapter 5 Self-Test (page 207)

1. $\left\{\left(\dfrac{12}{5}, \dfrac{3}{20}\right)\right\}$ **2.** $\{(44, -104)\}$ **3.** $\{(2, 3, 1)\}$ **4.** $\emptyset$ **5.** $\{(x, y, z): x = \frac{7}{5}z - \frac{2}{5};$

$y = \frac{6}{5}z - \frac{6}{5};\ z \in R\}$ **6.** $\{(2, -4)\}$ **7.** $\left\{(0, 4), \left(\dfrac{-12}{5}, \dfrac{16}{5}\right)\right\}$, (Figure 7 follows)

8. $\{(0, 3), (0, -3)\}$, (Figure 8 follows)   **9.** $\left\{(1, 1), (-1, -1), \left(\dfrac{1}{\sqrt{3}}, \dfrac{3}{\sqrt{3}}\right), \left(\dfrac{-1}{\sqrt{3}}, \dfrac{-3}{\sqrt{3}}\right)\right\}$

10. $\left\{\left(\dfrac{10}{\sqrt{3}}, \dfrac{5}{\sqrt{3}}\right), \left(\dfrac{-10}{\sqrt{3}}, \dfrac{-5}{\sqrt{3}}\right)\right\}$

7.

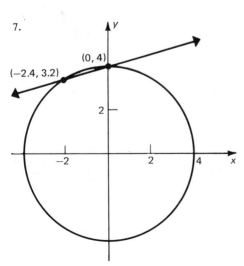

8.

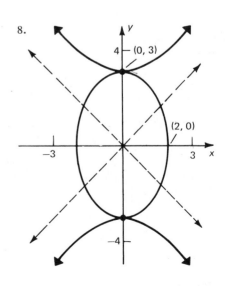

Exercise Set F   (page 209)

1. $\dfrac{5}{x-3} + \dfrac{6}{x+3}$ **3.** $\dfrac{3}{x-4} - \dfrac{6}{3x+2}$ **5.** $\dfrac{1/3}{5x-1} - \dfrac{2/3}{x+4}$ **7.** $\dfrac{2}{x-1} + \dfrac{4}{x+2} - \dfrac{5}{x-3}$

9. $\dfrac{5}{x} - \dfrac{1}{x^2} - \dfrac{3}{x-2}$ **11.** $\dfrac{-1}{x} + \dfrac{2}{x-1} + \dfrac{3}{(x-1)^2}$ **13.** $\dfrac{-x-2}{x^2+2} + \dfrac{2x-1}{x^2+3}$ **15.** $\dfrac{2}{x+2} - \dfrac{1}{x^2+2x+2}$

17. $\dfrac{1/2a}{x-a} - \dfrac{1/2a}{x+a}$

Exercise Set 6.1   (page 214)

7. (0, 0), (0, −3), (−6, 0)   **13.** (9, 0), (4, 8), (−2, 5), (−2, 0)

1.
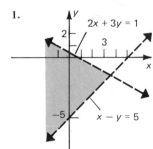

9.
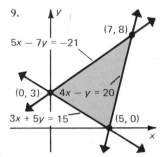

11.
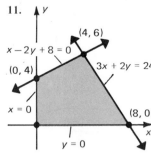

15.
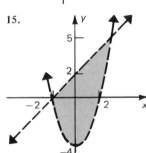

17.
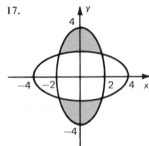

19.
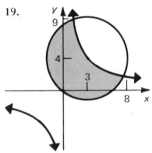

Exercise Set 6.2   (page 221)

1. $8; -\frac{5}{3}$ **3.** $7; -4$ **5.** 4 type A; 2 type B   **7.** 110 stock A; 90 stock B   **9.** 1500 gal. regular; 500 gal. premium   **11.** 500 lbs. of beef; 500 lbs. of pork   **13.** 15 from warehouse 1 to A; 60 from warehouse 2 to A; 50 from warehouse 1 to B

Chapter 6   Self-Test   (page 224)

1. (Figure 1 follows)   **3.** (Figure 3 follows)   **4.** (2, 0), (7, 1), (4, 7), (−8, 10)
5. (Figure 5 follows)   **7.** $15; -50$ **8.** $90; -69$ **9.** 4 canister, 4 tank; 0 canister, 8 tank
10. 3 of A, 2 of B

1.
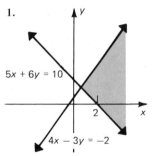

3.
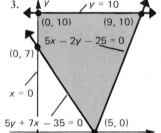

5.
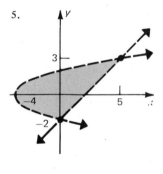

Exercise Set 7.1　　(page 231)

1. $\begin{bmatrix} 9 & -4 \\ 2 & -7 \end{bmatrix}$　**3.** $\begin{bmatrix} 6 & 9 \\ -4 & 5 \\ 9 & 9 \end{bmatrix}$　**5.** not possible　**7.** $\begin{bmatrix} 7 & 4 & 7 \\ -1 & -6 & -1 \end{bmatrix}$　**9.** $\begin{bmatrix} 2 & 4 & 6 \\ -8 & -10 & -12 \end{bmatrix}$

11. $\begin{bmatrix} 16 & 4 \\ -8 & -12 \\ 24 & 20 \end{bmatrix}$　**13.** $\begin{bmatrix} 30 & -11 \\ 8 & -25 \end{bmatrix}$　**15.** $\begin{bmatrix} 19 & 8 & 15 \\ 5 & -8 & 9 \end{bmatrix}$　**17.** $\begin{bmatrix} -\frac{2}{5} & \frac{22}{5} \\ -\frac{2}{5} & 6 \\ -\frac{3}{5} & \frac{2}{5} \end{bmatrix}$　**19.** $\begin{bmatrix} 18 & -18 \\ 12 & 2 \end{bmatrix}$

21. not possible　**23.** not possible　**25.** $\begin{bmatrix} 21 & 17 & -1 \\ -9 & 3 & -15 \end{bmatrix}$　**27.** $\begin{bmatrix} -18 & -10 \\ 42 & 19 \end{bmatrix}$

29. $\begin{bmatrix} 20 & 80 \\ 23 & 36 \end{bmatrix}$　**31.** $\begin{bmatrix} -5 \\ 7 \end{bmatrix}$　**33.** $\begin{bmatrix} -3 \\ 1 \\ 2 \end{bmatrix}$　**35.** Ranch 2 to Portland; ranch 3 to San Francisco

Exercise Set 7.2　　(page 238)

7. $\begin{bmatrix} -2 & 3 \\ 1 & -1 \end{bmatrix}$　**9.** singular　**11.** singular　**13.** $\begin{bmatrix} -\frac{2}{3} & -\frac{1}{3} & 2 \\ 1 & 0 & -2 \\ -\frac{1}{3} & \frac{1}{3} & 1 \end{bmatrix}$　**15.** $\{(3, 0)\}$　**17.** $\{(3, -2, 1)\}$

19. a. 1 lb. of type I, 5 lbs. of type II, no type III　**b.** no type I, 1 lb. of type II, 1 lb. of type III
21. 5 of I, 7 of II　**23. a.** 7 g. of I, 4 g. of II, 2 g. of III　**b.** 3 g. of I, 4 g. of II, 5 g. of III

Exercise Set 7.3　　(page 244)

1. -2　**3.** 0　**5.** 12　**7.** -10　**9.** 47　**11.** -20　**13.** 2　**15.** 48　**17.** -32
19. $7x - 2y - 4 = 0$　**21.** $4x - 7y - 9 = 0$　**23.** collinear　**25.** not collinear　**27.** $\{3\}$
29. $\{-5, -1\}$　**31.** 108　**33.** 4　**35.** $7x - 11y + 5z = 0$　**37.** $35x + 14y - 10z - 70 = 0$

Chapter 7　Self-Test　　(page 246)

1. $\begin{bmatrix} 13 & -10 \\ -5 & 9 \end{bmatrix}$　**2.** not possible　**3.** $\begin{bmatrix} 5 & -2 & -9 \\ 0 & -15 & 10 \end{bmatrix}$　**4.** $\begin{bmatrix} 9 & -6 & -21 \\ -4 & -35 & 14 \end{bmatrix}$

5. $\begin{bmatrix} -2 & 20 & -38 \\ 3 & -15 & 27 \end{bmatrix}$　**6.** $\begin{bmatrix} -19 & -20 & -21 \\ 1 & 2 & 3 \\ 3 & 0 & -3 \end{bmatrix}$　**7.** not possible　**8.** $\begin{bmatrix} -8 & 0 \\ -5 & -12 \end{bmatrix}$

9. $X = \begin{bmatrix} 1 & -1 & -3 \\ -1 & -5 & 1 \end{bmatrix}$　**10.** $\begin{bmatrix} -23 & 26 \\ 33 & -4 \end{bmatrix}$　**11.** $\{(5, -4)\}$　**12.** $\{(1, 3, -2)\}$　**14. a.** 3 of
type I, 2 of type II　**b.** 4 of type I, 4 of type II　**15.** -84　**16.** -16

Exercise Set G　　(page 247)

1. G-1　**3.** G-2　**5.** G-3　**7.** $4\begin{vmatrix} 3 & -2 & 5 \\ 1 & 0 & 7 \\ 1 & 3 & 2 \end{vmatrix}$　**9.** $\frac{1}{12}\begin{vmatrix} 1 & 3 & 0 \\ 8 & -2 & 5 \\ 2 & 4 & -5 \end{vmatrix}$　**11.** 41　**13.** $\frac{113}{15}$

15. 87　**17.** 348

Exercise Set H　　(page 250)

1. $\left\{\left(\frac{26}{3}, \frac{16}{3}\right)\right\}$　**3.** $\{(6, 4)\}$　**5.** not applicable　**7.** $\{(2, 2, -2)\}$　**9.** not applicable

11. $\left\{\left(\frac{1}{3}, 4, \frac{19}{3}\right)\right\}$　**13.** $\{(3, -1, 0, 3)\}$　**15.** $\left\{\left(\frac{1}{4}, -\frac{1}{6}\right)\right\}$　**17.** $y = \dfrac{a_1c_2 - a_2c_1}{a_1b_2 - a_2b_1} = \dfrac{\begin{vmatrix} a_1 & c_1 \\ a_2 & c_2 \end{vmatrix}}{\begin{vmatrix} a_1 & b_1 \\ a_2 & b_2 \end{vmatrix}}$

Exercise Set I　　(page 253)

1. $\begin{bmatrix} \frac{1}{3} & 0 \\ 0 & 1 \end{bmatrix}$　**3.** $\begin{bmatrix} 1 & 0 & 0 \\ 0 & 0 & 1 \\ 0 & 1 & 0 \end{bmatrix}$　**5.** $\begin{bmatrix} 1 & 0 & 0 \\ 0 & 1 & 0 \\ -3 & 0 & 1 \end{bmatrix}$　**9.** $T_1 = \begin{bmatrix} 1 & 0 \\ 2 & 1 \end{bmatrix}$; $T_2 = \begin{bmatrix} 1 & 0 \\ 0 & \frac{1}{4} \end{bmatrix}$; $T_3 = \begin{bmatrix} 1 & -5 \\ 0 & 1 \end{bmatrix}$

11. $C^{-1} = \begin{bmatrix} \dfrac{d}{ad - bc} & \dfrac{-b}{ad - bc} \\ \dfrac{-c}{ad - bc} & \dfrac{a}{ad - bc} \end{bmatrix}$; C^{-1} does not exist if $ad - bc = 0$.

Exercise Set 8.1 (page 260)

1. decreasing **3.** increasing **5.** neither

7.
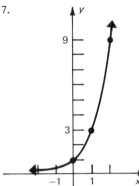

13.
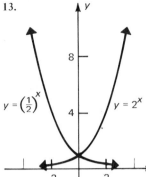

17.
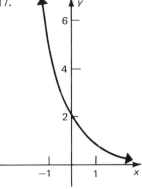

21.
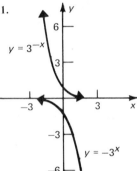

27.
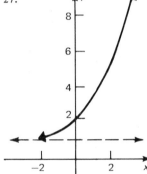

31.
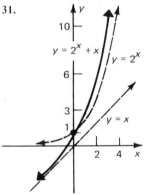

Exercise Set 8.2 (page 265)

1. $5 = \log_2 32$ **3.** $\frac{1}{5} = \log_{32} 2$ **5.** $-\frac{1}{2} = \log_{16} \frac{1}{4}$ **7.** 4 **9.** $\frac{1}{4}$ **11.** 4 **13.** 2 **15.** 1
17. $\frac{1}{3}$ **19.** $\{32\}$ **21.** $\{7\}$ **23.** $\{10\}$ **25.** (Figure 25 follows) (Figure 31 follows)
37. 9 **39.** -1 **41.** $\frac{6}{5}$ **43.** 2 **45.** 3 **47.** $\log_b x + \log_b y + \log_b z$

49. $2 \log_b x - \frac{1}{3} \log_b y - \frac{1}{2} \log_b z$ **51.** $\frac{1}{3} \log_b(x+y) - \frac{2}{3} \log_b(x-y)$ **53.** $\log_b xyz$ **55.** $\log_b \frac{x^{2/3}}{y^3 z^5}$

57. $\log_b \left[\frac{(x-y)(y-z)^2}{z-x} \right]^{-3}$ **61.** (Figure 61 follows)

25.
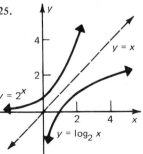

31.
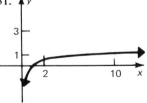

61.
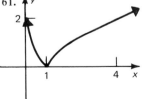

366

Exercise Set 8.3   (page 268)

1. 1.8405   **3.** 9.7769 − 10   **5.** 3.5553   **7.** 4.8676   **9.** 6.3099   **11.** 4.4189
13. 6.9634 − 10   **15.** 4.3039   **17.** 1.8469   **19.** 5.5013   **21.** 1.3653   **23.** 3.2620
25. 2.5936   **27.** 55   **29.** 8.8   **31.** 0.3

Exercise Set 8.4   (page 271)

1. {97}   **3.** $\left\{\frac{7}{2}\right\}$ **5.** {3}   **7.** $\varnothing$ **9.** $\{5+\sqrt{26}\}$ **11.** $\left\{\frac{\log 28}{\log 5}-2\right\}$ **13.** $\left\{\pm\sqrt{\frac{\log 9}{\log 2}}\right\}$
15. $\left\{-\frac{\log 0.52}{\log 25}\right\}$ **17.** $\left\{\frac{\log 49}{\log 9-\log 7}\right\}$ **19.** $\{-3\}$ **21.** 57.76   **23.** −57.76   **25.** 14.28
27. 1718   **29.** 9240   **31.** 2796   **33.** .07   **35.** $\{1,100\}$ **37.** $t=\dfrac{\log A-\log P}{n\,\log\left(1+\dfrac{r}{n}\right)}$

Exercise Set 8.5   (page 277)

1. \$7345; 7397; 7430; 7465   **3.** 11 yrs, 11 mos; 11 yrs, 10 mos; 11 yrs, 7 mos; 11 yrs, 5 mos
5. \$6485   **7.** \$2,996,400   **9.** \$3293   **11.** 2.4   **13.** 1.2×10^{-4}
15. 0.36 g   **17.** \$4298   **19.** 3,240,000   **21.** \$44,580   **23.** 847 yrs; no

Chapter 8   Self-Test   (page 279)

1. (Figure 1 follows)   **2.** (Figure 2 follows)   **3.** $4=\log_3 81$ **4.** $-\frac{1}{2}=\log_{25}\frac{1}{5}$ **5.** 4   **6.** $\frac{1}{4}$
7. 5   **8.** 15   **9.** 4   **10.** 4   **11.** (Figure 11 follows)   **12.** $\frac{1}{6}\log_b x+\frac{2}{3}\log_b y-\frac{1}{3}\log_b z$
13. $\log_b\dfrac{x^{1/3}z^2}{y^2}$ **14.** 1.7405   **15.** 9.6433 − 10   **16.** 4.8676   **17.** 9.5537 − 10   **18.** 4.2905
19. 5.8434   **20.** 1.5   **21.** 50   **22.** 0.4   **23.** {5}   **24.** $\left\{\dfrac{\log 64}{\log 3}\right\}$ **25.** $t=\dfrac{1}{k}\ln\dfrac{B}{B_0}$
26. \$1266; 1271; 1271   **27.** \$64.44   **28.** \$78,720

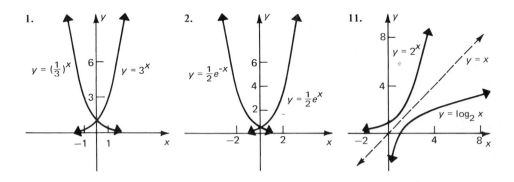

Exercise Set J   (page 283)

1. 2.4771   **3.** 3.0212   **5.** 8.7243 − 10   **7.** 8.3296 − 10   **9.** 0.8767   **11.** 7.0249 − 10
13. 246   **15.** 0.011   **17.** 6.46   **19.** 0.0001946   **21.** 91660   **23.** 175.5   **25.** 1551
27. 32,150,000   **29.** 0.08358   **31.** 3.247   **33.** 11.54

Exercise Set K   (page 285)

1. {1.1}. (Figure 1 follows)   **3.** {−0.8, 2}, (Figure 3 follows)   **5.** {0, 0.5}   **7.** (Figure 7 follows)   **a.** {$x: x \le 0$ or $x \ge 1$}   **b.** {$x: 0 \le x \le 1$}   **9. a.** {$x: 0 \le x \le 0.5$}   **b.** {$x: x \ge 0.5$}
11. $2^x > x$ **13.** $\sqrt{x} > \ln x$ **15.** {$x: -1 < x < 0$ or $x > 1$}

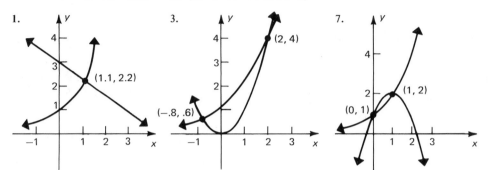

Exercise Set L   (page 287)

1. (Figure 1 follows)   **2.** (Figure 2 follows)   **3.** $e^x = y + \sqrt{y^2 + 1}$ **10.** (Figure 10 follows)

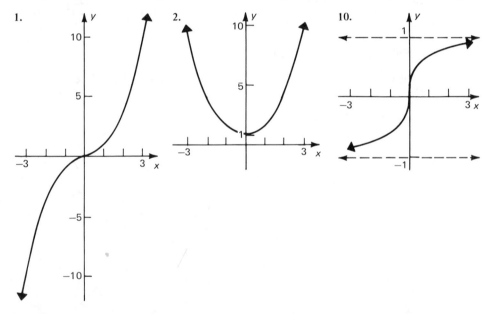

Exercise Set 9.1   (page 292)

1. $1, 3, 5, 7, \ldots, 2n - 1, \ldots$ **3.** $1, 2, 4, 8, \ldots, 2^{n-1}, \ldots$ **5.** $\dfrac{1}{3}, \dfrac{1}{9}, \dfrac{1}{27}, \dfrac{1}{81}, \ldots, \dfrac{1}{3^n}, \ldots$

7. $-1, 4, -36, 512, \ldots, (-1)^n(2n)^{n-1}, \ldots$ **9.** $2, \dfrac{3}{2}, \dfrac{4}{3}, \dfrac{5}{4}, \dfrac{6}{5}, \dfrac{7}{6}$ **11.** $1, 0, -\dfrac{1}{3}, \dfrac{1}{2}, -\dfrac{3}{5}, \dfrac{2}{3}$

13. $2, 7, 12, 17$ **15.** $\dfrac{1}{2}, \dfrac{1}{4}, \dfrac{1}{8}, \dfrac{1}{16}$ **17.** $\dfrac{x}{2}, \dfrac{x^3}{4}, \dfrac{x^5}{8}, \dfrac{x^7}{16}$ **19.** $7; 28, 35, 42; 7n$

21. $-9; 0, -9, -18; 36 - 9n$ **23.** $x + 2c; 4x + 7c, 5x + 9c, 6x + 11c; nx + (2n - 1)c$ **25.** -3

27. $12x + 11$ **29.** $2; 16, 32, 64; 2^n$ **31.** $-\dfrac{1}{3}; -\dfrac{1}{81}, \dfrac{1}{243}, -\dfrac{1}{729}; \dfrac{(-1)^{n+1}}{3^n}$

33. $\dfrac{x}{c}; \dfrac{x^2}{c^2}, \dfrac{x^3}{c^3}, \dfrac{x^4}{c^4}; \left(\dfrac{x}{c}\right)^{n-2}$ **35.** 768 **37.** $16x^{17}$ **39.** $2; 4; 50$ **41.** 6 **43.** 40
45. -13 **47.** $10, 17$ **49.** $\pm 6\sqrt{3}$

Exercise Set 9.2 (page 297)

1. Arith.; 468 **3.** Geom.; 121 **5.** Arith.; -198 **7.** Arith.; $-\frac{95}{2}$ **9.** Geom.; $\frac{127}{128}$
11. $7+9+11+13+15+17+19+21$; 112 **13.** $\frac{1}{9}+\frac{1}{27}+\frac{1}{81}+\frac{1}{243}+\frac{1}{729}$; $\frac{121}{729}$
15. $\frac{1}{4}-\frac{1}{8}+\frac{1}{16}-\frac{1}{32}+\frac{1}{64}-\frac{1}{128}$; $\frac{21}{128}$ **17.** $-\frac{5}{2}-2-\frac{3}{2}-1-\frac{1}{2}+0+\frac{1}{2}+1$; -6 **19.** 2132 **21.** 2060
23. \$512; \$1023 **25.** (1) \$46,500; (2) \$10,700,000 **29.** $\frac{3k}{2}(k+1)$

Exercise Set 9.3 (page 301)

1. $\frac{1}{2}+\frac{1}{4}+\frac{1}{8}+\frac{1}{16}+\cdots+\frac{1}{2^n}+\cdots$ **3.** $3-9+27-81+\cdots+(-1)^{n+1}3^n+\cdots$ **5.** $\frac{125}{3}$
7. no sum; $|r|=\left|\frac{5}{2}\right|>1$ **9.** 3 **11.** no sum; $|r|=\left|\frac{3}{2}\right|>1$ **13.** $\frac{9}{4}$ **15.** 3 **17.** $\frac{4}{9}$
19. $\frac{2}{11}$ **21.** $\frac{1}{2}$ **23.** $\frac{245}{37}$ **25.** $\frac{341}{27}$; 15 **27.** $\{x: x<-2 \text{ or } x>0\}$
29. $\{x: x<0 \text{ or } x>\frac{4}{3}\}$

Exercise Set 9.4 (page 305)

1. 7! **3.** $1\cdot2\cdot3\cdots17\cdot18\cdot19$ **5.** $1\cdot2\cdot3\cdots(n-3)(n-2)(n-1)$
7. $1\cdot2\cdot3\cdots(2n-2)(2n-1)2n$ **9.** $(9!)10\cdot11\cdot12$ **11.** $(n-1)!\cdot n\cdot(n+1)$ **13.** 56
15. 126 **17.** 35 **19.** $\frac{1}{13}$ **21.** n^2+n **23.** $2n^3-2n^2$
25. $x^6+12x^5+60x^4+160x^3+240x^2+192x+64$ **27.** $x^4+8x^3y+24x^2y^2+32xy^3+16y^4$
29. $81x^4-216x^3y+216x^2y^2-96xy^3+16y^4$
31. $32a^5-40a^4b+20a^3b^2-5a^2b^3+\frac{5}{8}ab^4-\frac{1}{16}b^5$
33. $x^{28}+56x^{27}y+1512x^{26}y^2+26208x^{25}y^3$ **35.** $x^{36}-18\sqrt{2}x^{34}+306x^{32}-1632\sqrt{2}x^{30}$ **37.** 1.20
39. \$1338

Chapter 9 Self-Test (page 306)

1. $-1, -4, -7, -10$; arith.; $d=-3$ **2.** $-6, 18, -54, 162$; geom.; $r=-3$
3. $\frac{3}{2}, \frac{3}{2}, \frac{1}{6}, \frac{1}{18}$; geom.; $r=\frac{1}{3}$ **4.** $\frac{3}{2}, \frac{11}{6}, \frac{13}{6}, \frac{5}{2}$; arith.; $d=\frac{1}{3}$
5. geom.; $4x^2, 2x, 1$; $s_n=(2x)^{6-n}$; $s_{10}=\frac{1}{16x^4}$ **6.** arith.; $3x-24, 3x-30, 3x-36$; $s_n=3x-6n$;
$s_{10}=3x-60$ **7.** 72; 3 **8.** 25; ±20 **9.** $\frac{1562}{3125}$ **10.** $\frac{155}{x}$
11. $x^2+2x^2+3x^2+\cdots+1000x^2$; $500,500x^2$
12. $\frac{1}{2^2}+\frac{1}{2^3}+\frac{1}{2^4}+\frac{1}{2^5}+\frac{1}{2^6}+\frac{1}{2^7}$; $\frac{63}{128}$ **13.** no sum **14.** 12 **15.** $-\frac{2}{9}$ **16.** no sum **17.** $\frac{2}{11}$
18. 462 **19.** $(n-1)n(n+1)(n+2)$ **20.** $x^5+15x^4+90x^3+270x^2+405x+243$
21. $16y^4-16y^3+6y^2-y+\frac{1}{16}$ **22.** 1.791

Exercise Set 10.1 (page 314)

1. a. {athletes on both football and track teams}; {athletes on either football or track team, or
both} **b.** 10; 7; 2; 15 **c.** 8 **3.** 94 **5.** 1 **7.** 50 **9.** 12 **11.** 15 **13.** 10
15. 6,760,000 **17.** 32 **19. a.** 1680 **b.** 5000

Exercise Set 10.2 (page 318)

1. 123, 132, 213, 231, 312, 321 **3.** 5040 **5.** 336 **7.** 72 **9.** 5040 **11. a.** 120
b. 3125 **13. a.** 27,216 **b.** 90,000 **15.** 12096; 6048 **17.** 720 **19.** 120 **21.** 151,200
23. 60,480 **25.** 48 **27.** 725,760

Exercise Set 10.3 (page 323)

1. 36 **3.** 6 **5.** 495 **7.** 190 **9.** 210 **11.** 2,598,960 **13.** 28; 56; 70 **15.** 350
17. a. 1044 **b.** 910 **19.** $13440x^4$ **21. a.** 80 **b.** 200 **25.** 286; 105

Exercise Set 10.4   (page 327)

1. {(1, 1), (2, 2), (3, 3), (4, 4), (5, 5), (6, 6)}; 6; "the numbers on the dice are different;" 30
3. {(1, 6), (2, 5), (3, 4), (4, 3), (5, 2), (6, 1)}; 6   **5.** 1; 2; 3; 4; 5; 5; 4; 3; 1   **7.** $\frac{1}{6}$; $\frac{5}{6}$ **9.** $\frac{1}{6}$
11. $\frac{1}{9}$ **13.** $\frac{1}{18}$ **15.** $\frac{5}{18}$ **17.** $\frac{13}{18}$ **19.** $\frac{7}{18}$ **21.** $\frac{10}{91}$ **23.** $\frac{3}{91}$ **25.** $\frac{15}{91}$ **27.** $\frac{36}{91}$ **29.** $\frac{51}{91}$
31. $\frac{25}{51}$ **33.** $\frac{1}{26}$ **35.** $\frac{2}{135}$ **37.** $\frac{17}{45}$

Exercise Set 10.5   (page 332)

1. A and B are independent; A and C are dependent; B and C are dependent.   **3.** $\frac{3}{8}$ **5.** $\frac{39}{80}$
7. $\frac{5}{17}$ **9.** $\frac{80}{153}$ **11.** $\frac{7}{17}$ **13.** $\frac{1}{2}$ **15.** $\frac{1}{4}$ **17.** $\frac{3}{8}$ **19.** $\frac{256}{270725}$

21. $1 - \dfrac{\binom{365}{5}}{(365)^5} \approx .0271$

Chapter 10   Self-Test   (page 333)

1. 335; 65   **2.** 5,760,000; 51,840,000   **3.** 720; 120   **4. a.** 24   **b.** 256   **5.** 151,200
6. $\frac{99}{425}$ **7. a.** 966   **b.** 1001   **c.** 637   **8.** $-3432y^7$ **9. a.** $\frac{44}{4165}$ **b.** $\frac{1}{20825}$ **10.** $\frac{3}{11}$ **11.** $\frac{2}{5}$
12. $\frac{1}{70}$

Exercise Set N   (page 337)

1. 0.1150   **3. a.** $\frac{1}{6561}$ **b.** $\frac{256}{6561}$ **c.** $\frac{6305}{6561}$ **d.** $\frac{129}{6561}$ **5. a.** $\frac{1}{2}$ **b.** $\frac{1}{32}$ **c.** $\frac{1}{2}$ **7.** $\frac{1161}{59049}$
9. $\frac{4528}{15625}$

INDEX

Family of lines, 103ff
Field, 14; ordered, 14
Fractions, partial, 207ff
Function, 76f; algebra of, 112ff; absolute value, 89; bracket, 91; composite, 113; composition of, 113; constant, 84; determinant, 240; decreasing, 256; exponential, 258, 260; graphing linear, 83; greatest integer, 91; hyperbolic, 286ff; increasing, 256; linear, 82ff; logarithmic, 262f; notation, 77; of a real variable, 77; one-to-one, 139; polynomial, 151ff; quadratic, 94ff; rational, 175ff; real-valued, 77; sequence, 288; signum, 92; step, 91
Fundamental theorem of algebra, 158

G
Geometric sequence, 291; common ratio, 291; general term, 291; sum of finite, 296; sum of infinite, 298ff
Graph, addition of ordinates, 148; discussing, 115ff; of a conic section, 120ff; of linear equation, 83ff; of linear inequality, 16, 64; of a linear system, 185; of a number, 15; of an ordered pair, 44; of a rational function, 177; of a relation, 77; of a set, 15; sign, 66; symmetry of, 116
Graphical solutions, of equations, 284ff; of inequalities, 66, 284ff; of linear systems, 185; of nonlinear systems, 202ff; of systems of inequalities, 211ff
Greater than relation, 9

H
Half-life, 275
Half-plane, 133; closed, 133; edge, 133; open, 133
Horizontal line test, 139
Hydrogen potential, 274
Hyperbola, asymptotes of, 125; branch of, 124; standard form of, 124, 130

I
Identity, 47, 207
Imaginary numbers, 22; pure, 22; unit, 20
Independent variable, 75
Inequality, 9; absolute, 63; absolute value, 69ff; conditional, 63; equivalent, 62ff; in one variable, 62; linear, 133; quadratic, 66, 134; sense of, 13; solution of, 63; solution set of, 63; solving graphically, 66, 284f
Intercepts, 83

Interest, compound, 272; continuous, 272; simple, 273
Interval, 16; closed, 16; endpoints, 16; half-closed, 16; half-open, 16; open, 16; unbounded, 16
Isomorphism, 21

L
Less than relation, 9
Linear combination, of equations, 187; method of, 188
Linear equations, double-intercept form, 87; point-slope form, 87; slope-intercept form, 87; standard form, 82
Linear interpolation, 281
Linear programming, 215ff
Location theorem, 163
Logarithmic equations, 269
Logarithmic function, graph of, 263; properties of, 263
Logarithms, characteristic of, 280; computations with, 280; mantissa of, 280; natural, 266

M
Mantissa, 280
Mapping, 78
Mathematical induction, 308ff
Matrix (matrices) 193ff; addition, 227; additive inverse, 228; augmented, 194; coefficient, 194; column of, 194; conformable (product), 230; conformable (sum), 227; diagonal, 232; elementary, 253ff; elements or entries, 225; equal, 226; equation, 228; identity, 233; inverse, 233; invertible, 234; nonsingular, 234; order of, 225; principal diagonal, 232; product, 227, 229; row-equivalent, 195; row of, 194; singular, 234; square, 232; zero, 228
Mean, arithmetic, 293; geometric, 293
Minor of determinant, 240
Monomial, coefficient, 32; degree, 32; numerical coefficient, 32
Multiple zero, 158
Multiplication, of real numbers, 7
Multiplication axioms, 8; associativity, 8; commutativity, 8; identity, 8; inverse, 8
Multiplicative inverse, 8, 31

N
Natural logarithms, 266ff
Negative, 7
Negative numbers, 9
Nonterminating decimal, 300

Number line, 15
Numerator, 12

O

One-to-one correspondence, 4
One-to-one function, 139
Opposite, 7
Order axioms, 9; closure, 9; transitivity, 9; trichotomy, 9
Ordered n-tuple, 184
Ordered pair, 43; component of, 43
Ordinate, 44

P

Parabola, 94; axis of symmetry, 94, 121; standard form equation, 95, 121; vertex, 94, 121
Parameter, 103
Partial sum, 294, 296
Permutation, 316
Polynomial, 32ff, 151ff; completely factored form of, 36; complex, 151; constant, 33; degree of, 33; differences, 33f; division algorithm for, 41; division of, 41f; divisor, 41; factoring, 35ff; function, 151ff; graph of, 170ff; identical, 207; leading coefficient of, 151; leading term of, 151; like terms of, 33; over J, 36; over specified sets, 37; products, 34; quotient, 41; real, 151; remainder, 41; sums, 33f; zero, 33; (see also zeros)
Positive numbers, 9
Power, 23
Probability, 324ff; binomial, 335f; conditional, 331
Product, 7
Pythagorean theorem, 85

Q

Quadrants, 43
Quadratic, equations, 51ff; inequalities, 133ff
Quadratic formula, 53; discriminant, 56
Quadratic functions, 94ff; graphs of, 94
Quadratic relations, 134ff; graphs of, 134
Quotient, 8

R

Radicals, 27ff; difference of, 28; index of, 27; of order n, 27; product of, 28; quotient of, 28; simplest form of, 27; sum of, 28; symbol for, 27
Radicand, 27
Radiocarbon dating, 276

Range, 75, 137, 259
Rate of change, average, 108ff; instantaneous, 110, 111
Rational expressions, 38ff; difference of, 39; least common denominator of, 39; lowest terms, 39; product of, 40; quotient of, 40; sum of, 39
Reciprocal, 8, 31
Rectangular coordinate system, 43f
Recursion formula, 289
Region, polygonal, 212; boundary of, 212; convex, 212; vertex of, 212
Relations, 74ff; domain, 75; graphing, 77ff; inverse, 137; quadratic, 134; range, 75
Remainder column, 156
Remainder theorem, 154
Roots, 23; complex n^{th}, 24; cube, 23; principal, 23f; square, 23, 30

S

Sample space, 324
Scientific notation, 280
Sentences, involving absolute value, 69ff; open, 7
Sequence, 288ff; arithmetic, 290; finite, 289; function, 288; general term, 288; geometric, 291; infinite, 288
Series, infinite, 299; infinite geometric, 299; sum of, 300
Set, 3ff; -builder notation, 4; Cartesian product, 74, 313; disjoint, 5; element of, 3; empty, 4; equal, 4; finite, 4; graph of, 15; infinite, 4; intersection of, 5; member of, 3; null, 4; of integers, 5; of irrationals, 5; of natural numbers, 3; of rational numbers, 5; of real numbers, 5; of whole numbers, 5; replacement, 3, 184, 289; union of, 5
Sigma notation, 296
Sign graphs, 66f
Slope, 85f; average, 110; instantaneous, 110
Solution, extraneous, 201; multiple, 168
Solution set, of an equation, 46; of an inequality, 63; of systems, 184, 211
Solution space, 211
Solving equations, by completing the square, 52; by factoring, 51; by formula, 53; by the square root method, 52
Solving inequalities, 62ff; absolute value, 69ff; sign graphs, 66f
Solving systems, by Cramer's rule, 250ff; by linear combinations, 188; by inverse matrices, 236; by matrices, 195; by substitution, 186

1 2 3 4 5 6 7 8 9 10 11 12 13 14 15 16 17 18 19 20 21 22 23 24 25 CP 82 81 80 79 78 77 76 75 74 73